Alternative Energy Systems and Applications

Alternative Energy Systems and Applications

B. K. Hodge

Mississippi State University

John Wiley & Sons, Inc.

EXECUTIVE PUBLISHER	Don Fowley
SENIOR ACQUISITIONS EDITOR	Michael McDonald
EXECUTIVE MARKETING MANAGER	Christopher Ruel
EDITORIAL PROGRAM ASSISTANT	Rachael Leblond
SENIOR PRODUCTION EDITOR	William A. Murray
SENIOR DESIGNER	Kevin Murphy
COVER DESIGN	David Levy
COVER PHOTO	Digital Vision

This book was set in Times Ten by Preparé and printed and bound by Malloy, Inc.

This book is printed on acid free paper. ∞

To order books or for customer service please call 1-800-CALL WILEY (225-5945).

ISBN: 978-0-470-14250-9

Printed in the United States of America

10 9 8 7 6 5 4 3 2 1

*To Gayle, my wife
and best friend*

Preface

In recent years much has been made of the impact of the myriad energy problems faced not only by the United States, but also the rest of the world. These impacts range from energy security issues (the dependence on imported energy sources) to economic issues (gasoline reached $4.00/gallon in the summer of 2008) to energy sustainability issues (minimum environmental and ecological impacts). Many in the engineering, corporate, and political communities advocate greater reliance on alternative energy sources. Because of the increased interests in such topics, a number of books on subjects ranging from sustainable energy to renewable energy to alternative energy have been offered in recent years. However, many of these books contain mostly qualitative information with little in the way of quantitative information or "engineering" calculations or procedures. Some also advocate specific alternative energy scenarios and some do not present balanced discussions. This textbook was written to address the above concerns.

Alternative Energy Systems and Applications is suitable for use at the senior or beginning graduate level for students in mechanical engineering or in energy-engineering related fields. Familiarity with the basic concepts of fluid mechanics, thermodynamics, and heat transfer is presumed in the development of the topics in the book, but maturity in these subjects is not needed in order to understand the developments.

The title, *Alternative Energy Systems and Applications*, is used to convey the idea that the topics covered encompass both alternative energy sources and alternative uses of existing energy sources. The solution to the current energy dilemma will contain features of both. The breadth of topics proposed for the book is delineated in the chapter headings. Chapter 1 critically examines energy usage in the United States. Although not explicitly subdivided into congruent topical areas, Chapters 2–5 treat turbomachine-based topics (hydro, wind, and combustion turbines), Chapters 6–9 consider solar-based topics (active, passive, and photovoltaic), Chapters 10–11 examine fuel cells and CHP (combined heating and power) applications, and Chapters 12–15 complete the review of alternative energy concepts (biomass, geothermal, ocean, and nuclear).

All chapters except chapter 1 broadly fit into one of two categories — (1) a review of the background information necessary for a topic or (2) an exploration of an alternative energy source or an alternative use of an existing energy source. Chapters 2 and 6, for example, are used to review the backgrounds necessary for turbomachines and solar energy, respectively.

Often alternative energy topics are equated to renewable energy resource discussions. In this book, Chapters 3–4, 7–9, and 12–14 consider topics usually associated with renewable energy resources. The chapters dealing with renewable energy topics present the physical principles involved in harvesting the renewable, review (in most cases) the amount of the renewable resource available, examine quantitative aspects of the harvesting, point out difficulties with utilizing the renewable resource, discuss limitations and economics aspects, and provide, if applicable, examples of commercial systems for harvesting the renewable. Where appropriate, website addresses are cited.

The chapters addressing alternative uses of existing energy sources are focused on applications. Combustion turbines, fuel cells, and CHP systems represent alternative uses of existing energy sources. The application chapters basically discuss the operation, the thermodynamic aspects, and the expected efficiencies of such systems and provide examples. As with the renewable energy topics, suitable websites are referenced.

All chapters, except Chapter 1, contain worked examples and review questions and exercise problems. The focus is on first-order engineering calculations. Mathcad® is used as the computational software system throughout the book. However, the examples/problems are fundamental, and many other computational systems (MATLAB®, EES, Mathematica®) could be readily adopted for use with little effort. The intent of the book is to provide students with a quantitative approach to alternative energy sources and alternate applications of existing energy sources. Since this is a survey textbook, it does not attempt to provide detailed engineering information on the topics discussed, but references are provided that do contain detailed engineering information.

This textbook is the outgrowth of several years of teaching ME 4353/6353 Alternate Energy Sources in the Bagley College of Engineering at Mississippi State University. The discretionary funds provided to me as holder of the Tennessee Valley Authority Professorship in Energy Systems and the Environment at Mississippi State University were very helpful in this endeavor and are acknowledged. Additionally, thanks are due to Professors Francis Kulacki, James Mathias, and David Ruzic who reviewed the manuscript. Their comments and insights were quite useful.

B. K. Hodge
Mississippi State University
January 2009

Contents

Energy Usage in the United States

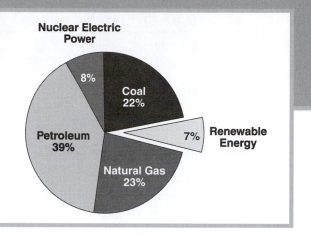

1.1 ENERGY AND POWER

A review of the customary units used for energy and power is appropriate to initiate a study of alternative energy sources and applications. Although much of the world uses the SI system (Système international d'unités), the United States, in addition to the SI system, uses the English Engineering and the British Gravitational systems of units. The unit of energy in the SI system is the N-m (newton-meter), which is also defined as the joule (J). Energy in the English Engineering (EE) system is defined as the Btu (British thermal unit), or alternatively, the ft-lbf (foot-pound force). Power is the rate of energy usage or transfer in J/sec, Btu/sec, or ft-lbf/sec. Power equal to 1 J/sec is defined as a watt (W). The most frequently used power unit is kilowatt (kW), or 1000 W. In the United States, power is sometimes expressed in terms of horsepower (hp), where 1 hp is 550 ft-lbf/sec or 0.7457 kW. The kilowatt-hour (kWh) is another frequently used unit of energy and represents a unit of energy (kW) multiplied by a unit of time (hour). The conversion is 3412.14 Btu = 1 kWh.

Tester et al. (2005) provide a sampling of power expenditures for various activities. Some of their results are reproduced as Table 1.1. The range of power expended is astonishing, about 9 orders of magnitude.

1.2 ENERGY USAGE AND STANDARD OF LIVING

An irrefutable fact is that the developed countries (the United States, Japan, the United Kingdom, etc.) use more energy per capita than the less developed countries (Mexico, Indonesia, etc.). Figure 1.1, taken from Tester et al. (2005), graphically presents energy consumption per capita as a function of gross national product (GNP) per capita for a number of countries. For the industrialized countries, the GNP per capita is from $15,000 to $25,000 while the energy consumption per capita is from

TABLE 1.1 Power expenditures for various activities

Activity	Power Expended
Pumping human heart	$1.5\,W = 1.5 \times 10^{-3}\,kW$
Household light bulb	$100\,W = 0.1\,kW$
Human, hard work	$0.1\,kW$
Draft horse	$1\,kW$
Portable floor heater	$1.5\,kW$
Compact automobile	$100\,kW$
SUV	$160\,kW$
Combustion turbine	$5000\,kW = 5\,MW$
Large ocean liner	$200{,}000\,kW = 200\,MW = 0.2\,GW$
Boeing 747 at cruise	$250{,}000\,kW = 250\,MW = 0.25\,GW$
Coal-fired power plant	$1 \times 10^{6}\,kW = 1000\,MW = 1\,GW$
Niagara Falls hydroelectric plant	$2 \times 10^{6}\,kW = 2000\,MW = 2\,GW$

150 GJ to 325 GJ. The United States and Canada have the highest energy consumption per capita. A number of reasons exist for the high energy consumption per capita in the United States, among them (1) historically cheap energy, (2) low population density, (3) large area, and (4) a history of abundant domestic energy.

As discussed later in this chapter, low energy costs and domestic energy abundance are a thing of the past, and the United States faces escalating energy costs and

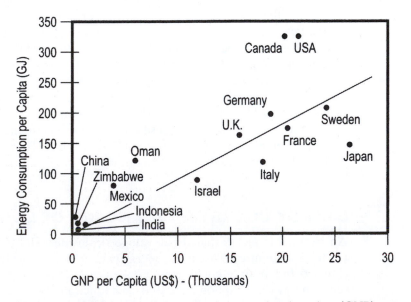

Figure 1.1 Per capita energy consumption versus gross national product (GNP) per capita for a number of countries (Tester et al. 2005).

the need for increasing energy imports (chiefly in the form of oil). The long-term implications of growing energy costs and imports will dramatically affect both the economy and the foreign policy posture of the United States. Indeed, the basis of this textbook is the need to consider both alternative energy sources and alternative (read more efficient) energy applications to address the energy problems facing the United States.

The energy problem in the United States is exacerbated by measures by other countries (India and China, for example) to increase the standard of living for their citizens. World energy consumption is rising faster than energy consumption in the United States.

1.3   A HISTORICAL PERSPECTIVE OF ENERGY USAGE IN THE UNITED STATES

The Energy Information Administration (EIA) of the United States Department of Energy (USDOE) provides a readily accessible and up-to-date source of energy statistics. The EIA website is www.eia.doe.gov. The EIA provides on a timely basis yearly energy statistics for the United States. These yearly energy summaries, titled *Annual Energy Review*, appear about eight months after the end of the calendar year and can be accessed at www.eia.doe.gov/aer. The information contained in this text is from calendar year 2007 and carries the USDOE accession number DOE/EIA-0384 (2007).

Figure 1.2, a mosaic of nighttime satellite photographs of the United States, is a rather dramatic illustration of the population density and dispersion and of the population of the United States, as well as the energy intensity distribution of night lighting (primarily from electricity usage).

Figure 1.2 Mosaic of nighttime satellite photographs of the United States (EIA 2007).

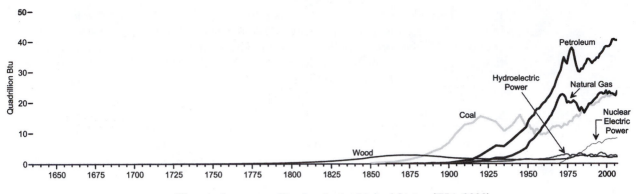

Figure 1.3 Historical energy utilization in the United States (EIA 2008).

Consider how the United States arrived at its current energy economy. Figure 1.3, taken from the EIA *Annual Energy Review 2007*, presents a graphical representation of historical energy utilization. The energy unit used is the quad (equal to 1 quadrillion Btu = 10^{15} Btu). Until the mid-1800s, energy utilization was mostly wood, with coal becoming increasingly important after 1850. By 1900 coal usage was much greater than wood usage, and petroleum was becoming more important as an energy source. In 1950 petroleum usage exceeded coal usage, and natural gas usage was dramatically rising. At the turn of the millennium, petroleum provided the most energy, with natural gas and coal vying for second and third place. Nuclear power was in fourth place, with hydroelectric and renewable energy (including wood) sources making the smallest contributions. In the next section, details of the energy utilization in 2007 will be explored.

The genesis of the energy problem is illustrated in Figure 1.4. Until about 1950, the United States had little dependence on energy imports. However, with the

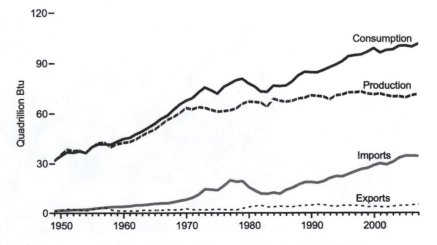

Figure 1.4 Energy consumption, imports, and exports for the United States.

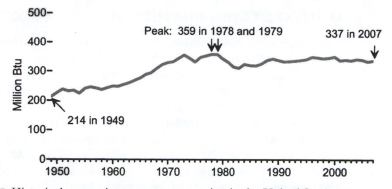

Figure 1.5 Historical per capita energy consumption in the United States.

post–World War II prosperity, energy exports began to increase as consumption out-paced domestic production. Since the 1980s, domestic production has increased, but at a rate slower than consumption has increased. The result has been a steady increase in energy imports.

Further explanation of how the United States arrived at its current rate of energy consumption is provided in Figures 1.5 and 1.6. Figure 1.5 tracks per capita energy consumption, and Figure 1.6 displays energy use per dollar of gross domestic product. Per capita energy consumption peaked at 360 million Btu in 1978 and declined during the "energy crisis" of the 1970s and 1980s. Some slight increases occurred during the 1990s, and in 2007 the per capita energy consumption was 337 million Btu. The energy crisis resulted in no dramatic decrease in per capita energy consumption in the United States; these results explain, in part, the current energy dilemma of the United States. In short, the United States failed to understand and heed the warnings of the first energy crisis. Figure 1.6 represents the energy usage per dollar of gross national product (GNP). Since the 1980s, the energy consumed per dollar of GNP has meaningfully declined from nearly 18,000 Btu to the 2007 value of 8,780 Btu. This decline is attributed to increased energy efficiency, especially in manufacturing, and to structural changes in the economy (the migration of much energy-intensive industry to other countries).

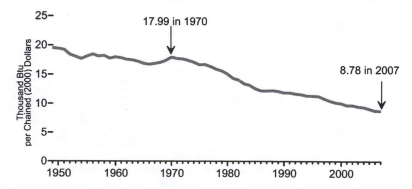

Figure 1.6 Historical energy use per dollar of gross domestic product in the United States.

1.4 | UNITED STATES ENERGY USAGE IN 2007

The EIA energy flow diagram, from the *Annual Energy Review 2007,* is arguably the most informative graphical representation in the *Review* and is reproduced as Figure 1.7. In this figure all energy usages are in quads (10^{15} Btu). Energy sources are delineated on the left-hand side of the diagram (coal at 23.48 quad, for example). The sources are then summed and expressed in terms of domestic production (71.71 quad) and imports (34.60 quad). The total supply is 106.96 quad with exports of 5.36 quad, which yields 101.6 quad for consumption. Thus, in 2007, the United States energy economy was 106.96 quad, of which 34.60 quad was imported.

The end-point energy usages (categorized as residential, commercial, industrial, and transportation) are shown on the right-hand side of the figure. They are also displayed on a pie chart in Figure 1.8. Industrial usage accounts for 32 percent of the total energy used, followed by 29 percent for transportation. The remainder is almost

(Quadrillion Btu)

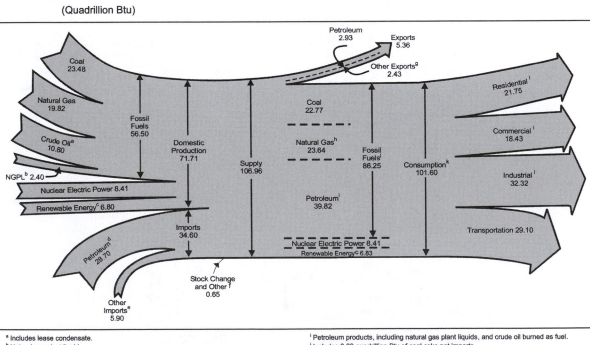

[a] Includes lease condensate.
[b] Natural gas plant liquids.
[c] Conventional hydroelectric power, biomass, geothermal, solar/photovoltaic, and wind.
[d] Crude oil and petroleum products. Includes imports into the Strategic Petroleum Reserve.
[e] Natural gas, coal, coal coke, fuel ethanol, and electricity.
[f] Adjustments, losses, and unaccounted for.
[g] Coal, natural gas, coal coke, and electricity.
[h] Natural gas only; excludes supplemental gaseous fuels.

[i] Petroleum products, including natural gas plant liquids, and crude oil burned as fuel.
[j] Includes 0.03 quadrillion Btu of coal coke net imports.
[k] Includes 0.11 quadrillion Btu of electricity net imports.
[l] Primary consumption, electricity retail sales, and electrical system energy losses, which are allocated to the end-use sectors in proportion to each sector's share of total electricity retail sales. See Note, "Electrical Systems Energy Losses," at end of Section 2.
Notes: • Data are preliminary. • Values are derived from source data prior to rounding for publication. • Totals may not equal sum of components due to independent rounding.
Sources: Tables 1.1, 1.2, 1.3, 1.4, and 2.1a.

Energy Information Administration / Annual Energy Review 2007

Figure 1.7 United States energy flow diagram for 2007.

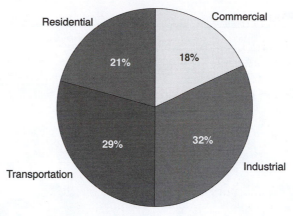

Figure 1.8 End-use energy utilization in 2007.

evenly split between residential and commercial. Because no end-use sector dominates the energy usage, alternative sources and applications are needed for all end-use sectors if significant reductions in energy usage are to be forthcoming.

In 2007 renewable energy from all sources amounted to about 7 percent of the total energy utilized in the United States. Figure 1.9 itemizes the contribution of

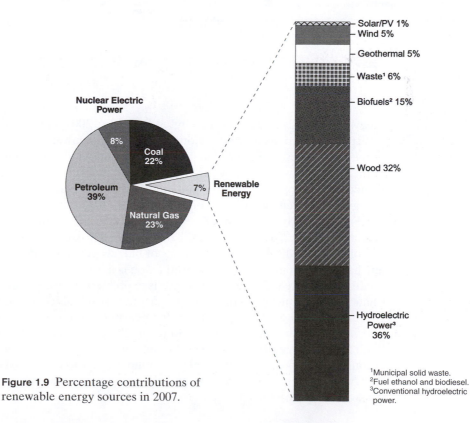

Figure 1.9 Percentage contributions of renewable energy sources in 2007.

[1]Municipal solid waste.
[2]Fuel ethanol and biodiesel.
[3]Conventional hydroelectric power.

(Million Barrels per Day)

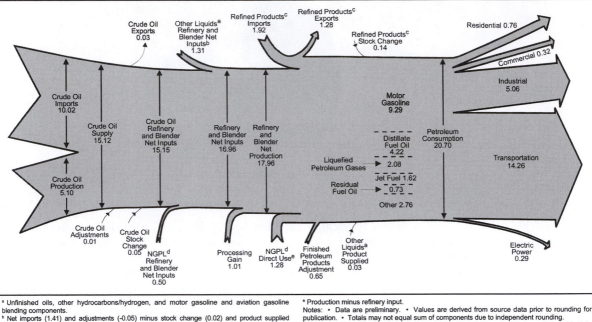

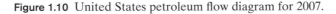

[a] Unfinished oils, other hydrocarbons/hydrogen, and motor gasoline and aviation gasoline blending components.
[b] Net imports (1.41) and adjustments (-0.05) minus stock change (0.02) and product supplied (0.03).
[c] Finished petroleum products, liquefied petroleum gases, and pentanes plus.
[d] Natural gas plant liquids.

[e] Production minus refinery input.
Notes: • Data are preliminary. • Values are derived from source data prior to rounding for publication. • Totals may not equal sum of components due to independent rounding.
Sources: Tables 5.1, 5.3, 5.5, 5.8, 5.11, 5.13a-5.13d, 5.16, and *Petroleum Supply Monthly*, February 2008, Table 4.

Figure 1.10 United States petroleum flow diagram for 2007.

renewable energy sources in the United States for 2007. Perhaps the most amazing statistic is that wood and conventional hydroelectric power accounted for 73 percent of the total renewable energy that year! Solar and wind contributed only 5 percent of the total renewable energy (or just 0.35 percent of the total consumption) in 2007. Hence, in spite of much interest and media hype, the penetration of solar and wind energy into the energy mix has not made much progress.

Figure 1.10 is a diagram of petroleum flow in the United States for 2007. The format of Figure 1.10 is similar to that of Figure 1.7 except that the numbers in the petroleum flow diagram are in millions of barrels per day (MMBD). Starting at the left-hand side, domestic crude oil production is a little more than half the amount of crude oil imported. The refinery output is expressed in terms of motor gasoline, distillate fuel oil, liquefied petroleum gases, jet fuel, residual fuel oil, and "other." Motor gasoline, at 9.29 MMBD, accounted for nearly one-half of the total utilization of petroleum products in the United States in 2007. The right-hand side of the petroleum flow diagram expresses the end-point petroleum energy usages. Transportation accounts for 69 percent of the total petroleum usage. Industrial usage is about 25 percent, with residential, commercial, and electric power generation responsible for the remaining 9 percent.

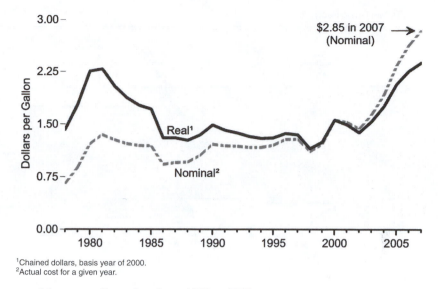

[1]Chained dollars, basis year of 2000.
[2]Actual cost for a given year.

Figure 1.11 Motor gasoline prices from 1978 to 2007.

Transportation, the dominant end-point petroleum energy usage, warrants further examination. Much insight can be gained by tracking the cost of a gallon of motor gasoline in terms of "real" and "nominal" dollars. Real dollars are the chained dollars based on the dollar in 2000, whereas nominal dollars represent the actual cost during a given year. Real dollars thus account for inflation. Figure 1.11 shows the real and nominal cost of a gallon of motor gasoline from 1978 to 2007. In real dollars gasoline was $2.25/gallon in 1980, a price not reached again until 2004. As of the summer of 2008, gasoline was more than $4.00/gallon, reflecting the disturbingly rapid increase since 2004. Indeed, only since 2000 has the rate of increase of the price of gasoline exceeded that of inflation. During the prosperous years of the 1990s, relative to inflation, gasoline prices declined! No wonder conservation, higher-gas-mileage vehicles, and alternative sources failed to arouse much interest in the public.

The natural gas flow diagram for 2007 is presented in Figure 1.12. Natural gas usage in this figure is expressed in trillions of cubic feet. As with the other energy flow diagrams (Figures 1.7 and 1.10), information proceeds from the left-hand side (sources) to the right-hand side (end-point usages). Imports account for about 18 percent of total consumption. Industrial usage and electric power generation account for 64 percent of the total natural gas utilization, with the remainder split between residential, commercial, and transportation.

The coal flow diagram for 2007 is shown in Figure 1.13, and usage is expressed in millions of short tons. In a fashion similar to the other energy flow diagrams (Figures 1.7, 1.10, and 1.12), information proceeds from the left-hand side (sources) to the right-hand side (end-point usages). All coal is produced domestically, with a small amount exported. Virtually all of the coal usage (93 percent) in the United States is for the generation of electricity, with some, about 8 percent, used in industrial

(Trillion Cubic Feet)

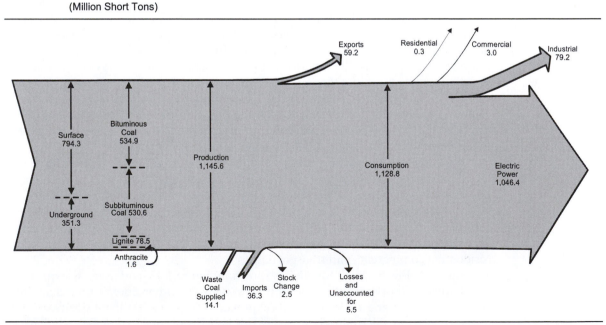

[a] Quantities lost and imbalances in data due to differences among data sources.
[b] Natural gas consumed in the operation of pipelines (primarily in compressors), and as fuel in the delivery of natural gas to consumers; plus a small quantity used as vehicle fuel.

Notes: • Data are preliminary. • Values are derived from source data prior to rounding for publication. • Totals may not equal sum of components due to independent rounding.
Sources: Tables 6.1, 6.2, and 6.5.

Figure 1.12 United States natural gas flow diagram for 2007.

(Million Short Tons)

[1] Includes fine coal, coal obtained from a refuse bank or slurry dam, anthracite culm, bituminous gob, and lignite waste that are consumed by the electric power industrial sectors.

Notes: • Production categories are estimated; other data are preliminary. • Values are derived from source data prior to rounding for publication. • Totals may not equal sum of components due to independent rounding.
Sources: Tables 7.1, 7.2, and 7.3.

Figure 1.13 United States coal flow diagram for 2007.

(Quadrillion Btu)

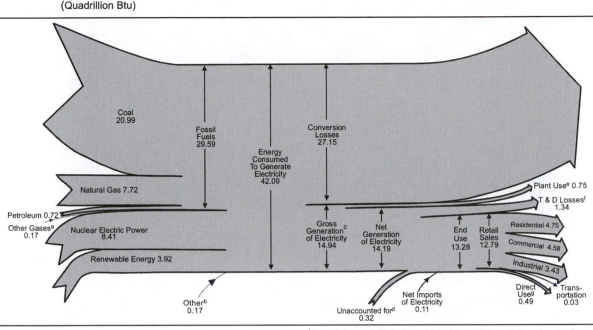

Figure 1.14 United States electricity flow diagram for 2007.

[a] Blast furnace gas, propane gas, and other manufactured and waste gases derived from fossil fuels.

[b] Batteries, chemicals, hydrogen, pitch, purchased steam, sulfur, miscellaneous technologies, and non-renewable waste (municipal solid waste from non-biogenic sources, and tire-derived fuels).

[c] Estimated as net generation divided by 0.95.

[d] Data collection frame differences and nonsampling error. Derived for the diagram by subtracting the "T & D Losses" estimate from "T & D Losses and Unaccounted for" derived from Table 8.1.

[e] Electric energy used in the operation of power plants, estimated as 5 percent of gross generation.

[f] Transmission and distribution losses (electricity losses that occur between the point of generation and delivery to the customer) are estimated as 9 percent of gross generation.

[g] Use of electricity that is 1) self-generated, 2) produced by either the same entity that consumes the power or an affiliate, and 3) used in direct support of a service or industrial process located within the same facility or group of facilities that house the generating equipment. Direct use is exclusive of station use.

Notes: • Data are preliminary. • See Note, "Electrical System Energy Losses," at the end of Section 2. • Values are derived from source data prior to rounding for publication. • Totals may not equal sum of components due to independent rounding.

Sources: Tables 8.1, 8.4a, 8.9, and A6 (column 4).

applications. Coal is the one energy source that the United States does not have to import. However, the extensive use of coal for electric generation poses significant environmental problems.

Although an end-point energy use rather than an energy source, an examination of the electricity flow in the United States is appropriate. Figure 1.14 presents the electricity flow diagram for 2007; the numbers in the figure are in quads (10^{15} Btu). The conversion factor is 3412 Btu = 1 kWh. The left-hand side delineates the input energy, including nuclear electric power. Coal is the dominant (71 percent) fossil fuel source of energy for electricity generation in the United States. The right-hand side of the diagram breaks down the end-point energy usages, including transmission and distribution losses (about 9 percent). With 42.09 quad consumed to generate 14.94 quad of electricity, the overall thermal efficiency of electricity generation is 35 percent. Hence, of the 42.09 quad of energy used to generate electricity in the United States in 2007, 27.15 quad represents conversion losses.

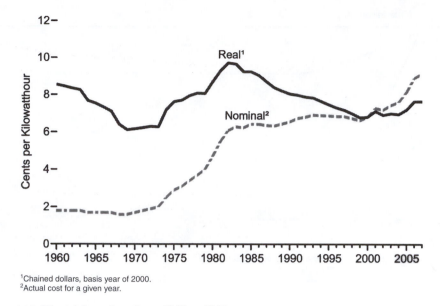

¹Chained dollars, basis year of 2000.
²Actual cost for a given year.

Figure 1.15 Electricity prices from 1960 to 2007.

In a fashion similar to Figure 1.11, Figure 1.15 tracks the real and nominal cost (in cents) of 1 kWh of electricity from 1960 until 2007. Real cents are the chained cents based on the cent in 2000, whereas nominal cents are the actual cost during a given year. In 1960 the real cost of 1 kWh of electricity was about 8.5 cents, compared to about 8 cents in 2007. Indeed, except for a few years in the 1980s, the real cost of 1 kWh has been less than in 1960. From 1980 until 1998, the real cost of 1 kWh of electricity monotonically declined. As with motor gasoline, the declining real cost of electricity during the prosperous years of the 1990s, relative to inflation, provided no economic impetus for conservation or exploration of alternative sources.

1.5   WORLDWIDE ENERGY USE

Although this chapter concentrates on the energy scenario in the United States, an examination of energy usage on a worldwide basis will enhance understanding of the global nature of the energy problem. Figure 1.16 shows the energy utilization for the world and for the countries with the highest energy consumption from 1996 to 2005. All data are presented in quads. The increases in energy use worldwide and by China are evident in the figure. The energy usage in Russia declined slightly, and energy usage by the United States increased, but not as rapidly as for China. The energy problems of the United States are exacerbated by the increasing demand for energy worldwide, and especially in countries with rapidly expanding economies.

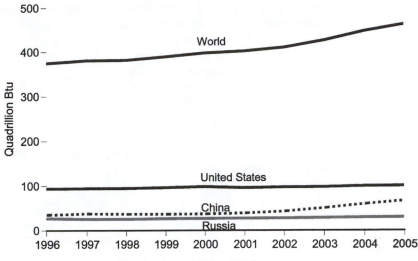

Figure 1.16 World energy utilization from 1996 to 2005.

EFFICIENCIES

The usual definition of the thermal efficiency of a heat engine is

$$\eta = \frac{\text{Power}_{\text{out}}}{\text{Power}_{\text{in}}} \tag{1-1}$$

The following three consequences of the Second Law of Thermodynamics are useful for assigning limits to the efficiency of a heat engine:

1. No heat engine can be more efficient than an externally reversible engine operating between the same temperature limits.

2. All externally reversible heat engines operating between the same temperature limits have the same efficiency.

3. The efficiency of any externally reversible heat engine operating between temperatures of T_H (high temperature) and T_L (low temperature) is given by the Carnot efficiency.

$$\eta_{\text{Carnot}} = \frac{T_H - T_L}{T_H} = 1 - \frac{T_L}{T_H} \tag{1-2}$$

The Carnot efficiency, η_{Carnot}, thus provides a bound on the maximum efficiency that can be obtained by any heat engine.

Tester et al. (2005) provide a useful summary of typical efficiencies for a number of components and devices associated with energy use. Table 1.2, adapted from Tester et al. (2005), presents the efficiency ranges for a number of energy system components, lighting sources, and biological systems. Additionally, Table 1.2 indicates the energy conversion path for each entry. For example, for large gas turbines, the energy conversion path is chemical to mechanical (c→m), and the nominal efficiency range is 35–40 percent. Some large components possess efficiencies above 90 percent, but many widely used components—internal combustion engines, for example—have efficiencies less than 30 percent. From a historical perspective, the reason for the demise of the steam locomotive is evident.

Much of the electricity generated is used for lighting. As Table 1.2 demonstrates, even the most efficient lighting source, high-pressure sodium, is only 15–20 percent efficient. Incandescent lighting is a woeful 2–5 percent efficient. Biological systems used for food production—milk or beef, for example—also exhibit low efficiencies. Locally, photosynthesis efficiency does not exceed 5 percent of the incident sunlight, but the global mean is much lower.

TABLE 1.2 Efficiencies of selected components and biological systems

Component	Energy Conversion Path	Efficiency (percent)
Large electric generators	m → e	98–99
Large power plant boilers	c → t	90–98
Large electric motors	e → m	90–97
Home natural gas furnaces	c → t	90–96
Drycell batteries	c → e	85–95
Waterwheels (overshot)	m → m	60–85
Small electric motors	e → m	60–75
Large steam turbines	t → m	40–45
Wood stoves	c → t	25–45
Large gas turbines	c → m	35–40
Diesel engines	c → m	30–35
Photovoltaic cells	r → e	20–30
Large steam engines	c → m	20–25
Internal combustion engines	c → m	15–25
Steam locomotives	c → m	3–6
Light Sources		
High-pressure sodium lamps	e → r	15–20
Fluorescent lights	e → r	10–12
Incandescent light bulbs	e → r	2–5
Paraffin candles	c → r	1–2
Biological Systems		
Milk production	c → c	15–20
Broiler production	c → c	10–15
Beef production	c → c	5–10
Local photosynthesis	r → c	4–5
Global photosynthesis	r → c	0.3

Energy conversion path labels: c = chemical, e = electrical, m = mechanical, r = radiant, t = thermal.

What Table 1.2 ultimately indicates is that much engineering effort is needed to reduce energy utilization by improving the efficiencies of various components.

1.7 CLOSURE

The information in this chapter introduces the purpose of this book. If we are to meet the increasing energy demands of the United States and the world, the use of alternative energy sources and alternative uses of existing energy sources must be considered. The remaining chapters examine fundamental principles and facts concerning a wide variety of alternative energy sources and alternative energy utilization schemes. The level of the coverage for a topic is, in most instances, fundamental. To perform detailed engineering work on most topics represented in this book, additional technical information will be needed. However, the material presented herein provides an introduction to and overview of many alternative energy scenarios.

REVIEW QUESTIONS

1. What is a quad?
2. What is the difference between energy and power?
3. What was the total energy usage in the United States in 2007?
4. In 2007 in the United States, what percentage of coal production was used to generate electricity?
5. How much renewable energy was used in the United States in 2007? What were the dominant sources of renewable energy?
6. How much energy used in the United States in 2007 was imported?
7. What has happened to United States energy imports since the 1970s?
8. What does EIA stand for?
9. In 2007, what percentages of U.S. energy were used in residential, commercial, industrial, and transportation applications?
10. From 1980 through 1995, in terms of inflation-adjusted dollars, what happened to the prices of gasoline and electricity?
11. What has happened to world energy usage since the 1970s?
12. What is the Carnot efficiency? What is its importance?
13. In comparison to most mechanical devices, how efficient is photosynthesis?

REFERENCES

Energy Information Administration (EIA). 2008. *Annual Energy Review 2007*. Washington, DC: U.S. Dept. of Energy.

Tester, J. W., Drake, E. M., Driscoll, M. J., Golay, M. W., and Peters, W. A. 2005. *Sustainable Energy*. Cambridge, MA: MIT Press.

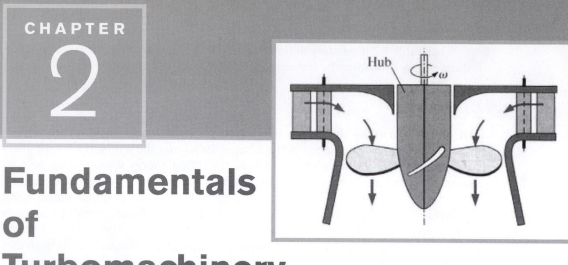

CHAPTER 2

Fundamentals of Turbomachinery

2.1 DEFINITION OF A TURBOMACHINE

Two of the important categories of renewable, alternative energy sources are wind energy and hydroelectric energy, and one frequently used alternative application for fossil fuels is the gas or combustion turbine. All of these devices are classified as turbomachines, and the operating principle is similar for all. The purpose of this chapter is to develop a basic understanding of and nomenclature for turbomachines. The definition of a turbomachine is a good place to start:

Turbomachine A device in which *energy is transferred* either to or from a *continuously flowing fluid* by the *action* of one or more *moving blade rows*.

The italicized words in the definition are important. Energy transfer, not energy transformation, takes place in a turbomachine. Since the fluid is continuously flowing, a turbomachine is not a positive-displacement device. The energy transfer that takes place is caused by the action of moving blade rows.

2.2 TURBOMACHINE CLASSIFICATIONS

Turbomachines are often classified as compressors (or pumps) and turbines. A compressor, pump, or fan is a turbomachine in which energy is added to the fluid as the fluid traverses the rotor. In a turbine, energy is extracted from the fluid as the fluid traverses the rotor.

Another classification is based on the path of the fluid as it traverses the rotor. A turbomachine can be a radial device, an axial device, or a mixed device. As the names imply, in a radial device the path of the fluid is in the radial direction, in an axial device the path of the fluid is in the axial direction, and in a mixed device the

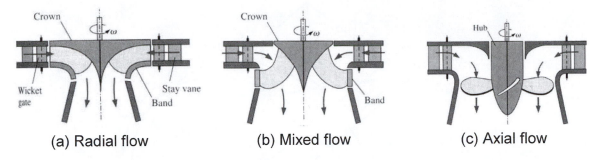

Figure 2.1 Types of turbomachines according to the fluid path (Cengel and Cimbala, 2006, used with permission).

fluid path has both radial and axial components. Figure 2.1 illustrates some common turbomachines exhibiting radial, axial, and mixed characteristics. In the radial device, Figure 2.1(a), the flow enters the rotor radially, and in the mixed device, Figure 2.1(b), the flow enters the rotor with both a radial and an axial component of velocity. The axial device, Figure 2.1(c), a propeller-type turbine, exhibits axial flow, and the flow both enters and leaves the rotor with axial velocity but little radial velocity.

2.3 TURBOMACHINE ANALYSIS

Figure 2.2 is the control volume that will be used to develop fundamental relations for a turbomachine. The control volume is rotating at an angular velocity, ω, about the shaft. Face 1 represents the inlet control surface, and face 2 represents the control surface at the exit. At an arbitrary location on the inlet control surface the absolute velocity vector, V_1, is shown. V_2 represents the absolute velocity at the exit. In this context, the absolute velocity is measured with respect to a stationary observer. The arbitrary points are located at radial distances r_1 and r_2 from the axis of rotation. Three mutually perpendicular velocity components are defined: V_a, the axial component; V_u, the tangential component; and V_m, the radial (or meridional) component. These components completely resolve the absolute velocity vectors. The components' names are also descriptive of their directions: V_a is in the axial direction and is collinear with the axis of rotation. Turbomachinery analysis is greatly facilitated by using the velocity components.

Conservation of angular momentum is the fundamental principle used in analyzing a turbomachine. For a control volume, conservation of angular momentum appears as

$$\sum \overline{M} = \frac{\partial}{\partial t} \int_{CV} \rho \, (\overline{r} \times \overline{V}) \, d\,\text{Vol} + \int_{CS} \rho \, (\overline{r} \times \overline{V}) \, \overline{V} \cdot \overline{dA} \tag{2-1}$$

In words, Eq. (2-1) states that the vector sum of moments acting on the control volume is equal to the time rate of change of angular momentum within the control volume plus the net efflux of angular momentum across the control surface. However,

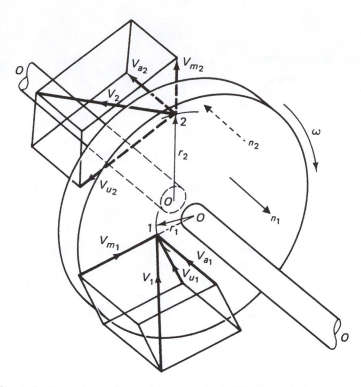

Figure 2.2 Control volume for turbomachinery analysis (Hodge and Taylor, 1999, used with permission).

what is of interest for a turbomachine is the moment about the axis of rotation. Consider, as in Figure 2.3, a schematic representation of the velocity components at an arbitrary point located a distance r from the axis of rotation. The x, y, and z axes as well as the unit vectors i, j, and k are indicated in the figure.

The vectors $\bar{r}$ and $\overline{V}$ then become

$$\bar{r} = r\,i \tag{2-2}$$

$$\overline{V} = V_m i + V_u j + V_a k \tag{2-3}$$

and the cross product, $\bar{r} \times \overline{V}$, is defined as

$$\bar{r} \times \overline{V} = \begin{vmatrix} i & j & k \\ r & 0 & 0 \\ V_m & V_u & V_a \end{vmatrix} = r\,V_u\,k - r\,V_a\,j \tag{2-4}$$

The moment or torque of interest is that about the z axis, the axis of rotation. As Eq. (2-4) indicates, the only component of the cross product that produces angu-

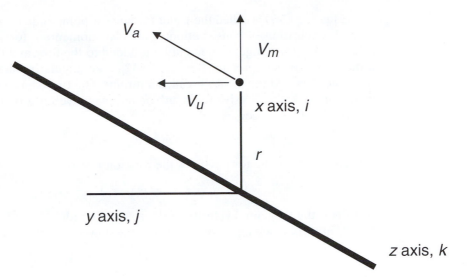

Figure 2.3 Schematic representation of the velocity components at a point.

lar momentum about the z axis is $r\,V_u$. Using τ as the summation of moments about the axis of rotation and invoking the steady state, conservation of angular momentum reduces to

$$\tau = \int_{CS} \rho\, r V_u\, \overline{V} \cdot \overline{dA} \tag{2-5}$$

For a control volume, as in Figure 2.2, with an inlet and an exit, Eq. (2-5) becomes

$$\tau = -\int_{inter} \rho\, r V_u\, V_a\, dA + \int_{exit} \rho\, r V_u\, V_a\, dA \tag{2-6}$$

Since n_1, the unit outward normal for the inlet (as indicated in Figure 2.2), is collinear with the z axis, $\overline{V} \cdot \overline{dA}$ is equal to $-V_a\, dA$. The mean value of $r\,V_u$ is defined as $\overline{r\,V_u}$ such that

$$\int \rho r\, V_u V_a\, dA = \overline{r\,V_u} \int \rho\, V_a\, dA = \dot{m}\,\overline{r V_u} \tag{2-7}$$

If the assertion is made that no flow rate bypasses the rotor, then the mass flow rates entering and leaving the rotor are the same: $\dot{m} = \dot{m}_1 = \dot{m}_2$. The expression for torque about the axis of rotation takes the form

$$\tau = -\dot{m}\left(\overline{rV_u}\right)_1 + \dot{m}\left(\overline{rV_u}\right)_2 = \dot{m}\left[\left(\overline{rV_u}\right)_2 - \left(\overline{rV_u}\right)_3\right] \tag{2-8}$$

For many turbomachines, either $r\,V_u$ is constant across a control surface, or r or V_u is constant and the "bar" notation is dropped, so that

$$\tau = \dot{m}\left(r_2 V_{u_2} - r_1 V_{u_1}\right) \tag{2-9}$$

Equation (2-9) is called the Euler turbine (or pump) equation and is the fundamental application of conservation of angular momentum for a turbomachine. If $r_2 V_{u_2} > r_1 V_{u_1}$, then angular momentum is added to the flow, and the device is called a pump, fan, or compressor. If $r_2 V_{u_2} < r_1 V_{u_1}$, then angular momentum is extracted from the flow, and the device is called a turbine. The turbomachine sign convention is for torque to be positive for a turbine and negative for a pump. Thus, the usual form of Eq. (2-9) becomes

$$\tau = \dot{m}\left(r_1 V_{u_1} - r_2 V_{u_2}\right) \tag{2-10}$$

The power is the torque times the rotation rate or

$$\text{Power} = \tau \cdot \omega = \dot{m}\left(\omega\, r_1\, V_{u_1} - \omega\, r_2\, V_{u_2}\right) \tag{2-11}$$

But $r\,\omega$ is the tangential velocity of the rotor at a radius r and is assigned the variable U in turbomachinery notation. Thus, the power is expressed as

$$\text{Power} = \dot{m}\left(U_1\, V_{u_1} - U_2\, V_{u_2}\right) \tag{2-12}$$

The "head" change is the power per unit mass flow rate or

$$\frac{\text{Power}}{\dot{m}} = U_1\, V_{u_1} - U_2\, V_u \tag{2-13}$$

Equation (2-13) follows the turbomachinery protocol of defining head as the energy per unit mass (or power per unit mass flow rate).

Two velocities have been introduced: the absolute velocity, V, and the tangential rotor velocity, U. The relative velocity, V_r, is defined by the vector equation

$$\overline{V} = \overline{U} + \overline{V}_r \tag{2-14}$$

The graphical interpretation of Eq. (2-14), the velocity triangle, is provided in Figure 2.4. Since V_u is the tangential component of the absolute velocity, V_u is identified as the projection of V in the direction of U, that is, $V \cos\alpha$.

The angle V makes with U, α, is often called the "nozzle angle." The angle V_r makes with U, β, is the blade angle since V_r represents the velocity with respect to the rotor. The law of cosines for the inlet and exit velocity triangles can be written as

$$V_{r_1}^2 = U_1^2 + V_1^2 - 2\,U_1\, V_1 \cos(\alpha_1) = U_1^2 + V_1^2 - 2\,U_1\, V_{u_1} \tag{2-15}$$

$$V_{r_2}^2 = U_2^2 + V_2^2 - 2\,U_2\, V_2 \cos(\alpha_2) = U_2^2 + V_2^2 - 2\,U_2\, V_{u_2} \tag{2-16}$$

Subtracting Eq. (2-16) from Eq. (2-15) and dividing by 2 yields

$$U_1\, V_{u_1} - U_2\, V_{u_2} = \frac{1}{2}\left[\left(V_1^2 - V_2^2\right) + \left(U_1^2 - U_2^2\right) + \left(V_{r_2}^2 - V_{r_1}^2\right)\right] = H \tag{2-17}$$

which represents the change in head across a turbomachine expressed in terms of the absolute, relative, and rotor velocities—the components of the velocity triangle.

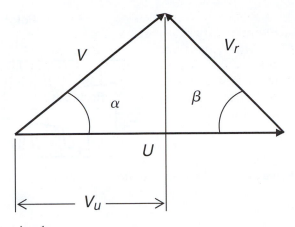

Figure 2.4 Velocity triangle.

An examination of each of the terms in the expression between the equal signs in Eq. (2-17) yields some insight into energy transfer in turbomachines. Consider the following:

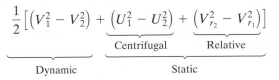

$$\frac{1}{2}\left[\left(V_1^2 - V_2^2\right) + \underbrace{\left(U_1^2 - U_2^2\right)}_{\text{Centrifugal}} + \underbrace{\left(V_{r_2}^2 - V_{r_1}^2\right)}_{\text{Relative}}\right]$$

The first term represents the change in absolute velocity and is called the dynamic effect. The remaining two terms, the centrifugal and the relative, form the static effect. The total change in head is the sum of the static and dynamic effects. Some useful physics can be deduced from Eq. (2-17) and the classification of dynamic, centrifugal, and relative. In a radial device the centrifugal effect is often dominant. For a radial-flow compressor, the head, H, is negative; thus, the centrifugal effect aids the energy transfer process if $U_2 > U_1$—meaning the exit radius should be greater than the inlet radius. A turbine is just the opposite, as the centrifugal effect should be positive, meaning that the inlet radius should be larger than the exit radius. This principle is explored in Example 3.3.

The ratio of the static effect to the total change in head is called the reaction and is defined as

$$R = \frac{\left(U_1^2 - U_2^2\right) + \left(V_{r_2}^2 - V_{r_1}^2\right)}{\left(V_1^2 - V_2^2\right) + \left(U_1^2 - U_2^2\right) + \left(V_{r_2}^2 - V_{r_1}^2\right)} = \frac{\frac{1}{2}\left[\left(U_1^2 - U_2^2\right) + \left(V_{r_2}^2 - V_{r_1}^2\right)\right]}{U_1 V_{u_1} - U_2 V_{u_2}} \quad (2\text{-}18)$$

If the reaction, R, is zero, then the change in head is due just to the dynamic effect— the change in absolute velocity across the rotor. Hence, the static change in head is zero, and the rotor can be open (no casing needed). $R = 0$ devices are often referred to as impulse machines since the total head change is due only to the dynamic effect.

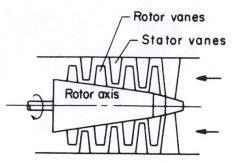

Figure 2.5 Staged axial-flow device schematic (Logan, 1993, used with permission).

Axial-flow turbomachines are devices in which motion of the fluid as it traverses the rotor is predominantly in the axial direction. A schematic of an axial flow compressor is shown in Figure 2.5. In an axial turbomachine with stages, rotating blade rows (rotors) are followed by stators, stationary blade rows in which energy transformation, not energy transfer, takes place. The stationary blades ahead of the first rotor are often called inlet guide vanes. For a given rotor or stator, the inlet and exit radii are essentially equal, so that $r = r_1 = r_2, U = U_1 = U_2$, and $V_a = V_{a_1} = V_{a_2}$. For axial flow, the expressions for torque, power, head, and reaction reduce to the following:

$$\tau = \dot{m}\,r(V_{u_1} - V_{u_2}) \tag{2-19}$$

$$\text{Power} = \dot{m}\,U(V_{u_1} - V_{u_2}) \tag{2-20}$$

$$H = U\,(V_{u_1} - V_{u_2}) \tag{2-21}$$

$$R = \frac{\left(V_{r_2}^2 - V_{r_1}^2\right)}{\left(V_1^2 - V_2^2\right) + \left(V_{r_2}^2 - V_{r_1}^2\right)} = \frac{\frac{1}{2}\left(V_{r_2}^2 - V_{r_1}^2\right)}{U\left(V_{u_1} - V_{u_2}\right)} \tag{2-22}$$

2.4   EXAMPLE PROBLEMS

Some examples exercising the concepts developed about turbomachines are in order.

Example 2.1

An axial-flow turbomachine possesses symmetrical velocity triangles as shown in Figure 2.6. For these conditions answer the following questions.

- How are the velocity components related?
- What is the reaction?
- Is this device a turbine or a compressor? Explain.

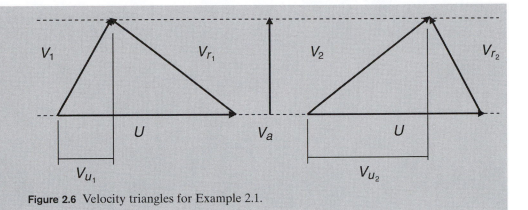

Figure 2.6 Velocity triangles for Example 2.1.

Solution The velocity triangles in Figure 2.6 can be envisioned as appropriate for the rotor arrangement of Figure 2.2. In an axial-flow device, the radial component of velocity is, by definition, zero, and the planes of the velocity triangles are collinear with the axial direction. The axial velocity is then the projection of the absolute velocity perpendicular to the tangential direction. For symmetrical velocity triangles, $V_1 = V_{r_2}$ and $V_{r_1} = V_2$. Substituting these relations into Eq. (2-22) results in

$$R = \frac{\left(V_1^2 - V_2^2\right)}{\left(V_1^2 - V_2^2\right) + \left(V_1^2 - V_2^2\right)} = \frac{1}{2}$$

Symmetrical velocity triangles in an axial-flow turbomachine result in $R = 0.5$. For many gas turbines or combustion turbines with staged axial-flow compressors or turbines, a design condition is frequently $R = 0.5$.

Since $V_{u_2} > V_{u_1}$ the device is a compressor (or pump) because the angular momentum is increased as the fluid traverses the rotor (i.e., energy is added to the fluid).

Although they are given here as part of the problem statement, the first step in solving any turbomachinery problem is to construct the velocity triangles. We will consider a more complex problem next.

Example 2.2

A centrifugal-flow turbomachine has the relative velocity components as indicated in Figure 2.7. Additionally, the following are specified:

$$U_1 = r_1 \omega = 3.75 \frac{m}{sec}$$

$$U_2 = r_2 \omega = 7.50 \frac{m}{sec}$$

$$\dot{m} = 57 \frac{kg}{sec}$$

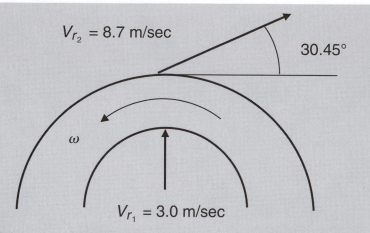

Figure 2.7 Schematic for Example 2.2.

Determine the power, classify the device as a turbine or compressor, and find the reaction.

Solution   The first step is to complete the velocity triangles. Consider the inlet velocity triangle. The values of U_1 and V_{r_1}, as well as the fact that V_{r_1} is radial, are known. Three pieces of information must be known in order to uniquely specify a velocity triangle. The inlet velocity triangle is a right triangle and, thus, is completely specified by the angle and the two velocity components. Likewise, two velocity components (U_2 and V_{r_2}) as well as an angle are known for the exit velocity triangle. Since $V_{r_2} \cos(30.45) = 7.5$ m/sec $= U_2$, the exit velocity triangle is also a right triangle. The velocity triangles appear as in Figure 2.8.

The velocity triangles must obey the vector expression $\overline{V} = \overline{U} + \overline{V}_r$. Since V_2 is perpendicular to U_2, $V_{u_2} = 0$; and with V_{r_1} perpendicular to U_1, $V_{u_1} = U_1$. The power can be computed as

$$\text{Power} = \dot{m}\,(U_1 V_{u_1} - U_2 V_{u_2}) = \dot{m}\,(U_1 U_1 - U_2 \cdot 0) = \dot{m}\,U_1^2$$

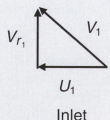

Inlet

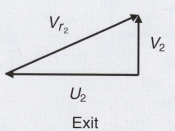

Exit

Figure 2.8 Velocity triangles for Example 2.2.

With numbers and units included, the power becomes

$$\text{Power} = 57 \, \frac{\text{kg}}{\text{sec}} \left(3.75 \, \frac{\text{m}}{\text{sec}} \right)^2 = 802 \, \frac{\text{kg m}^2}{\text{sec}^3} \frac{\text{N sec}^2}{\text{kg m}} = 802 \, \frac{\text{N m}}{\text{sec}} = 802 \, \frac{\text{J}}{\text{sec}} = 802 \, \text{W}$$

The device is a turbine since $r_1 V_{u_1} > r_2 V_{u_2}$. With U and V_r known for both velocity triangles, the absolute velocities can be computed and are $V_1 = 4.80$ m/sec and $V_2 = 4.41$ m/sec.

An alternative, but illuminating, approach to computing the power is to use an extension of Eq. (2-17) for the power:

$$\text{Power} = \dot{m}H = \frac{\dot{m}}{2} \left[\left(V_1^2 - V_2^2 \right) + \left(U_1^2 - U_2^2 \right) + \left(V_{r_2}^2 - V_{r_1}^2 \right) \right]$$

which becomes for this example

$$\text{Power} = \dot{m} \, H = \frac{57}{2} \, \frac{\text{kg}}{\text{sec}} \left[\left(4.8^2 - 4.41^2 \right) + \left(3.75^2 - 7.5^2 \right) + \left(8.7^2 - 3^2 \right) \right] \frac{\text{m}^2}{\text{sec}^2}$$

$$= 28.5(3.59 - 42.2 + 66.7) \, \text{W} = 802 \, \text{W}$$

Since the power from a turbine is positive, the dynamic contribution and the relative velocity contribution are positive, but the centrifugal effect is negative and reduces the power extracted from the fluid. This device is a radial outflow turbine that must overcome the negative contribution of the centrifugal effect in order to extract power. As discussed earlier, a radial inflow turbine would take advantage of the centrifugal effect.

With the velocity components known, the reaction, R, can be calculated directly as

$$R = \frac{\left(U_1^2 - U_2^2 \right) + \left(V_{r_2}^2 - V_{r_1}^2 \right)}{\left(V_1^2 - V_2^2 \right) + \left(U_1^2 - U_2^2 \right) + \left(V_{r_2}^2 - V_{r_1}^2 \right)}$$

$$= \frac{\left(3.75^2 - 7.5^2 \right) + \left(8.7^2 - 3^2 \right)}{\left(4.8^2 - 4.41^2 \right) + \left(3.75^2 - 7.5^2 \right) + \left(8.7^2 - 3^2 \right)} = 0.87$$

With little change in the absolute velocity, the dynamic effect is small. This value of R shows that in spite of the relative and centrifugal effects acting opposite, the static change is the dominant effect in this device.

Example 2.3

A radial inflow turbine, with radial entry and radial exit, possesses a mass flow rate of 2 kg/s. At the inlet to the turbine, the wheel speed, U, is 350 m/s and the rotor diameter is 0.3 m.

Find:

1. The power extracted from the stream
2. The rotational speed of the rotor
3. The torque on the shaft

Solution   In turbine nomenclature, "radial entry" means that the inlet relative velocity, V_r, is in the radial direction ($V_{r_1} \perp$ to U_1); "radial exit" means that the absolute exit velocity, V, is radial ($V_2 \perp$ to U_2). A schematic of the turbine with the velocity triangles is presented in Figure 2.9.

Because of the centrifugal effect, a radial turbine is more effective if the inlet is at the outer diameter of the rotor and the exit at the inner diameter. The inlet and outlet velocity triangles are sketched in the figure. The Euler equation for the power is as follows:

$$P = \dot{m}\left(U_1 V_{u_1} - U_2 V_{u_2}\right)$$

Since V_2 is perpendicular to U_2, $V_{u_2} = 0$, and since V_{r_1} is perpendicular to U_1, $V_{u_1} = U_1$. Examine the velocity triangles to verify these conditions. Hence, the power becomes

$$P = \dot{m}\, U_1 V_{u_1} = \dot{m}\, U_1^2$$

so that

$$P = 2\,\frac{\text{kg}}{\text{sec}} \left(350\,\frac{\text{m}}{\text{sec}}\right)^2 \frac{\text{sec}^2\,\text{N}}{\text{kg m}}\,\frac{\text{W sec}}{\text{N m}} = 245{,}000\,\text{W} = 245\,\text{kW} \qquad (2\text{-}34)$$

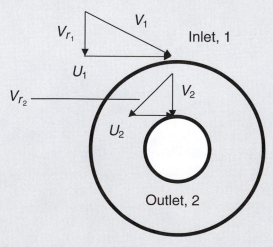

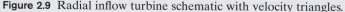

Figure 2.9 Radial inflow turbine schematic with velocity triangles.

The wheel speed, U, is the linear speed of the rotor, $r\omega$. Then

$$U = r\omega = \frac{D}{2}\omega$$

The angular velocity then becomes

$$\omega = \frac{2U}{D} = 2 \cdot 350 \frac{m}{sec} \cdot \frac{1}{0.3\ m} = 2333\ \frac{1}{sec}\ \frac{rev}{2\pi}\ \frac{60\ sec}{m} = 22{,}282\ RPM$$

The Euler form of the expression for the shaft torque is

$$\tau = \dot{m}(r_1 V_{u_1} - r_2 V_{u_2}) = \dot{m}\,r_1 V_{u_1} = \dot{m}\,r_1 U_1$$

$$= 2\ \frac{kg}{sec} \cdot \frac{0.3m}{2} \cdot 350 \frac{m}{sec}\ \frac{N\ sec^2}{kg\ m} = 105\ Nm \qquad (2\text{-}37)$$

The signs for both the power and the torque are positive, indicating that energy is extracted from the stream by the device; thus the device is as indicated in the problem statement, a turbine.

2.5   CLOSURE

This introduction to turbomachinery analysis and nomenclature is sufficient for the purposes of this book. For additional information, consult the references for this chapter. The next three chapters treat hydroelectric power, wind energy, and combustion turbines, all topics for which a fundamental understanding of turbomachinery is needed.

REVIEW QUESTIONS

1. What is the vector form of the steady-state conservation of angular momentum expression?
2. What is a turbomachine?
3. What velocity component contributes to torque about the shaft of a turbomachine?
4. What is the mean radius assumption?
5. What is the Euler pump or turbine equation?
6. How does a pump differ from a turbine?
7. What are the components of a velocity triangle? Explain the physical meaning of each.
8. What is reaction?
9. What is an impulse device?

EXERCISES

1. A centrifugal pump running at 3500 RPM pumps water at a flow rate of 0.01 m³/sec. The water enters axially and leaves the rotor at 5 m/sec relative to the blades, which are radial at the exit. If the pump requires 5 kW and is 67 percent efficient, estimate the basic dimensions (rotor exit diameter and width).

2. The dimensions of a centrifugal pump are provided in the table.

	Inlet	Outlet
Radius, mm	175	500
Blade width, mm	50	30
Blade angle, degrees	65	70

The pump handles water and is driven at 750 RPM. Calculate the increase in head and the power input if the flow rate is 0.75 m³/sec.

3. Velocity components are given for various turbomachines in the sketches. The following information is the same for all of the sketches: outer radius = 300 mm, inner radius = 150 mm, Q = 2 m³/sec, ρ = 1000 kg/m³, and ω = 25 rad/sec. Determine the torque, power, change in head, and reaction for each set of conditions.

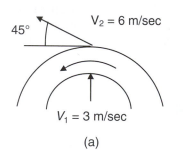

(a)

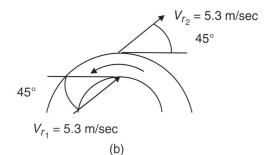

(b)

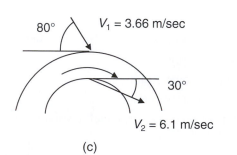

(c)

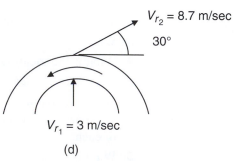

(d)

Figure P2.3

(continued on next page)

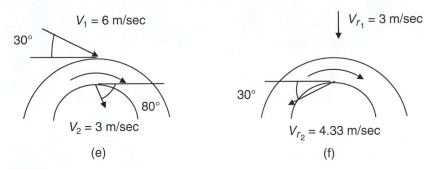

Figure P2.3 *(cont.)*
(Potter and Wiggert, 1997, *Mechanics of Fluid*, Prentice-Hall, used with permission)

REFERENCES

Cengel, Y. A., and Cimbala, J. M. 2006. *Fluid Mechanics Fundamentals and Applications*. New York: McGraw-Hill.

Dixon, S. L. 1998. *Fluid Mechanics and Thermodynamics of Turbomachinery*, 4th ed. Elsevier.

Logan, E. 1993. *Turbomachinery*, 2nd ed. New York: Dekker.

Shepherd, D. G. 1956 *Principles of Turbomachinery*. New York: Macmillan. This is the classic turbomachinery textbook.

Wright, T. 1999. *Fluid Machinery: Performance, Analysis, and Design.* Boca Raton, FL: CRC Press.

CHAPTER 3

Hydropower

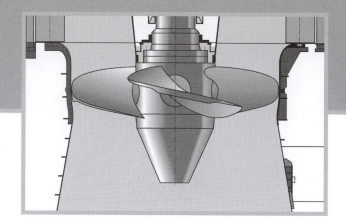

3.1   INTRODUCTION

Hydroelectric dams are among the best known and largest of man-made structures. Such dams have a long and successful history in the United States. In the early decades of the 1900s, many hydroelectric dams were constructed. By the 1930s, the United States was generating about 40 percent of its electricity by hydropower. In 2007, as indicated in Figures 1.9 and 1.14, that number declined to about 3 percent. Norway, Nepal, New Zealand, and Canada generate more than 50 percent of their electricity by hydropower.

For years hydroelectric power has been viewed as the most environmentally friendly of all electricity power sources. However, in recent times the environmental impacts of hydroelectric power have been viewed as less benign and more intrusive. Recent books by Leslie (2005) and Scudder (2005) address the environmental problems associated with large dams.

Large hydroelectric dams require large capital investments and a time scale of years to build and complete. Once complete, hydroelectric power has the lowest operating costs of any electricity source. Figure 3.1, taken from www.wvic.com/hydrofacts, provides a comparison of the total fuel, maintenance, and operating costs of elec-

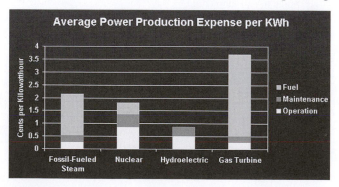

Figure 3.1 Costs of generating electricity from various energy sources (www.wvic.com, used with permission).

tricity generated from various sources. With a few exceptions, hydroelectric dams have long lifetimes and permit electrical generating output to be quickly varied. The United States has about 2400 power-producing dams.

3.2   EXAMPLES OF HYDROELECTRIC DAMS

Perhaps the most famous dam in the United States is the Hoover Dam on the Colorado River. Hoover Dam, 726 ft (221 m) high and 1244 ft (379 m) long, is one of the world's largest dams and was built between 1931 and 1936 (see Table 3.1).

Hoover Dam is a major supplier of hydroelectric power and allows for flood control and improved navigation. The electricity is distributed to California (56 percent), Nevada (25 percent), and Arizona (19 percent). Lake Mead, the largest reservoir in the United States, is impounded by the dam.

Figure 3.2 is a reproduction of an original planform for Hoover Dam and is included here for historical interest. Figure 3.3 presents a view of the dam from the tailrace looking upstream at the main structure.

TABLE 3.1 Hoover Dam statistics

726.4 feet (221 m) high
1244 feet (379 m) wide
660 feet (203 m) thick at the base
45 feet (13 m) thick at the top
4.5 years construction time
4.4 million yards of concrete used for construction
Construction started March 1931
17 generators
>4 billion kWh produced each year

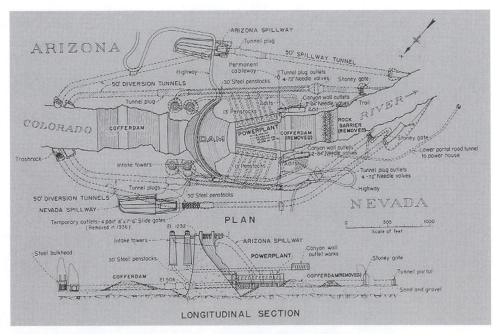

Figure 3.2 Reproduction of a planform drawing of Hoover Dam (U.S. Bureau of Reclamation).

Figure 3.3 View of Hoover Dam (U.S. Bureau of Reclamation).

The Itaipú hydroelectric power plant on the Parana River in South America is the largest operating hydroelectric dam in the world. Built from 1975 to 1991, the plant has a total production capacity of 12,600 MW (18 generating units of 700 MW each) and a reliable output of 75 million MWh a year. This output, in a single powerhouse, is enough to power most of California. The Itaipú dam contributes 25% of the energy supply for Brazil and 78% of that for Paraguay.

The materials list for constructing the Itaipú dam is impressive. The volume of iron and steel would be enough to build 380 Eiffel Towers, and the volume of concrete used represents 15 times the volume utilized to build the Channel Tunnel between France and England. Figure 3.4 provides an overview of the dam and reservoir.

Perhaps the most famous contemporary dam in the world is the Three Gorges Dam, being built in the Republic of China. As of the summer of 2008, most of the

Figure 3.4 Aerial view of the Itaipú hydroelectric power plant.

TABLE 3.2 Three Gorges Dam statistics

606 feet (185 m) high
6500 feet (1983 m) wide
377 ft (115 m) thick at the base
1993–2009 construction period
10.8 million tons of concrete used for construction
84 billion kWh to be produced each year

construction was complete, although much of the work of installing generators remains. Because of the number of people who had to be relocated and the area of land that will be flooded to form the reservoir, the Three Gorges Dam has generated much worldwide controversy. Nonetheless, the government of China has continued with its plans. Table 3.2 lists some statistics of the Three Gorges Dam.

The following websites are recommended for additional information on hydroelectric dams:

http://www.pbs.org/wgbh/buildingbig/wonder/structure/three_gorges.html
http://www.pbs.org/wgbh/buildingbig/wonder/structure/itaipunew_dam.html
http://www.pbs.org/wgbh/buildingbig/dam/index.html
http://www.c1b.com/dam2.jpg

3.3 HYDRAULIC ANALYSIS

A quantitative description of a hydroelectric dam involves two components: (1) a hydraulic analysis of the penstock, the fluid's path between the upper and lower water levels; and (2) an overview of the characteristics of turbines suitable for different combinations of flow rate and elevation change. This section develops the hydraulic analysis of the penstock flow, and the next section examines the different types of turbines suitable for use in hydroelectric installations.

Figure 3.2 illustrates the penstock arrangement of Hoover Dam. The intake towers feed the primary penstocks, each 30 feet in diameter. The primary penstocks then branch into several parallel penstocks that feed the turbines located in the power plant. The water traverses the turbines and is discharged downstream of the dam. The flow arrangement thus consists of an inlet, lengths of penstock, a turbine, and an exit. A general arrangement is shown schematically in Figure 3.5. The arrangement shows several valves, several fittings, the turbine, and penstock lengths.

The hydraulic analysis starts with the application of the conservation of energy equation, typically just called the energy equation, to the system. The general energy equation from station A to station B appears as

$$\frac{P_A}{\gamma} + \frac{V_A^2}{2g} + Z_A = \frac{P_B}{\gamma} + \frac{V_B^2}{2g} + Z_B + \sum_{n=1}^{N} \frac{V_n^2}{2g}\left(f_n \frac{L_n}{D_n} + K_n + C_n \cdot f_{T_n}\right) + W_t \frac{g_c}{g} \quad (3\text{-}1)$$

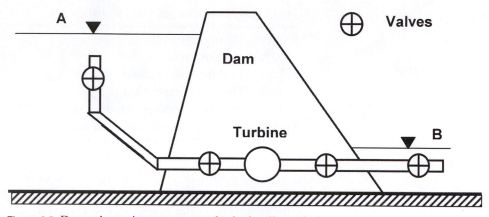

Figure 3.5 Dam schematic arrangement for hydraulic analysis.

where the variables are defined as follows:

P = pressure $\qquad$ V = velocity $\qquad$ Z = elevation

γ = specific weight $\quad$ f = friction factor $\quad$ L = pipe length

D = diameter $\qquad$ K = minor loss $\qquad$ C = minor loss

f_T = fully-rough friction factor $\qquad$ W_t = change in head across the turbine

N = number of pipe segments

g_c is used when the English Engineering system is employed and is defined as

$$g_c = 32.174 \; \frac{\text{ft lbm}}{\text{lbf sec}^2} = 1 \; \frac{\text{m kg}}{\text{N sec}^2} \qquad (3\text{-}2)$$

g_c can be used with the SI system, but it is an identity defined just in terms of the newton (N). In Eq. (3-1), the units of change in head of the turbines are in energy per unit mass (N m/kg or ft-lbf/lbm), in conformance with turbomachinery practice.

$\quad$ The first questions to be answered are where to initiate application of the energy equation and where to terminate application of the energy equation. For the case of a hydroelectric dam as illustrated in Figure 3.5, the free surfaces of the water levels (A and B in the figure) upstream and downstream are taken as the range of application of the energy equation. Since both surfaces are at atmospheric pressure, $P_A = P_B$; and since the velocities of the surfaces of the upstream and downstream reservoir are low, $V_A = V_B$. The change in head across the turbine becomes

$$W_t \frac{g_c}{g} = Z_A - Z_B - \sum_{n=1}^{N} \frac{V_n^2}{2g}\left(f_n \frac{L_n}{D_n} + K_n + C_n \cdot f_{T_n} \right) \qquad (3\text{-}3)$$

The change in head across the turbine is the difference in elevation minus the major and minor losses. A more useful expression results if the velocity is expressed in terms of the flow rate as

$$\frac{V^2}{2g} = \left(\frac{Q}{A}\right)^2 \frac{1}{2g} = \left(Q\frac{4}{\pi D^2}\right)^2 \frac{1}{2g} = \frac{8Q^2}{g\pi^2 D^4} \tag{3-4}$$

So that Eq. (3-3) becomes

$$W_t \frac{g_c}{g} = Z_A - Z_B - \sum_{n=1}^{N} \frac{8Q^2}{g\pi^2 D_n^4}\left(f_n \frac{L_n}{D_n} + K_n + C_n \cdot f_{T_n}\right) \tag{3-5}$$

For a penstock system such as the one illustrated in Figure 3.5, the maximum flow rate (in the absence of a pump) occurs when the turbines extract no energy, that is, when the change in head across the turbine is zero. Thus, the major and minor losses are equal to the elevation difference. The maximum change in head across the turbine occurs when the flow rate vanishes. Since the power extracted by the turbine is

$$\text{Power}_{\text{extracted}} = \rho Q \cdot W_t \tag{3-6}$$

the power extracted by either the maximum flow rate or the maximum change in head is zero. So between these extremes there exists, for a penstock arrangement, a flow rate that will yield the maximum power extracted from the fluid. In general, the power extracted for penstock arrangement will be double valued with respect to flow rate.

In Eq. (3-5), values for the friction factor and fully-rough friction factor are needed. In introductory fluid mechanics courses, the Moody diagram is often used to present the functional dependence of the friction factor, f, on the Reynolds number, $\text{Re}_D = \rho V D/\mu$, and the relative roughness, ε/D. However, the Moody diagram is unwieldy for computer-based solutions, and a closed-form expression is desired. In the laminar regime, the usual expression is

$$f = \frac{64}{\text{Re}_D} \tag{3-7}$$

Several different representations are available for turbulent flow. In this textbook the Haaland equation, from Haaland (1983), is used.

$$f = \frac{0.3086}{\left[\log\left(\left(\frac{\varepsilon}{3.7D}\right)^{1.11} + \frac{6.9}{\text{Re}_D}\right)\right]^2} \tag{3-8}$$

Minor loss terms are sometimes expressed as equivalent lengths using the fully-rough friction factor, f_T, the asymptotic value of the friction factor for a given relative roughness. From the Haaland equation, the fully-rough friction factor becomes

$$f_T = \frac{0.3086}{\left[\log\left(\left(\frac{\varepsilon}{3.7D}\right)^{1.11}\right)\right]^2} \tag{3-9}$$

With the foregoing description as the basis for piping system problem solution formulation, some examples of the approach will be examined and discussed.

The minor loss representations, K and C in Eq. (3-5), are available in virtually any textbook on fluid mechanics and in specialized textbooks such as Hodge and Taylor (1999) and Cengel and Cimbala (2006). "*Flow of Fluids,*" Crane Company Technical Report No. 410 (1957), is frequently cited as a consistent source of major and minor loss information. Consider the following example.

Example 3.1

A proposed design for a hydroelectric project is based on a discharge of 0.25 m³/sec through the penstock and turbine as shown in Figure 3.6. The minor losses are negligible.

Determine the power in kW that can be expected from the facility, if the turbine efficiency is 0.85.

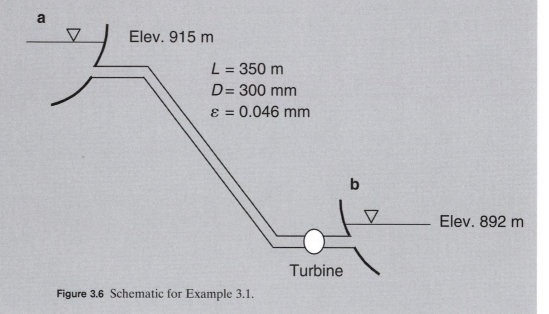

Figure 3.6 Schematic for Example 3.1.

Solution

Application of the general energy equation would yield Eq. (3-5), except that $K = C = 0$, since the minor losses are specified to be zero. In the classification of fluid mechanics hydraulics problems, finding the solution would be a category I problem since the flow rate is given and the head change is to be determined. Any fluid mechanics textbook has extensive information on the solution of problems such as this one. In this textbook, Mathcad will be used as the arithmetic engine. Figure 3.7 shows the Mathcad worksheet with the solution to this problem.

Mathcad is used as the arithmetic engine for the solution of Example 3.1. The first part of the worksheet defines the systems with inputs such as pressures, lengths, diameters, minor losses, and physical properties. The Mathcad solve block procedure is used to secure the solution. The solve block is initiated by the Given statement and ends with the Find command in which the unknown is specified. The result is that 25 kW is delivered from the turbine.

With some modifications, the procedure of Example 3.1, as illustrated in Figure 3-7, can be used for virtually any penstock problem. Additional details are available in Hodge and Taylor (1999) and Hodge (2006).

$\text{ORIGIN} \equiv 1$ Set origin for counters to 1 from the default value of 0.

Input the pipe geometry:

Diameter in mm   Length in m   Roughness in mm:

$$D := \begin{pmatrix} 0.3 \\ 0.3 \end{pmatrix} \cdot m \qquad L := \begin{pmatrix} 175 \\ 175 \end{pmatrix} \cdot m \qquad \varepsilon := \begin{pmatrix} 0.046 \\ 0.046 \end{pmatrix} \cdot mm \qquad \text{(commercial pipe)}$$

Input the system boundary (initial and end) conditions:

Pressures in Pa   Elevations in m:

$$\begin{pmatrix} P_a \\ P_b \end{pmatrix} := \begin{pmatrix} 0 \\ 0 \end{pmatrix} \cdot Pa \qquad \begin{pmatrix} Z_a \\ Z_b \end{pmatrix} := \begin{pmatrix} 915 \\ 892 \end{pmatrix} \cdot m$$

Input the loss coefficients:

K factor   Equivalent length   Number of pipes

$$K := \begin{pmatrix} .78 \\ 0 \end{pmatrix} \qquad C := \begin{pmatrix} 0 \\ 0 \end{pmatrix} \qquad N := \text{length}(D)$$

Input the fluid properties:

Density in kg/m^3 Kinematic Viscosity in m^2/s

$$\rho := 1000 \frac{kg}{m^3} \qquad \nu := 1.14 \, 10^{-6} \cdot \frac{m^2}{sec}$$

Input the flow rate in cms: Initial guess on turbine change in head:

$$Q := 0.25 \frac{m^3}{sec} \text{ Flow rate.} \qquad W_t := 15 \cdot newton \cdot \frac{m}{kg}$$

Define constants and adjust units for consistency:

$$g := 9.806 \frac{m}{sec^2} \qquad g_c := 1 \cdot \frac{m \cdot kg}{newton \cdot sec^2}$$

Define the functions for Reynolds number, fully-rough friction factor, and friction factor:

$$Re(q,d) := \frac{4 \cdot q}{\pi \cdot d \cdot \nu} \qquad f_T(d,\varepsilon) := \frac{0.3086}{\log\left[\left(\dfrac{\varepsilon}{3.7 \cdot d}\right)^{1.11}\right]^2}$$

$$f(q,d,\varepsilon) := \left| \begin{array}{l} \dfrac{0.3086}{\log\left[\dfrac{6.9}{Re(q,d)} + \left(\dfrac{\varepsilon}{3.7 \cdot d}\right)^{1.11}\right]^2} \quad \text{if } Re(q,d) > 2300 \\[2em] \dfrac{64}{Re(q,d)} \quad \text{otherwise} \end{array} \right.$$

(continued on next page)

Figure 3.7 Mathcad worksheet for Example 3.1.

The generalized energy equation is:

Given

$$W_t \cdot \frac{g_c}{g} = Z_a - Z_b - \sum_{i=1}^{N} \frac{8}{\pi^2} \cdot \frac{Q^2}{(D_i)^4 \cdot g} \cdot \left(f(Q, D_i, \varepsilon_i) \cdot \frac{L_i}{D_i} + K_i + C_i \cdot f_T(D_i, \varepsilon_i) \right)$$

$$W_t := Find(W_t)$$

$$W_t = 117.688 m^2 \sec^{-2} \qquad W_t = 117.688 newton \cdot \frac{m}{kg}$$

The head change across the turbine can be used to find the power generated.

$$Power := Q \cdot \rho \cdot W_t \qquad Power = 29.422 kW \qquad \text{Power extracted from fluid.}$$

$$Power_{gen} := 0.85 \cdot Power \qquad Power_{gen} = 25.009 kW \qquad \text{Power delivered by turbine.}$$

Figure 3.7 (cont.)

3.4 TURBINE SPECIFIC SPEED CONSIDERATIONS

The hydraulic solution addressed in Section 3.3 does not deal with the details of energy transfer in the turbine or with any details of turbine specification. Because of the wide range of elevation differences and flow rates associated with hydropower, additional information related to turbines is needed.

The starting point for more detailed considerations of turbines suitable for hydroelectric application is dimensional analysis. A complete set of dimensional variables for a turbomachine is presented in Table 3.3.

With ω, ρ, and D as repeating variables, the complete set of dimensionless parameters is as follows:

$$\text{Power coefficient:} \qquad C_W = \frac{W}{\rho \omega^3 D^5} \qquad (3\text{-}10)$$

TABLE 3.3 Turbomachine dimensionless analysis parameters

Parameter	Symbol	Dimensions
Power	W	ML^2/T^3
Speed	ω	$1/T$
Rotor diameter	D	L
Flow rate (volumetric)	Q	L^3/T
Pressure change	ΔP	M/LT^2
Density	ρ	M/L^3
Viscosity	μ	M/LT
Efficiency	η	—

$$\text{Pressure coefficient:} \qquad C_P = \frac{\Delta P}{\rho \omega^2 D^2} \qquad (3\text{-}11)$$

$$\text{Flow coefficient:} \qquad C_Q = \frac{Q}{\omega D^3} \qquad (3\text{-}12)$$

$$\text{Reynolds number:} \qquad \text{Re} = \frac{\rho \omega D^2}{\mu} \qquad (3\text{-}13)$$

If $\Delta P = \gamma H$, then

$$\text{Head coefficient:} \qquad C_H = \frac{gH}{\omega^2 D^2} \qquad (3\text{-}14)$$

Eliminating D and evaluating the resulting dimensionless parameter at the point of maximum efficiency result in the specific speed. Since the specific speed is derived from dimensionless products by eliminating the diameter, the specific speed is also dimensionless. General usage is to define the specific speed for a pump in terms of the flow rate and increase in head; for a turbine, the usual practice is to define its specific speed in terms of the power and the available head. Thus, the specific speeds of the two devices differ. The traditional definitions are

$$\text{Pump:} \qquad N_{\text{SP}} = \frac{\omega \sqrt{Q}}{(gH)^{3/4}} \qquad \text{at maximum efficiency} \qquad (3\text{-}15)$$

$$\text{Turbine:} \quad N_{\text{SP}} = \frac{\omega \sqrt{W/\rho}}{(gH)^{5/4}} \qquad \text{at maximum efficiency} \qquad (3\text{-}16)$$

The customary procedure in the United States is not to include the density, ρ, and the acceleration of gravity, g, and to express the specific speeds using the following units:

$$\omega \equiv \text{rpm} \qquad Q \equiv \text{GPM} \qquad H \equiv \text{ft} \qquad W \equiv \text{hp}$$

A consequence of these practices is that the specific speed values are no longer dimensionless. The customary *dimensional* versions of the specific speeds are

$$\text{Pump:} \qquad N_{\text{SP}} = \frac{\omega \sqrt{Q}}{(H)^{3/4}} \qquad (3\text{-}17)$$

$$\text{Turbine:} \quad N_{\text{SP}} = \frac{\omega \sqrt{W}}{(H)^{5/4}} \qquad (3\text{-}18)$$

The utility of using the specific speeds is illustrated in the following composite graphs. When expressed at the maximum-efficiency operating point, the specific speeds for the pump and the turbine define global machine geometries for optimal operating conditions. In both of these plots, the customary dimensional specific speeds are used. Although this chapter primarily treats turbines used in hydropower, the pump specific speed information is included for completeness and because some hydropower systems use pumps as well as turbines.

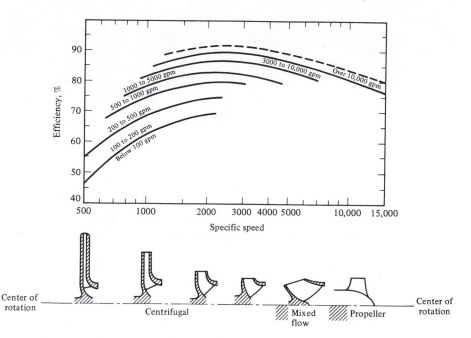

Figure 3.8 Pump specific speed relationship (Worthington Pumps, used with permission).

Figure 3.8 demonstrates why the specific speed is so useful in classifying turbo-machines such as pumps. The specific speed in this figure is based on the traditional United States procedure and uses Eq. (3-16). Below the abscissa, on which the pump specific speed is indicated, are cross-sectional schematics illustrating the salient features of pumps as a function of specific speed. At low values of N_{SP}, the flow rates are low and the increases in head are high—leading to centrifugal (radial) flow pumps, in which the centrifugal effect is dominant and the inlet area is relatively small. As the specific speed increases, the centrifugal effects diminish and the inlet areas increase—with the result that the pump geometry progresses from radial to mixed to axial (propeller). Hence, the specific speed defines the general geometry of a pump for a given speed, flow rate, and increase in head across the pump. Also presented in this figure are the nominal efficiencies expected from a well-designed pump as a function of flow rate and specific speed. Generally, as the size of a pump increases, the efficiencies increase since larger pumps are associated with higher flow rates. The reason is that viscous effects become more important as the pump size (and, hence, the flow rate) is decreased and as more of the flow is affected by viscous effects near surfaces (or walls).

Figure 3.9 is the specific speed relationship for a turbine. For hydroelectric applications, three types of turbines are in common use: the impulse turbine (or Pelton wheel) for low values of turbine specific speed, the Francis turbine for moderate specific speeds, and the Kaplan turbine for high specific speeds. The general cross-sectional shape of each type of device as well as expected efficiencies are included in the figure. Figure 3.10 shows a schematic of a Pelton wheel as well as details of the splitter

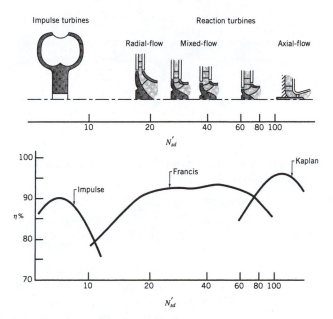

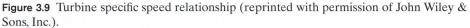

Figure 3.9 Turbine specific speed relationship (reprinted with permission of John Wiley & Sons, Inc.).

ridge in the "bucket." Flow is accelerated through a nozzle and impacts the moving buckets, the moving blade row, along the periphery of the Pelton wheel. Angular momentum is extracted at the splitter ridge, and the fluid is discharged. Figure 3.11 presents schematics of a radial-flow Francis turbine and a Kaplan turbine. Later in this section, additional information on these types of turbines will be presented.

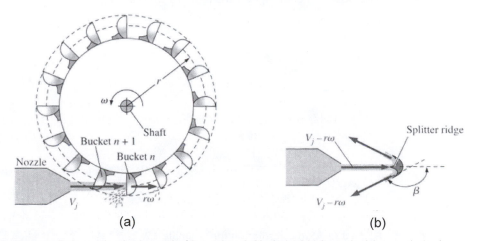

Figure 3.10 Pelton wheel schematic (Cengel and Cimbala, 2006, used with permission).

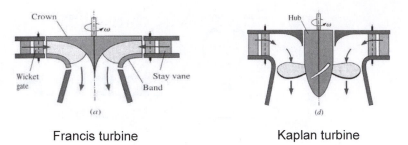

Francis turbine   Kaplan turbine

Figure 3.11 Francis turbine and Kaplan turbine schematics (Cengel and Cimbala, 2006, used with permission).

Example 3.2	For the conditions of Example 3.1, determine the type of turbine to be installed if the desired rotational speed is 1200 RPM.

Solution   The specific speed will be calculated using the United States customary approach with Eq. (3-17). The Mathcad solution is reproduced in Figure 3.12. Mathcad was chosen because it easily handles units. In Figure 3.12 N_S is the dimensionless version of the specific speed, and the results are shown in unitless (dimensionless) form in Mathcad. Using the customary definition, with units of RPM, hp, and ft, the specific speed is computed to be 31.247. In Mathcad, dividing a variable by a set of units results in a unitless (nondimensional) number. So using customary units and Eq. (3-17) yields the appropriate results, shown here as N_{cus}. For a customary value of 31.247, an examination of Figure 3.9 shows that the appropriate turbine type for this application is a Francis turbine and that an efficiency of 0.85 is reasonable.

$$\text{Power} := Q \cdot p \cdot W_s \qquad\qquad \text{Power} = -29.422 \text{ kw}$$
$$\text{Power gen} := 0.85 \cdot \text{Power} \qquad \text{Power gen} = -25.009 \text{ kw}$$

The power specific speed for this example is

$$\omega := 1200 \cdot \frac{2 \cdot \pi}{min}$$

$$N_S := \omega \cdot \frac{\left(\dfrac{-\text{Power}_{gen}}{\rho}\right)^{0.5}}{\left[g \cdot (Z_a - Z_b)\right]^{\frac{5}{4}}} \qquad\qquad N_S = 0.719 \qquad \text{(This is the "dimensionless" version.)}$$

$$N_{cus} := \frac{\omega}{2 \cdot \dfrac{\pi}{min}} \cdot \frac{\left(\dfrac{-\text{Power}_{gen}}{hp}\right)^{0.5}}{\left[\dfrac{(Z_a - Z_b)}{ft}\right]^{\frac{5}{4}}} \qquad N_{cus} = 31.247 \quad \text{(This is the customary version.)}$$

Figure 3.12 Mathcad solution for Example 3.2.

3.5   ENERGY TRANSFER IN TURBINES

The previous section introduced the turbine specific speed and discussed its importance in classifying the type of turbine required for specific applications. This section examines some details of energy transfer in various turbines. The Pelton wheel will be examined first.

Pelton wheels are appropriate for low values of turbine specific speed. Low values of turbine specific speed are associated with large head (elevation difference) availability and relatively small flow rates. Figure 3.13 presents a photograph of a Pelton wheel.

The water jet to the splitter ridge is supplied from a nozzle. In large Pelton wheels more than one nozzle may be used. Use of multiple nozzles permits the use of a smaller wheel for the same amount of power extracted, but multiple nozzles complicate the physical arrangement. Voith-Siemens markets a Pelton wheel with six nozzles. The complexity of a six-nozzle arrangement is illustrated in Figure 3.14. In this figure, four nozzles are drawn and two are indicated by dashes.

The analysis of a Pelton wheel is initiated by defining the velocity triangles. Figure 3.15 illustrates the splitter ridge and the inlet and outlet velocity triangles for a Pelton wheel. Since the jet enters and leaves the splitter ridge at the same radial location, $U_1 = U_2 = U$. A Pelton wheel is an impulse device with no casing; therefore, the reaction is zero. For an impulse device with $U_1 = U_2 = U$, Eq. (2-18) requires that $V_{r_1} = V_{r_2}$. The velocity triangles appear as in Figure 3.15 with the inlet velocity V_1 in the tangential direction and, thus, collinear with the wheel speed, U. The relative

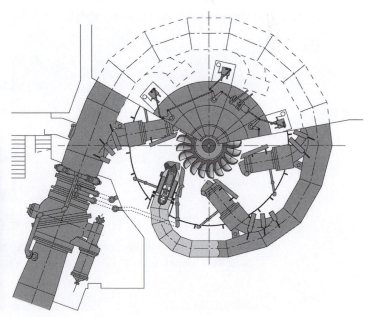

Figure 3.13 Photograph of a commercial Pelton wheel (Voith-Siemens Hydro).

Figure 3.14 A Pelton wheel with six nozzles (Voith-Siemens Hydro).

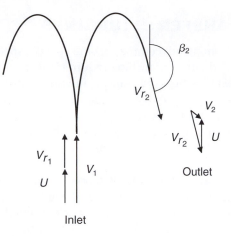

Figure 3.15 Splitter ridge schematic with velocity triangles.

velocity component V_{r_1} completes the inlet velocity triangle. The exit splitter ridge makes an angle β_2 ($< 180°$) with respect to V_1 so that the fluid can exit the splitter ridge without interfering with the incoming stream. With V_{r_2} and U known, the exit velocity triangle is defined.

The power extracted from the fluid is

$$P = \dot{m}\left(U_1 V_{u_1} - U_2 V_{u_2}\right) = \dot{m}U\left(V_{u_1} - V_{u_2}\right) \tag{3-19}$$

And from the velocity triangles,

$$V_{u_1} = V_1 = U + V_{r_1}$$
$$V_{u_2} = U + V_{r_2}\cos(\beta_2) \tag{3-20}$$

Substituting Eqs. (3-20) into Eq. (3-19) with $V_{r_1} = V_{r_2}$ yields

$$P = \dot{m}U\left(U + V_{r_1} - U - V_{r_2}\cos(\beta_2)\right)$$
$$= \dot{m}U\left(V_{r_1}\left(1 - \cos(\beta_2)\right)\right) = \dot{m}U\left(V_1 - U\right)\left(1 - \cos(\beta_2)\right) \tag{3-21}$$

The wheel speed, U, for the maximum power extracted is obtained from

$$\frac{dP}{dU} = 0 = \dot{m}\left(V_1 - U\right)\left(1 - \cos(\beta_2)\right) - \dot{m}U\left(1 - \cos(\beta_2)\right)$$
$$= \dot{m}\left(1 - \cos(\beta_2)\right)\left(\left(V_1 - U\right) - U\right) = V_1 - 2U \tag{3-22}$$

Equation (3-22) yields the important result that for maximum power extraction from a Pelton wheel, $U = V_1/2$. The maximum power extracted can be cast as

$$P_{\max} = \dot{m}\frac{V_1}{2}\left(V_1 - \frac{V_1}{2}\right)\left(1 - \cos(\beta_2)\right)$$
$$= \dot{m}\left(\frac{V_1^2}{4}\right)\left(1 - \cos(\beta_2)\right) \tag{3-23}$$

Since β_2 is typically very near 180°, $(1 - \cos(\beta_2))$ is very close to 2 and

$$P_{max} = \dot{m}\frac{V_1^2}{2} \tag{3-24}$$

Equation (3-24) is a simple expression, but it contains many implications for Pelton turbines. Implicit in the expression is the relationship between the velocity available at the exit of the nozzle, the inlet velocity, and the wheel speed. Also, the relationship between power and the inlet velocity is quadratic. The inlet velocity depends on the elevation difference available: the larger the difference, the higher the velocity. High inlet velocities arise only because of large elevation differences— hence the stipulation that Pelton wheels are appropriate for dams with high available heads. An example will prove useful.

Example 3.3

What power, in kilowatts, can be developed by the Pelton wheel shown in Figure 3.16 if the generator efficiency is 0.85 percent? The total minor losses are given as $K = 5.25$ and $C = 1500$. What should be the angular speed of the wheel for maximum efficiency? What is the power specific speed? What is the torque?

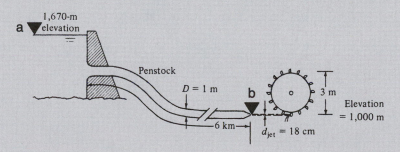

Figure 3.16 Schematic for Example 3.3 (reprinted with permission of John Wiley & Sons).

Solution For an arrangement such as in Figure 3.16, the energy equation must be applied between the free surface of the upper reservoir and the exit plane of the nozzle that feeds the Pelton turbine. The energy equation becomes

$$\frac{P_a}{\gamma} + \frac{V_a^2}{2g} + Z_a = \frac{P_b}{\gamma} + \frac{V_b^2}{2g} + Z_b + \frac{V^2}{2g}\left(f\frac{L}{D} + K + C \cdot f_T\right)$$

As with the earlier hydraulic example, $P_a = P_b$ and $V_a = 0$. Unlike Example 3.1, V_b is not zero but rather the nozzle exit velocity, V_{noz}, and since the Pelton turbine is downstream of b, no energy is extracted by the turbine between a and b. The reduced form of the energy equation becomes

$$Z_a = \frac{V_b^2}{2g} + Z_b + \frac{V^2}{2g}\left(f\frac{L}{D} + K + C \cdot f_T\right)$$

or

$$\frac{V_b^2}{2g} = Z_a - Z_a - \frac{V^2}{2g}\left(f\,\frac{L}{D} + K + C\cdot f_T\right)$$

This equation is interesting in that it states that the kinetic energy available for the Pelton turbine is equal to the elevation difference (potential energy) minus the major and minor losses. However, a more useful expression for solving the problem is obtained if the velocities are cast in terms of the flow rate using Eq. (3-4); the result is

$$\frac{8Q^2}{g\pi^2 D_b^4} = Z_a - Z_b - \frac{8Q^2}{g\pi^2 D^4}\left(f\,\frac{L}{D} + K + C\cdot f_T\right)$$

Put in the form of Example 3.1, the equation becomes

$$0 = Z_a - Z_b - \frac{8Q^2}{g\pi^2 D_B^4} - \frac{8Q^2}{g\pi^2 D^4}\left(f\,\frac{L}{D} + K + C\cdot f_T\right)$$

The Mathcad worksheet used to solve Example 3.1 was modified for the solution to this problem. The solution is presented in Figure 3.17. The first part of the worksheet presents the solution for the flow rate. The flow rate solution is a category II pipe flow problem, and the result is 2.79 m³/sec. The remainder of the worksheet is concerned with the Pelton turbine computations. The power extracted, 16,780 kW, is computed using Eq. (3-23), and the wheel rotational speed, 349 rpm, is calculated from $V_1/2$.

$ORIGIN \equiv 1$ Set origin for counters to 1 from the default value of 0.

Input the pipe geometry:

Diameter in m   Length in m   Roughness in mm   Nozzle Diameter

$D := 1\cdot m$ $L := 6000\ \cdot m$ $\varepsilon := 0.046\cdot mm$ (commercial pipe) $D_{noz} := 18\cdot cm$

Input the system boundary (initial and end) conditions:

Pressures in Pa   Elevations in m:

$$\begin{pmatrix} P_a \\ P_b \end{pmatrix} := \begin{pmatrix} 0 \\ 0 \end{pmatrix}\cdot Pa \qquad \begin{pmatrix} Z_a \\ Z_b \end{pmatrix} := \begin{pmatrix} 1670 \\ 1000 \end{pmatrix}\cdot m$$

Input the loss coefficients:

K factor   Equivalent length

$K := 5.25$ $C := 1500$

Input the fluid properties:

Density in kg/m³   Kinematic viscosity in m²/s

$$\rho := 1000\cdot\frac{kg}{m^3} \qquad\qquad \nu := 1.14\cdot 10^{-6}\,\frac{m^2}{sec}$$

Input the flow rate in cms:

$$Q := 1\cdot\frac{m^3}{sec} \qquad\qquad \text{Initial guess on flow rate.}$$

(continued on next page)

Figure 3.17 Mathcad solution for Example 3.3.

Define the functions for Reynolds number, fully-rough friction factor, and friction factor:

$$Re(q,d) := \frac{4 \cdot q}{\pi \cdot d \cdot \nu} \qquad\qquad f_T(d,\varepsilon) := \frac{0.3086}{\left[\log\left[\left(\frac{\varepsilon}{3.7 \cdot d} \right)^{1.11} \right] \right]^2}$$

$$f(q,d,\varepsilon) := \left| \begin{array}{l} \dfrac{0.3086}{\left[\log\left[\dfrac{6.9}{Re(q,d)} + \left(\dfrac{\varepsilon}{3.7 \cdot d} \right)^{1.11} \right] \right]^2} \quad \text{if } Re(q,d) > 2300 \\[20pt] \dfrac{64}{Re(q,d)} \quad \text{otherwise} \end{array} \right.$$

The generalized energy equation is:

Given

$$0 \cdot m = Z_a - Z_b - \left(\frac{8}{\pi^2} \cdot \frac{Q^2}{D_{noz}^4 \cdot g} \right) - \left[\frac{8}{\pi^2} \cdot \frac{Q^2}{D^4 \cdot g} \cdot \left(f(Q, D, \varepsilon) \cdot \frac{L}{D} + K + C \cdot f_T(D, \varepsilon) \right) \right]$$

$$q := Find(Q)$$

$$q = 2.79 m^3 \cdot s^{-1} \qquad\qquad q = 1.674 \times 10^5 \frac{liter}{min} \qquad\qquad q = 737.167 \frac{gal}{sec}$$

The jet velocity is:

$$V_{jet} := \frac{q}{0.25 \cdot \pi \cdot D_{noz}^2} \qquad\qquad V_{jet} = 109.659 m \cdot s^{-1}$$

The kinetic energy available for the Pelton wheel is 0.5 V_{jet}^2; the available power is the available kinetic energy times flow rate. This assumes $(1 + \cos \beta_2 = 2)$.

$$Power := q \cdot \rho \cdot \frac{V_{jet}^2}{2} \qquad\qquad Power = 1.678 \times 10^4 \, kW$$

$$Power_{gen} := 0.85 \cdot Power \qquad\qquad Power_{gen} = 1.426 \times 10^4 \, kW$$

For maximum efficiency of a Pelton wheel, the blade speed is 0.5 of the jet speed.

$$U := 0.5 \cdot V_{jet} \qquad\qquad U = 54.83 \, m \cdot s^{-1} \qquad\qquad RPM := \frac{2 \cdot \pi \cdot rad}{min}$$

$$D_{wheel} := 3.0 \cdot m \qquad\qquad \omega := \frac{U}{0.5 \cdot D_{wheel}} \qquad\qquad \omega = 36.553 s^{-1}$$

$$Torque := \frac{Power_{gen}}{\omega} \qquad\qquad Torque = 3.902 \times 10^5 \, newton \cdot m \qquad\qquad \omega = 349.056 RPM$$

The power specific speed for this example is

$$N_{cus} := \frac{\omega}{RPM} \cdot \frac{\left(\dfrac{Power_{gen}}{hp} \right)^{0.5}}{\left(\dfrac{Z_a - Z_b}{ft} \right)^{\frac{5}{4}}} \qquad\qquad N_{cus} = 3.207 \quad \text{This is the customary version.}$$

Figure 3.17 *(cont.)*

The power specific speed is 3.207. An examination of Figure 3.9 confirms that the power specific speed is in the range expected for a Pelton turbine and that the specified efficiency is appropriate for this device.

For moderate values of turbine specific speed, the Francis turbine is the preferred type. Figure 3.18 is an illustration of a Francis turbine installation.

Francis turbines are the most frequently used of the three types. With the advent of large dams, Francis turbines have grown in size. In Figure 3.19, Voith-Siemens Hydro has documented the increase in size and output of the Francis turbines they have manufactured over the years.

The Kaplan turbine is the third type of device frequently used in hydropower applications. Figure 3.20 is a reproduction of a Kaplan turbine, and Figure 3.21 schematically illustrates a typical installation.

Both Francis and Kaplan turbines are reaction rather than impulse devices. The general approach to solving problems involving Francis or Kaplan turbines is to construct the velocity triangles and apply the Euler equation. The next example is typical of the approach used for these devices.

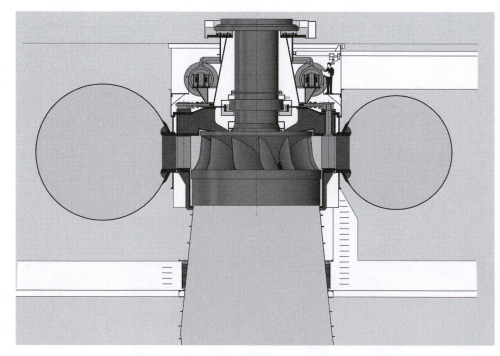

Figure 3.18 Francis turbine installation (Voith-Siemens Hydro).

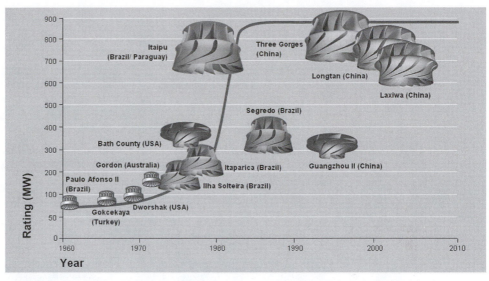

Figure 3.19 Increases in size of Voith-Siemens Francis turbines since 1960 (Voith-Siemens Hydro).

Figure 3.20 Kaplan turbine (Voith-Siemens Hydro).

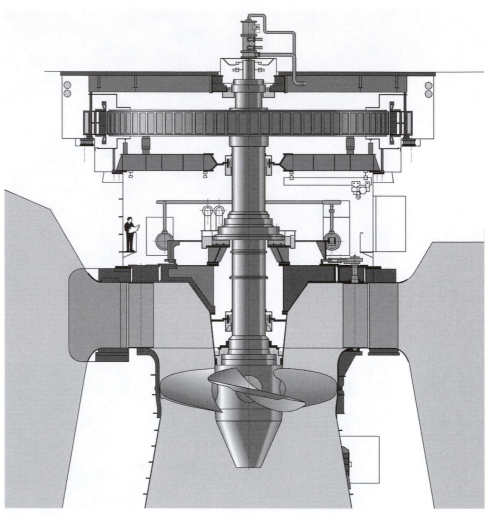

Figure 3.21 Kaplan turbine installation (Voith-Siemens Hydro).

Example 3.4	A 90 percent efficient radial-flow Francis turbine produces 111 MW at a speed of 167 RPM. The blade height is 0.732 m, and the inlet radius is 5.52 m. The flow exits the turbine in a radial direction.

(a) If the angle between the radial direction and the absolute velocity at the inlet is 30º, determine the volume flow rate.

(b) What is the change in head across the turbine?

Solution	This turbine has radial flow at the inlet and radial flow at the exit; thus, the flow is radial throughout the turbine. Figure 3.22 illustrates the general arrangement and the velocity triangles. Figure 3.23 shows a side view with the blade geometry defined.

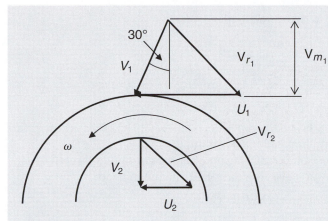

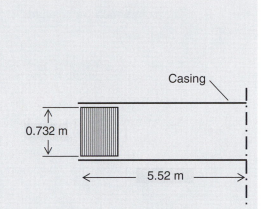

Figure 3.22 Schematic and velocity triangles for Example 3.4.

Figure 3.23 Turbine blade details.

The inlet radius and the speed of the rotor permit U_1 to be computed directly as

$$U_1 = r_1\omega = 5.52 \text{ m} \cdot 167 \frac{\text{rev}}{\text{min}} \frac{2\pi}{\text{rev}} \frac{\text{min}}{60 \text{ sec}} = 95.53 \frac{\text{m}}{\text{sec}}$$

The power extracted from the fluid is the same as the power input to the turbine and can be expressed as

$$P_{\text{in}} = \frac{P_{\text{out}}}{\eta} = \frac{111 \text{ MW}}{0.9} = 123.33 \text{ MW}$$

Since V_2 is radial, $V_{u_2} = 0$, and the expression for the power becomes

$$P_{\text{in}} = \dot{m} U_1 V_{u_1}$$

The flow rate into the turbine is the radial velocity times the free flow area, or

$$Q = 2\pi r_1 h_1 V_{m_1} = 2\pi \cdot 5.52 \text{ m} \cdot 0.732 \text{ m} \cdot V_{m_1} = 25.39 \text{ m}^2 \cdot V_{m_1}$$

Combining the previous two expressions yields

$$P_{\text{in}} = 123.33 \text{ MW} = \rho \cdot 25.39 \text{ m}^2 \cdot V_{m_1} U_1 V_{u_1} \tag{3-25}$$

But from the inlet velocity triangle, $V_{m_1} = V_{u_1} \cot(30°)$, so

$$123.33 \text{ MW} = \rho \cdot 25.39 \text{ m}^2 \cdot V_{m_1} U_1 V_{u_1} = 1000 \frac{\text{kg}}{\text{m}^3} 25.39 \text{ m}^2 \cdot V_{u_1} \cot(30°) 96.53 \frac{\text{m}}{\text{sec}} V_{u_1}$$

$$= 4.245 \cdot 10^6 \frac{\text{kg}}{\text{sec}} V u_1^2 \tag{3-26}$$

from which we conclude $V_{u_1} = 5.39$ m/sec and

$$V_{m_1} = 5.39\frac{m}{sec}\cos(30°) = 9.336\frac{m}{sec}$$

$$Q = 25.39\ m^2 \cdot 9.336\frac{m}{sec} = 237\frac{m^3}{sec} = 3.757 \cdot 10^6\frac{gal}{min} \qquad (3\text{-}27)$$

As might be expected from a Francis turbine that delivers 111 MW, the flow rate is high: 237 m³/sec or 3.757 million gallons per minute!

The change in head across the Francis turbine can be computed directly from the relationship between power, mass flow rate, and change in head:

$$\text{Power} = Q\Delta P = Q\rho g\,\Delta h$$

$$\Delta h = \frac{\text{Power}_{in}}{Q\rho g} = \frac{123.33\ \text{MW}\dfrac{10^6\,\text{W}}{\text{MW}}\dfrac{\text{J}}{\text{sec}\,W}\dfrac{\text{N m}}{\text{J}}}{237\dfrac{m^3}{sec}\,1000\dfrac{kg}{m^3}\,9.807\dfrac{m}{sec^2}}\dfrac{\text{kg m}}{\text{N sec}^2}$$

$$= 53.06\ m$$

Thus, the change in head across the Francis turbine is 53.06 m.

Example 3.4 is typical of the process used to work problems associated with turbomachinery devices. Much of the basis for the solution was developed in Chapter 2.

3.6   CLOSURE

This chapter has explored hydropower by examining some large hydroelectric sites, developing and utilizing the concept of specific speed, considering how a Pelton turbine (impulse device) works, and demonstrating how to use velocity triangles to solve turbine problems.

REVIEW QUESTIONS

1. Write the energy equation between points a and b of a pipe segment. Explain the meaning of each term.

2. What is the dimensional (traditional) representation of the specific speed for a turbine?

3. Why is the specific speed such an important parameter?

4. Why is the Pelton wheel considered an impulse device?

5. Why is the Francis turbine classified as a radial-flow device?

6. The Mississippi River has an enormous flow rate driven by small elevation differences. Would a Pelton wheel (impulse device), a Francis turbine, or a Kaplan turbine be the best choice for generating power from the river? Why?

7. Using Figure 3.9, explain the general turbine geometry as the specific speed varies from small values to large values.

EXERCISES

1. A turbine develops 15,500 hp with a decrease in head of 37 ft and a rotational speed of 106 RPM. What type of turbine is best suited for this application?

2. A Francis turbine is used in an installation to generate power. The power output is 115,000 hp under a head of 480 ft. The outer radius of the rotor is 6.75 ft; the inner radius is 5.75 ft. The rotational speed is 180 RPM, and the blade height is 1.5 ft. The turbine efficiency is 0.95 percent. The absolute velocity leaving the rotor is in the radial direction. Determine the angle that the absolute velocity entering the rotor makes with the radial direction.

3. A hydroelectric facility operates with an elevation difference of 50 m with a flow rate of 500 m³/sec. If the rotational speed is 90 RPM, determine the most suitable type of turbine and estimate the power output of the arrangement.

4. A Pelton wheel, located at an elevation of 200 m, is used to extract power from a reservoir at an elevation of 800 m. A penstock 1500 m long and 0.8 m in diameter connects the reservoir with the nozzle of the Pelton wheel. Minor losses are $K = 3$ and $C = 1500$. The Pelton wheel is 3 m in diameter, and the nozzle exit is 0.2 m in diameter. Determine the maximum power output of the Pelton wheel. Is a Pelton wheel the appropriate turbine type for this application?

5. A proposed design for a hydroelectric project is based on a discharge of 0.25 m³/sec through the penstock and turbine as illustrated in the accompanying figure. The minor losses are considered negligible.

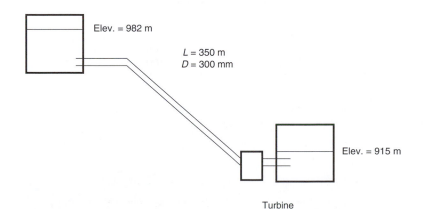

Figure P3.5

(a) Determine the power in kW that can be expected from the facility, if the turbine efficiency is 0.85.

(b) What type of turbine should be installed if the desired rotational speed is 1200 RPM?

6. A hydroelectric facility extracts power from a reservoir at an elevation of 800 m, discharging into a tailrace at an elevation of 200 m. A penstock 1500 m long and 0.8 m in diameter connects the reservoir to the tailrace. Minor losses are $K = 3$ and $C = 1500$. Determine the power extracted by the turbine if the flow rate is 2 m³/sec. If a Francis turbine is to be used, specify an appropriate speed.

7. A Pelton wheel is used to produce electricity in a hydroelectric facility. The radius of the wheel is 1.83 m, and velocity of the fluid exiting the 10-cm diameter nozzle is 102 m/sec. The exit blade angle, β, is 165°.

(a) Sketch the "bucket" of the Pelton wheel and the inlet and outlet velocity triangles.

(b) Calculate the flow rate.

(c) What is the speed in RPM for maximum power extraction?

(d) If the efficiency is 0.82, estimate the output shaft power.

(e) Calculate the power specific speed. Based on the result, are these conditions suitable for a Pelton turbine?

8. A 0.9 efficient Francis turbine produces 200 MW with a speed of 150 RPM. The blade height is 1 m, and the inlet radius of the rotor is 7 m. The flow exits the turbine in a radial direction.

(a) If the angle between the radial direction and the absolute velocity at the inlet is 30°, determine the volume flow rate.

(b) What is the change in head across the turbine?

9. A developer has constructed an elevated reservoir, as shown in the sketch, and estimates that a flow rate of 200 L/min is available on a continuous basis.

(a) Determine the power in kW that can be expected from the arrangement, if the turbine efficiency is 0.85.

(b) Discuss the implications of the results for the individual home owner interested in a small hydroelectric system.

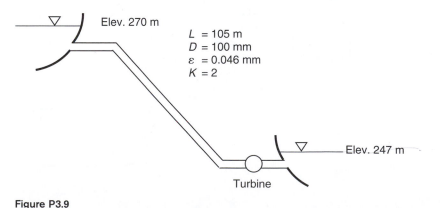

Elev. 270 m

$L = 105$ m
$D = 100$ mm
$\varepsilon = 0.046$ mm
$K = 2$

Elev. 247 m

Turbine

Figure P3.9

10. A 1-m diameter penstock is 10 km long and carries water to an impulse turbine. If the turbine/generator is 83 percent efficient, what power can be extracted if the elevation difference between the reservoir and the nozzle is 650 m? The jet (nozzle) diameter is 16 cm. What should the Pelton wheel diameter be for maximum efficiency if 360 RPM is to be maintained? What is the power specific speed in EE (English Engineering) units? Would this be a good application for an impulse turbine? Explain.

11. Turbines at the Conowingo plant on the Susquehanna River each develop 54,000 hp at 82 RPM under a head of 89 feet. What is the turbine type? Estimate the flow rate of each turbine.

12. Many areas of the country have reservoirs or waterways with locks and dams. An interesting problem is to evaluate the potential for hydroelectric generation at a local facility with a spillway. For example, would a hydroelectric facility be feasible at the Columbus (Mississippi) lock and dam on the Tennessee-Tombigbee waterway?

13. If a hydroelectric facility is to produce 100,000 hp with an elevation difference of 300 feet, what is a suitable speed (in RPM) if a Kaplan turbine, operating at high efficiency, is used?

REFERENCES

Cengel, Y. A., and Cimbala, J. M. 2006. *Fluid Mechanics: Fundamentals and Applications*. New York: McGraw-Hill.

"Flow of Fluids." 1957. Technical Paper No. 410. Chicago: Crane Company.

Haaland, S. E. 1983. "Simple and Explicit Formulas for the Friction Factor in Turbulent Flow," *Journal of Fluids Engineering* 103(5), 89–90.

Hodge, B. K. 2006. "A Unified Approach to Piping System Problems," *Computers in Education Journal* 16(2), 68–79.

Hodge, B. K., and Taylor, R. P. 1999. *Analysis and Design of Energy Systems*, 3rd ed. Upper Saddle River, NJ: Prentice-Hall.

Leslie, J. 2005. *Deep Water: The Epic Struggle over Dams, Displaced People, and the Environment*. New York: Farrar, Straus and Giroux.

Munson, B. R., Young, D. F., and Okishii, T. H. 2006. *Fundamentals of Fluid Mechanics*, 4th ed. New York: Wiley.

Scudder, T. 2005. *The Future of Large Dams: Dealing with Social, Environmental, Institutional, and Political Costs*. London: Earthscan.

CHAPTER 4

Wind Energy

4.1 INTRODUCTION

Wind energy, like solar energy, has captured much media attention recently. Indeed, wind energy has exhibited the most rapid growth of all renewable energy sources the last few years. As shown in Figure 1.9, wind energy accounted for 4 percent of the renewable energy used in the United States in 2007. Wind energy is also an increasingly important part of the energy mix in western Europe. The purposes of this chapter are to develop the fundamental principles of wind energy and to impart a quantitative understanding of wind energy.

Devices to harvest wind energy are available in many different configurations. A number of possible configurations are illustrated in Figures 4.1 and 4.2. Fundamental designations of a wind energy device include the horizontal-axis wind turbine (HAWT) shown in Figure 4.1 and the vertical-axis wind turbine (VAWT) in Figure 4.2. The designation depends simply on the axis of rotation; HAWT devices rotate in the horizontal plane and VAWT devices rotate in the vertical plane. HAWTs are more common than VAWTs, but horizontal devices must have a mechanism—a yaw control—to keep them pointed into the wind. VAWTs, on the other hand, do not need a yaw control.

Many of the wind energy devices illustrated in Figures 4.1 and 4.2 are speculative; they have little or no demonstrated functionality and are not available commercially. The most common configurations for the HAWT are the two- and three-bladed turbines, the windmill, and the sail wing. The two- and three-bladed wind turbines can be either upwind or downwind , with upwind being the most common. In terms of installed kilowatts, the total capacity of the HAWTs greatly exceeds that of the VAWTs. Figure 4.3 presents photographs of horizontal- and vertical-axis wind turbines; Figure 4.3(a) shows a large HAWT suitable for commercial power generation, and the turbine in Figure 4.3(b) is a much smaller device, suitable for a residence. A detailed examination of the characteristics of commercially available horizontal wind turbines will be provided later in this chapter.

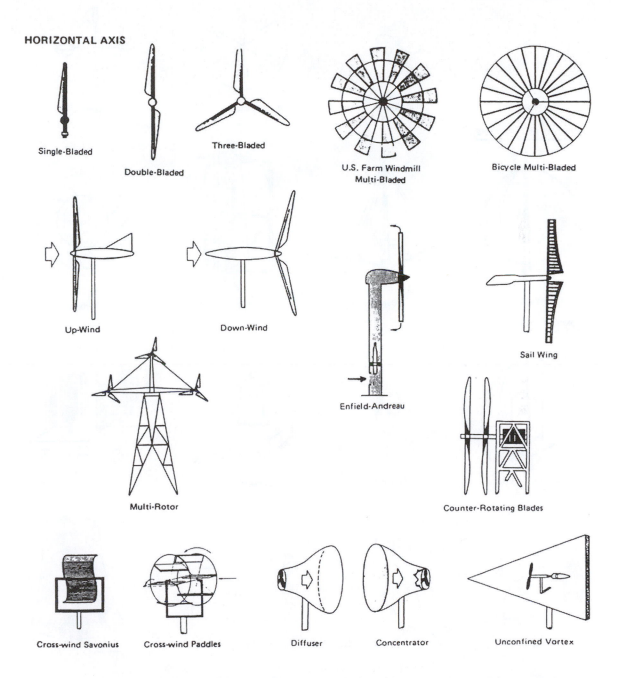

HORIZONTAL AXIS

Single-Bladed

Double-Bladed

Three-Bladed

U.S. Farm Windmill
Multi-Bladed

Bicycle Multi-Bladed

Up-Wind

Down-Wind

Enfield-Andreau

Sail Wing

Multi-Rotor

Counter-Rotating Blades

Cross-wind Savonius

Cross-wind Paddles

Diffuser

Concentrator

Unconfined Vortex

Figure 4.1 Horizontal-axis wind turbine taxonomy (Kreith and West 1997).

PRIMARILY DRAG-TYPE

Savonius

Multi-Bladed
Savonius

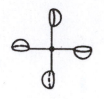

Plates

Shield

Cupped

PRIMARILY LIFT-TYPE

ϕ-Darrieus

$\triangle$-Darrieus

Giromill

Turbine

COMBINATIONS

Savonius/ϕ-Darrieus

Split Savonius

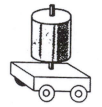

Magnus

Airfoil

OTHERS

Deflector

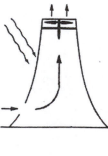

Sunlight

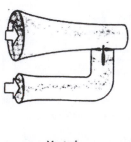

Venturi

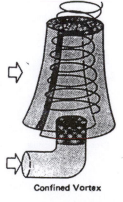

Confined Vortex

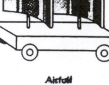

Figure 4.2 Vertical-axis wind turbine taxonomy (Kreith and West 1997).

(a) HAWT (600-kW Mitsubishi) (b) HAWT (1.5-kW Bergey) (c) HAWT (FloWind Corporation 17 EHD)

Figure 4.3 Examples of wind turbines (NREL).

The Savonius and the Darrieus are the most common versions of the VAWT, with most of the remaining vertical configurations either in the experimental/developmental or speculative stage. However, even the Savonius and the Darrieus configurations have not been extensively commercialized. VAWTs are generally classified as drag or lift devices, based on their operating principle. The Savonius rotor is a drag device since the drag of the wind on the "cups" generates the torque on the axis. The Darrieus is classified as a lift device since the shaft torque results primarily from lift on the blades. One indication of the dominance of the horizontal-configuration wind turbines is that on the National Renewable Energy Laboratory (NREL) website, only one photograph (depicting a Darrieus) of a vertical-axis wind turbine is presented in the renewable-energy photographic section. Figure 4-3(c), reproduced from the NREL website, is a photograph of a Darrieus wind turbine, a FloWind Corporation 17 EHD, taken in 1995.

4.2 FUNDAMENTAL CONCEPTS

As a prelude to discussing wind turbine operation, some details on the components of a wind turbine are needed. Figure 4.4 is a schematic illustrating the important parts of a typical HAWT. The tower is mounted to a base, and on top of the tower is the nacelle, which contains the gearbox, controls, and generator subsystems. The rotor is attached to the gearbox and generator by a shaft. The rotor diameter and the swept area of the blades are indicated, as is the hub height.

The power available from a wind of speed V with mass flow rate $\dot{m}$ sweeping an area A is

$$\text{Power} = \frac{1}{2}\dot{m}V^2 \qquad (4\text{-}1)$$

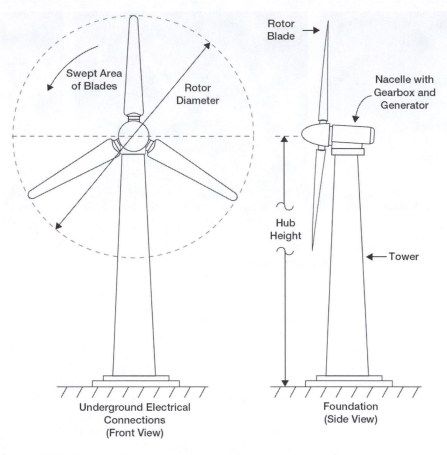

Figure 4.4 HAWT schematic and nomenclature (www.retscreen.net).

But the mass flow rate is $\dot{m} = \rho AV$, so the power available becomes

$$\text{Power} = \frac{1}{2}\rho AV^3 \qquad (4\text{-}2)$$

Equation (4-2) is very important, as it demonstrates that the power available from the wind is proportional to the area swept by a wind turbine and the cube of the wind speed. With a cubic power relationship, it is not surprising that the ideal location for wind turbines is where wind speeds are high. Wind resource information will be presented in Section 4.3. The fundamental question relating to Eq. (4-2) is, how much of the available power can be extracted from the wind? The analysis of Betz is the traditional approach used to answer this question.

The Betz analysis uses an actuator disk approach. In actuator theory, all energy transfer takes place in the plane of the actuator, and only energy transformation takes place upstream and downstream of the actuator. Actuator disk analysis is illustrated in Figure 4.5. The figure depicts three distributions: pressure, velocity, and cross-sectional

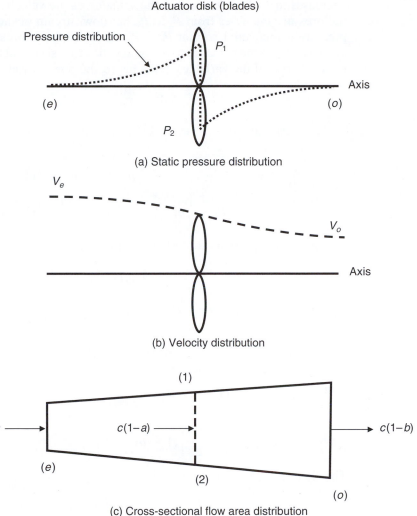

Figure 4.5 Actuator disk schematic with pressure, velocity, and area distributions illustrated.

flow area. Each will be examined in turn. The actuator disk is represented by the blades in the pressure and velocity distributions and by a dotted line in the cross-sectional area distribution. The actuator area, the area swept by the blades, is $A = (\pi/4)D^2$, where D is the rotor diameter. Far upstream from the actuator, at the point denoted as (e), for entrance, in Figure 4.5(c), the velocity is the wind velocity— c in this analysis. At the actuator disk, the velocity has been reduced to $c\,(1 - a)$, and at the outlet, indicated as (o), the velocity is $c\,(1 - b)$. As the velocity is reduced from c to $c\,(1 - a)$ upstream of the actuator, the static pressure increases from P_e to P_1. When energy is extracted from the flow at the actuator, the pressure discontinuously decreases to P_2. Downstream of the actuator, as the velocity continues to decrease, the pressure increases from P_2 to P_o. Far downstream of the actuator, atmospheric pressure is recovered, so that $P_o = P_e$, the atmospheric pressure. Since the velocity decreases from the entrance to the outlet, the cross-sectional flow area must increase. Cast in terms of the variables in the figure, the power available, Eq. (4-2), becomes

$$\text{Power}_{\text{avail}} = \frac{1}{2}\rho A c^3 \tag{4-3}$$

Upstream and downstream of the actuator, the Bernoulli equation applies and can be written as

$$P_e + \frac{1}{2}\rho c^2 = P_1 + \frac{1}{2}\rho c^2 (1 - a)^2$$

$$P_2 + \frac{1}{2}\rho c^2 (1 - a)^2 = P_0 + \frac{1}{2}\rho c^2 (1 - b)^2 \tag{4-4}$$

Solving for the pressure differences yields

$$\frac{P_e - P_1}{\rho} = \frac{1}{2}c^2\left[(1 - a)^2 - 1\right]$$

$$\frac{P_2 - P_0}{\rho} = \frac{1}{2}c^2\left[(1 - b)^2 - (1 - a)^2\right] \tag{4-5}$$

Adding the two expressions and using $P_e = P_o$, the expression for the change in pressure across the actuator becomes

$$\frac{P_2 - P_o + P_e - P_1}{\rho} = \frac{P_2 - P_1}{\rho} = \frac{1}{2}c^2\left[(1 - b)^2 - (1 - a)^2 + (1 - a)^2 - 1\right]$$

$$= \frac{1}{2}c^2\left[(1 - b)^2 - 1\right] \tag{4-6}$$

The change in pressure, $P_1 - P_2$, is thus

$$P_1 - P_2 = \frac{1}{2}\rho c^2\left[1 - (1 - b)^2\right] \tag{4-7}$$

The axial thrust across the actuator is

$$T = (P_1 - P_2)A = \frac{1}{2}\rho A c^2\left[1 - (1 - b)^2\right] \tag{4-8}$$

But the axial thrust is equal to the change in axial momentum,

$$T = \dot{m}\left[c - c(1 - b)\right] = \rho A c(1 - a)\left[c - c(1 - b)\right] = \rho A c^2(1 - a)b \tag{4-9}$$

Equating Eqs. (4-8) and (4-9) and solving yields

$$\frac{1}{2}\rho A c^2\left[1 - (1 - b)^2\right] = \rho A c^2(1 - a)b$$

$$a = \frac{b}{2} \tag{4-10}$$

Equation (4-10) demonstrates that the change in velocity upstream of the actuator plane is equal to the change in velocity downstream of the actuator plane. In many actuator disk analyses, the velocity at the actuator plane is taken as the average of the far upstream velocity, c, and far downstream velocity, $c(1-b)$—an equivalent result.

The rate of kinetic energy change, expressed in terms of c and a, is

$$E_k = \frac{1}{2}\rho Ac(1-a)\left[c^2 - c^2(1-b)^2\right] = \frac{1}{2}\rho Ac(1-a)\left[c^2 - c^2(1-2a)^2\right] \quad (4\text{-}11)$$

The maximum rate of change of kinetic energy, the maximum power extracted, will occur when the derivative of E_k with respect to a is set equal to zero, or

$$\frac{dE_k}{da} = 0 = (1)(1-a)^2 + 2a(1-a)(-1)$$

$$a = \frac{1}{3} \quad (4\text{-}12)$$

Then, using $a = \frac{1}{3}$ in Eq. (4-11) yields the expression for maximum power extracted:

$$\text{Power}_{\text{max}} = \frac{8}{27}\rho Ac^3 \quad (4\text{-}13)$$

The power coefficient is defined as the power extracted divided by the available power of the wind stream or

$$Cp = \frac{\text{Power}_{\text{ext}}}{\frac{1}{2}\rho AV_{\text{wind}}^3} \quad (4\text{-}14)$$

The maximum value of the power coefficient, the Betz limit, then becomes

$$Cp_{max} = \frac{\frac{8}{27}\rho Ac^3}{\frac{1}{2}\rho Ac^3} = \frac{16}{27} = 0.5926 \quad (4\text{-}15)$$

The Betz limit represents the maximum value of the power coefficient and defines the maximum power that can be extracted from a given wind stream. The power coefficient is perhaps the most important single metric used in characterizing a wind turbine. Equation (4-14) is usually rewritten to specify the power extracted in terms of the power coefficient:

$$\text{Power}_{\text{ext}} = \frac{1}{2}Cp\rho AV_{\text{wind}}^3 \quad (4\text{-}16)$$

If the power coefficient is the most important metric for wind turbines, then Figure 4.6 is arguably the most referenced figure in wind turbine engineering. This figure presents the expected range of power coefficients for well-designed wind

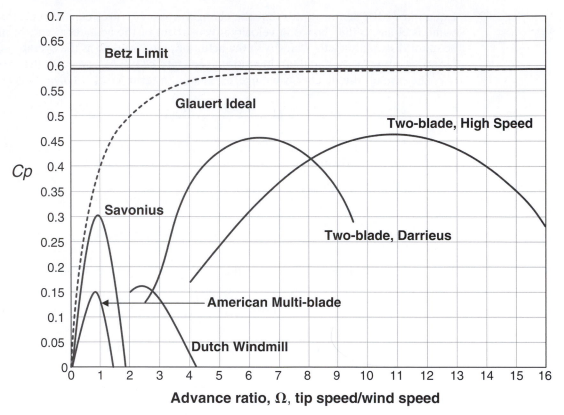

Figure 4.6 Power coefficient versus advance ratio for different wind turbine configurations.

turbines of different configurations. The abscissa, the x axis, is the advance ratio, which is the rotor tip speed divided by the wind speed, or

$$\Omega = \frac{r\omega}{V_{\text{wind}}} \tag{4-17}$$

where r is the rotor radius and ω is the rotor rotational rate.

The Betz limit is shown in the figure. The Glauert ideal Cp is the result of a more detailed analysis that explicitly involves the advance ratio. Lying below and to the right of the Glauert ideal are the expected Cp values for different wind turbine configurations. The American multi-blade and the Dutch windmill have the lowest Cp values. In many recent publications, the values for the American multi-blade and the Savonius configurations have been interchanged, with the result that the Savonius Cp values are smaller than those for the American multi-blade. The values presented in Figure 4.6 are correct. As pointed out by Johnson (2001), such an error will take years to correct. In addition to the Savonius and the American multi-blade, typical power coefficient values for the Darrieus and the modern, high-speed two-blade HAWT are presented. The HAWT shows relatively high Cp values over a rather wide range of advance ratios.

Example 4.1

A 27-mph wind at 14.7 psia and a temperature of 70°F enters a two-bladed wind turbine with a diameter of 36 ft. Calculate (a) the power of the incoming wind, (b) the theoretical maximum power that could be extracted, (c) a reasonable value for attainable power, (d) the rotor speed, in RPM, required for part (c), and (e) the torque for part (c).

Solution

The power of the incoming wind is defined by Eq. (4-2):

$$\text{Power} = \frac{1}{2}\rho A V^3$$

The thermal equation of state is

$$P = \rho R T$$

where R is the gas constant for air and, in English Engineering units, has the value

$$R = 53.35\frac{\text{ft lbf}}{\text{lbm R}}$$

The absolute temperature is required for the equation of state:

$$70°F = (460° + 70°)R = 530°R$$

Using the thermal equation of state, the density becomes

$$\rho = \frac{P}{RT} = 14.7\frac{\text{lbf}}{\text{in}^2}144\frac{\text{in}^2}{\text{ft}^2}\frac{\text{lbm R}}{53.35\text{ ft lbf}}\frac{1}{530°R} = 0.075\frac{\text{lbm}}{\text{ft}^3}$$

The area swept by the rotor is

$$A = \frac{\pi}{4}D^2 = \frac{\pi}{4}(36\text{ ft})^2 = 1018\text{ ft}^2$$

The power available from the wind is thus

$$\text{Power}_{\text{avail}} = \frac{1}{2}\rho A V^3 = \frac{1}{2}0.075\frac{\text{lbm}}{\text{ft}^3}1018\text{ ft}^2\left(27\frac{\text{mi}}{\text{h}}\frac{5280\text{ ft}}{\text{mi}}\frac{\text{h}}{3600\text{ sec}}\right)^3$$

$$= 2.371\cdot10^6\frac{\text{lbm ft}^2}{\text{sec}^3}\frac{\text{lbf sec}^2}{32.174\text{ ft lbm}} = 73,693\frac{\text{ft lbf}}{\text{sec}}\frac{\text{sec hp}}{550\text{ ft lbf}}$$

$$= 134\text{ hp} = 100\text{ kW}$$

The power available is 134 hp or 100 kW. The theoretical maximum power that could be extracted is the Betz limit or

$$\text{Power}_{\text{max}} = 0.5926\,\text{Power}_{\text{avail}} = 0.5926\cdot134\text{ hp} = 79.4\text{ hp} = 59.3\text{ kW}$$

For a two-bladed HAWT, Figure 4.6 suggests $Cp = 0.45$ for an advance ratio, $\Omega = r\omega/V_{\text{wind}}$, of 11. A reasonable value for attainable power from such a wind turbine is

$$\text{Power}_{\text{act}} = 0.45\cdot134\text{ hp} = 60.3\text{ hp} = 45\text{ kW}$$

Using the definition of advance ratio, the rotor speed can be cast as

$$\omega = \frac{\Omega \cdot V_{wind}}{0.5D} = 11 \cdot 39.6 \frac{ft}{sec} \frac{1}{0.5 \cdot 36\ ft} = 24.2 \frac{1}{sec} = 24.2\ Hz$$

$$= 24.2 \frac{1}{sec} \frac{Rev}{2\pi} \frac{60\ sec}{min} = 231\ RPM$$

In order to attain a Cp value of 0.45, a speed of 231 RPM is needed. From Eq. (2-11), the torque becomes

$$\tau = \frac{Power_{act}}{\omega} = 60.3\ hp\ \frac{550\ ft\ lbf}{sec\ hp} \frac{sec}{24.2} = 1370\ ft\ lbf$$

4.3 | WIND ENERGY RESOURCES

The previous section treated wind turbine performance in terms of a specified wind speed and pointed out the cubic relationship between speed and power from the wind. Since the wind is variable in terms of speed, direction, and altitude, the results of the previous section need to be extended to account for site-specific wind conditions. Extensive wind data are available at the NREL website, www.nrel.gov. The primary wind resources document is the *Wind Energy Resource Atlas of the United States*. This atlas contains annual average wind resource data as well as regional summaries and extensive explanations and references for the statistics of wind. A unique feature is an assessment of the "certainty" of the wind data. The wind data are rated from 1 (lowest degree) to 4 (the highest degree of certainty).

The wind power density distribution across the United States is very useful for first-order evaluations of candidate locations to determine if wind energy harvesting is feasible. The wind power density is the available average wind power per m^2 of wind turbine area. The basis of the distribution classification are the wind power density classes. The wind power density is measured from class 1 (lowest) to class 7 (highest) and is specified at nominal 10-m and 50-m elevations. Table 4.1 presents the wind power density classes and their associated wind speeds.

TABLE 4.1 Wind power density classes

Wind Power Class	10 m (33 ft)		50 m (164 ft)	
	Wind Power Density (W/m²)	Speed m/sec (mph)	Wind Power Density (W/m²)	Speed m/sec (mph)
1	0	0	0	
2	100	4.4 (9.8)	200	5.6 (12.5)
3	150	5.1 (11.5)	300	6.4 (14.3)
4	200	5.6 (12.5)	400	7.0 (15.7)
5	250	6.0 (13.4)	500	7.5 (16.8)
6	300	6.4 (14.3)	600	8.0 (17.9)
7	400	7.0 (15.7)	800	8.8 (19.7)
	1000	9.4 (21.1)	2000	11.9 (26.6)

Each wind class is bounded by a minimum wind power density and a maximum wind power density. The wind speeds in m/sec and mph corresponding to the minimum and maximum wind power density values are also indicated in Table 4.1. For example, class 4 is defined as a wind power density between 200 W/m^2 and 250 W/m^2 or a wind speed range between 5.6 m/sec and 6.0 m/sec, all at a 10-m elevation. At a 50-m elevation, the wind power classes are defined using different power density values than at a 10-m elevation.

The most useful summary figure from the *Wind Energy Resource Atlas of the United States* is the annual average wind power density distribution, reproduced as Figure 4.7. In this figure, the darker the color, the higher the wind class. Much of the Midwest is assessed as wind class 3 or 4, with isolated portions of the mountain west containing embedded regions of class 5 or 6. Except for areas near the coast, the Southeast is not generally suitable for wind energy production. NREL states that wind energy is appropriate only for wind class 3 and above.

The wind power density as illustrated in Figure 4.7 accounts for the yearly variation in wind speed at given locations. To understand the basis of development of information such as that in Figure 4.7, the statistics of wind energy need to be explored.

The probability of occurrence of a given wind speed is expressed by the Weibull distribution,

$$h(v, k, c) = \frac{k}{c}\left(\frac{v}{c}\right)^{k-1} \exp\left[-\left(\frac{v}{c}\right)^k\right] \qquad (4\text{-}18)$$

In the Weibull distribution, c is the scale parameter and k is the shape parameter. The shape parameter controls the shape of the distribution; the larger the shape factor, the closer the distribution comes to being Gaussian. The scale parameter controls the value of the mode (the most probable speed). The larger the scale parameter, the higher the mode, and the lower the probability of a given speed less than the mode. The shape parameter is dimensionless, and the scale parameter must have the same units as the speed. For wind distributions, the shape parameter value is usually near 2 (Patel 2005).

Figure 4.8 presents the probability distribution, expressed in percentages, for shape parameters of 1, 2, and 3 for a constant value of $c = 10\ mph$. The shape parameter of 1 yields an exponential distribution that has the highest percentage of hours at zero speed; the shape parameter of 2 results in a distribution skewed toward a higher percentage of occurrence at the lower speeds. The shape parameter of 3 yields a more symmetrical distribution and starts to resemble the Gaussian. The shape factor of $k = 2$ provides a generally acceptable match for the wind speed distribution at most sites.

Figure 4.9 illustrates the probability distribution, expressed in hours per year per mph, for scale parameter values of 10 mph, 15 mph, and 20 mph for a constant value of $k = 2$. The smaller the value of the scale parameter, the more hours at lower wind speeds. As the value of c increases, the mode wind speed values increase and the number of hours per year at wind speed higher than the mode increases.

Figures 4.8 and 4.9 provide a visual interpretation of the behavior of the Weibull distribution. However, the real question is, how does the Weibull distribution relate

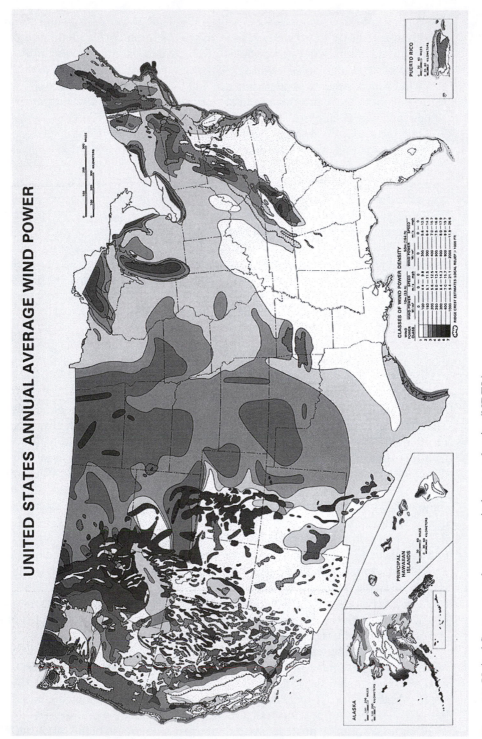

Figure 4.7 United States annual average wind power density (NREL).

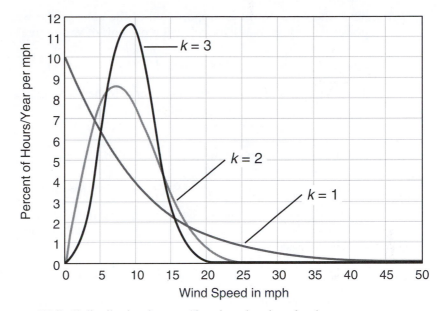

Figure 4.8 Weibull distribution for $c = 10$ mph and various k values.

to assessing the metrics of wind energy? The mode speed represents the most probable speed in a distribution. The mean speed is defined as

$$V_{\text{mean}} = \int_0^\infty h(v, k, c) \cdot v \cdot dv \qquad (4\text{-}19)$$

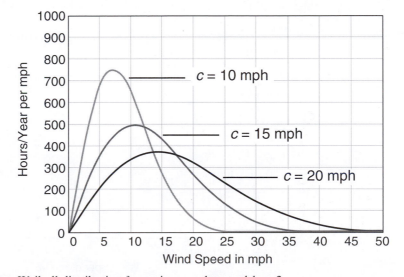

Figure 4.9 Weibull distribution for various c values and $k = 2$.

Since the wind power is proportional to the cube of the wind speed, the average power density available for collection per unit of swept area is

$$\text{Power}_{\text{avail}} = \int_0^\infty \frac{1}{2} \cdot \rho \cdot h(v, k, c) \cdot v^3 \cdot dv \tag{4-20}$$

Thus, the speed of interest for wind energy is the root-mean-cube speed

$$V_{\text{rmc}} = \sqrt[3]{\int_0^\infty h(v, k, c) \cdot v^3 \cdot dv} \tag{4-21}$$

and the annual average power density available becomes

$$\text{Power}_{\text{avail}} = \frac{1}{2} \rho V_{\text{rmc}}^3 \tag{4-22}$$

Based on the results of Figure 4.6, a reasonable power coefficient value for a modern, well-designed wind turbine is 0.5. The annual average extraction power density can be cast as

$$\text{Power}_{\text{ext}} = \frac{1}{4} \rho V_{\text{rmc}}^3 \tag{4-23}$$

The total energy that can be extracted per year for a given distribution is the integral of Eq. (4-23) for each velocity over all possible velocities, or

$$\text{Energy}_{\text{rmc}} = 0.25 \cdot \rho \int_0^\infty h(v, k, c) \cdot 8760 \cdot v^3 \cdot dv \tag{4-24}$$

The best way to assimilate all this information is via an example problem.

Example 4.2

Find V_{mode}, V_{mean}, V_{rmc}, the power density available distribution, and the power extracted per m^2 for a wind turbine at a site corresponding to a Weibull wind distribution with $c = 15$ m/sec and $k = 1.5$. The air density is 1.225 kg/m^3.

Solution

A graphical representation of the Weibull distribution for $k = 1.5$ and $c = 15$ m/sec is presented in Figure 4.10.

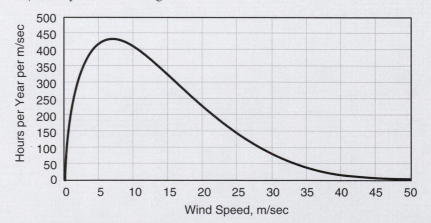

Figure 4.10 Weibull distribution for $k = 1.5$ and $c = 15$ m/sec.

The mode, the most probable wind speed, occurs at 7.21 m/sec. The mean wind speed and the root-mean-cube speed are defined in Eqs. (4-19) and (4-21), respectively. The arithmetic for this example is accomplished in Mathcad, and the Mathcad worksheet is reproduced in Figure 4.11.

$$mph := \frac{mi}{hr}$$

define mph as miles per hour

$$kW := 1000 \cdot watt$$

define kW

$$\rho := 1.225 \cdot \frac{kg}{m^3}$$

density

$$h(v, k, c) := \frac{k}{c} \cdot \left(\frac{v}{c}\right)^{k-1} \cdot e^{-\left(\frac{v}{c}\right)^k}$$

Weibull distribution function definition.

$$PowerDen\ (V) := 0.5 \cdot \rho \cdot V^3$$

Power density function definition.

$$c := 15 \cdot \frac{m}{sec} \qquad k := 1.5$$

specify values of the scale parameter and shape parameter

$$V_{mode} := 7.21 \cdot \frac{m}{sec} \qquad PowerDen\left(V_{mode}\right) = 229.563\ \frac{watt}{m^2}$$

$$V_{mean} := \int_{0 \cdot \frac{m}{sec}}^{\infty \cdot \frac{m}{sec}} \frac{k}{c} \cdot \left(\frac{v}{c}\right)^{k-1} \cdot e^{-\left(\frac{v}{c}\right)^k} \cdot v\ dv$$

Definition of mean speed, Equation (4-19).

$$V_{mean} = 13.541\ \frac{m}{sec} \qquad PowerDen\left(V_{mean}\right) = 1.521 \times 10^3\ \frac{watt}{m^2}$$

$$V_{rmc} := \sqrt[3]{\int_{0 \cdot \frac{m}{sec}}^{1000 \cdot \frac{m}{sec}} \frac{k}{c} \cdot \left(\frac{v}{c}\right)^{k-1} \cdot e^{-\left(\frac{v}{c}\right)^k} \cdot v^3\ dv}$$

Definition of rmc speed, Equation (4-21).

$$V_{rmc} = 18.899\ \frac{m}{sec} \qquad PowerDen\left(V_{rmc}\right) = 4.134 \times 10^3\ \frac{watt}{m^2}$$

(continued on next page)

Figure 4.11 Mathcad solution for Example 4.2.

$$\text{Power (v)} := 0.5 \cdot \rho \cdot h(v, k, c) \cdot v^3 \qquad \text{Power density available with Cp} = 1.0.$$

$$v := 0 .. 50 \cdot \frac{m}{\text{sec}} \qquad\qquad \text{wind velocity range}$$

$$\text{Energy} := \int_{0 \cdot \frac{m}{\text{sec}}}^{1000 \cdot \frac{m}{\text{sec}}} 0.25 \cdot \rho \cdot \left[\frac{k}{c} \cdot \left(\frac{v}{c}\right)^{k-1} \cdot e^{-\left(\frac{v}{c}\right)^k} \right] \cdot 8760 \frac{\text{hr}}{\text{yr}} \cdot v^3 \, dv$$

$$\text{Energy} = 1.811 \times 10^4 \, \text{kW} \cdot \frac{\text{hr}}{\text{yr} \cdot m^2}$$

Figure 4.11 *(cont.)*

At the start of the worksheet, mph and kW are defined in Mathcad variables, and the density (specified in the problem statement) is indicated. Functions are defined for the Weibull distribution, $h(v, k, c)$, and the power density, PowerDen(V). The shape and scale parameters from the problem statement are inserted into the worksheet. The mode speed is 7.21 m/sec, and the power density is calculated as 230 W/m². The mean speed is computed using Eq. (4-19) to be 13.54 m/sec with a power density of 1521 W/m². The root-mean-cube (rmc) speed is 18.9 m/sec [using Eq. (4-21)], and the power density at the rmc speed is 4134 W/m². The use of Mathcad, with its ability to carry units in computations, simplifies the arithmetic in this example. Since the cubic power of the wind speed is specified in the average power density expression, care should always be exercised to differentiate between the mode, mean, and rmc speeds, as the power density manifests significant variation depending on which value is used.

The energy density available for collection over a year per unit area of wind turbine is specified in Eq. (4-24) since $h(v, k, c) \cdot 8760$ represents the hours per year at a given wind speed from the distribution. Figure 4.12 presents the energy density per year per m² per m/sec for the wind speed distribution specified in the problem statement. The mode for the annual energy available is near 27 m/sec. Although the number of hours per year at 27 m/sec is small (far from the mode, as illustrated in Figure 4.10), the product of this wind speed and the number of hours per year is a maximum for the distribution because of the cubic functional dependence on wind speed.

An estimate of the annual energy extracted for $Cp = 0.5$ for a given wind speed distribution is provided by Eq. (4-20). Under the assumption of $Cp = 0.5$, the annual energy extracted for this distribution is 18,110 $\frac{\text{kW hr}}{\text{yr m}^2}$.

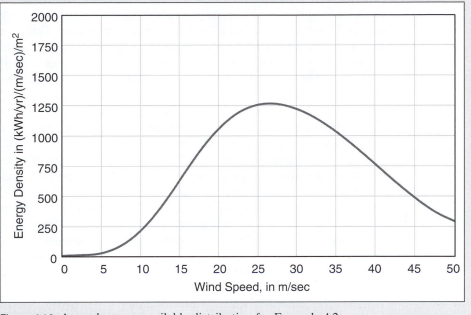

Figure 4.12 Annual energy-available distribution for Example 4.2.

Section 4.2 delineated the performance characteristics of wind turbines at specified speeds. This section has described the wind speed distribution and presented the statistics of the wind speed distribution. The next section assimilates the material in these two sections and addresses the operation and control of wind turbines as a function of wind speed.

4.4   WIND TURBINE OPERATION

What should the operating strategy be for a wind turbine as a function of wind speed? The answer is *not* to operate the wind turbine at the maximum power coefficient. Why not operate the wind turbine at the maximum Cp? Operation at the maximum Cp for all wind speeds would maximize the energy extracted, but factors such as generator capacity, structural requirements, and safety preclude such operation. The maximum-speed range will occur for only a few hours for a given wind speed distribution. Thus, sizing a generator for an input corresponding to the maximum-speed range would result in an oversized generator that would operate at maximum output only a few hours per year. And if the maximum Cp were used, the advance ratio, Ω, would have to be maintained constant. As the wind speed increased, the rotor rotation rate would have to increase to maintain a constant Cp. Since the radial stresses in a rotor are proportional to the rotation speed, operating at high wind speeds and a constant Ω would require a structurally robust wind turbine. Additionally, as the wind speed increases, safety and structural integrity become of increasing concern.

But no matter what operating strategy is used, a wind turbine must contain a controller to implement the strategy and mechanical elements to respond to the controller. Figure 4.13, from the NREL website, graphically illustrates typical HAWT features. The cutaway of the nacelle shows the gearbox and generator as well as other elements needed for operation. The blades and tower are also shown. The yaw motor and yaw drive are used to keep the plane of the blade oriented into the wind. The pitch mechanism on the blades adjusts the angle of the blades (the pitch) with respect to the wind direction in order to control the power extracted from the wind. The purpose of the brake is to slow down or completely stop the rotor. All of these elements are needed to implement the control strategy.

The ultimate purpose of a wind turbine control strategy is to regulate the power output of the turbine as a function of wind speed and direction. Additionally, the control protocol must ensure safe operation over all wind conditions. Patel (2005) suggests that the power output versus wind speed characteristic of a wind turbine can be viewed as being composed of several regions. Figure 4.14, adapted from Patel, illustrates a typical power output as a function of wind speed and delineates the various operating regimes and conditions. The ordinate variable is the percentage of generator output.

The first condition is the cut-in speed of the system. Below the cut-in wind speed, the system component efficiencies are so low that running the system is not worthwhile. Once the cut-in speed is reached, the system is operated in a constant-Cp region. In the constant-Cp region, the turbine extracts the maximum power from the wind, but the power extracted is less than the rated input to the generator. The rotor

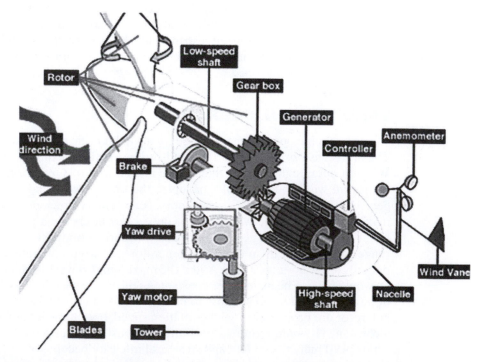

Figure 4.13 HAWT nacelle components and features (NREL).

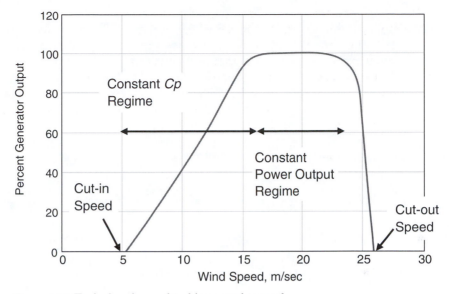

Figure 4.14 Typical regimes of turbine speed control.

speed is varied so that the advance ratio is maintained near the maximum Cp value. When the wind speed is sufficiently high, the power extracted by the rotor exceeds the rated input of the generator. In this regime, the constant power output regime, the system is made to produce the rated output of the generator by operating the turbine at a Cp lower than the maximum Cp. The cut-out speed is the wind speed beyond which operation would damage the system. For speed in excess of the cut-out, the rotor pitch is set to unload the rotor, and the rotor is locked with the brake. The total energy that can be extracted from a given wind distribution is reduced by the rated input of the generator and the cut-in and cut-out speeds. Example 4.2 will be extended to illustrate these effects.

Example 4.3	The system described in Example 4.2 is specified to have a cut-in speed of 5 m/sec, a cut-out speed of 35 m/sec, and a rated generator input of 7.5 kW/m². The maximum power coefficient, Cp, is 0.5. Determine and plot the following for both the system with no controls and the system controlled to meet the constraints: (a) the power density of the system, (b) the Cp versus wind speed required, (c) the energy extraction, and (d) the total energy extracted by the system.

Solution	Much of the information needed for the "no controls" part of this problem was developed in Example 4.2. However, since the results are more meaningful if the "no controls" and the controlled versions are compared, both versions will be examined in the solution. Mathcad will be utilized for all the calculations. The Mathcad worksheet for the solution to this problem is given in Figure 4.15.

The conditions and constraints defined in the problem statement are entered in the worksheet. The range of the wind speed is designated by a Mathcad range

$Cp_{nom} := 0.5$ Nominal value of the power coefficient.

$v := 0..50 \cdot \dfrac{m}{sec}$ wind velocity range

$V_{cutin} := 5 \cdot \dfrac{m}{sec}$ cut-in speed

$V_{cutout} := 35 \cdot \dfrac{m}{sec}$ cut-off speed

$Powerin_{max} := 7.5 \cdot \dfrac{kW}{m^2}$ rated input of generator

$PowerDen(v) := 0.5 \cdot Cp_{nom} \cdot \rho \cdot v^3$ Power density available with nominal Cp.

$$PowerDenCon\,(v) := \begin{vmatrix} 0 \cdot \dfrac{kW}{m^2} & \text{if } v < V_{cutin} \\[2ex] Powerin_{max} & \text{if } PowerDen\,(v) > Powerin_{max} \\[2ex] 0 \cdot \dfrac{kW}{m^2} & \text{if } v > V_{cutout} \\[2ex] PowerDen\,(v) & \text{otherwise} \end{vmatrix}$$

Piece-wise continuous function defined to implement the cut-in, cut-out, and rated power constraints.

$CpV(v) := \dfrac{PowerDenCon(v) \cdot Cp_{nom}}{PowerDen(v)}$

$Energy\,(v) := PowerDen\,(v)\,h(v,k,c) \cdot 8760 \cdot hr$ Energy density available with nominal Cp.

$EnergyCon\,(v) := PowerDenCon\,(v) \cdot h(v,k,c) \cdot 8760 \cdot hr$ Energy density with controls.

$$EnergyCon := \int_{0 \cdot \frac{m}{sec}}^{1000 \cdot \frac{m}{sec}} EnergyCon(v) \cdot \dfrac{1}{yr}\, dv$$ Energy extracted per year per m2 for system with controls.

$EnergyCon = 1.171 \times 10^4\, kW \cdot \dfrac{hr}{yr \cdot m^2}$

$$Energy_{max} := \int_{0 \cdot \frac{m}{sec}}^{1000 \cdot \frac{m}{sec}} Energy\,(v) \cdot \dfrac{1}{yr}\, dv$$ Energy extracted per year per m2 with no controls.

$Energy_{max} = 1.811 \times 10^4\, kW \cdot \dfrac{hr}{yr \cdot m^2}$

$CaptureRatio := \dfrac{EnergyCon}{Energy_{max}}$ $CaptureRatio = 0.647$

Figure 4.15 Mathcad worksheet for Example 4.3.

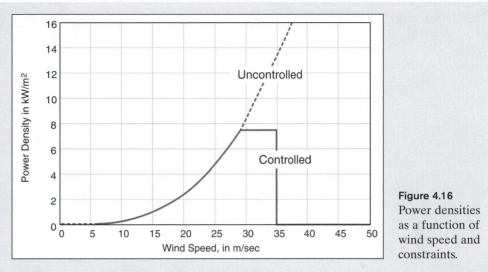

Figure 4.16 Power densities as a function of wind speed and constraints.

variable, $v := 0.50 \cdot$ m/sec. This range variable specifies that the speed, v, is to "range" from 0 to 50 m/sec. Any time v is specified, all of the values in the range are automatically included. The power density for the speed range specified is

$$\text{PowerDen}(v) = 0.5 \cdot Cp_{\text{nom}} \rho v^3 \qquad (4\text{-}25)$$

The PowerDenCon(v) function is a piecewise continuous function in Mathcad that calculates the power density subject to the cut-in speed, the cut-out speed, and the rated input power of the generator. The first line sets the power density to zero for wind speeds less than the cut-in speed; the second line constrains the power extracted not to exceed the generator input power. The third line implements the cut-out speed constraint, and the last line ensures the constant-Cp regime results. Figure 4.16 presents the power densities for the controlled (constrained) and uncontrolled conditions. The power coefficient, Cp, can be expressed as

$$Cp(v) = \frac{\text{PowerDenCon}(v)}{\text{PowerDen}(v)} \cdot Cp_{\text{nom}} \qquad (4\text{-}26)$$

and is graphically illustrated in Figure 4.17. The Cp is zero until the cut-in speed is reached and zero for speeds greater than the cut-out speed. The constant-Cp regime is present as is the variable-Cp regime.

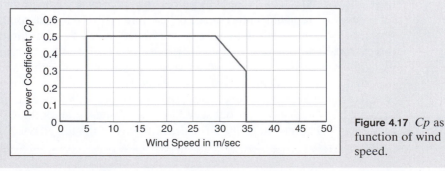

Figure 4.17 Cp as function of wind speed.

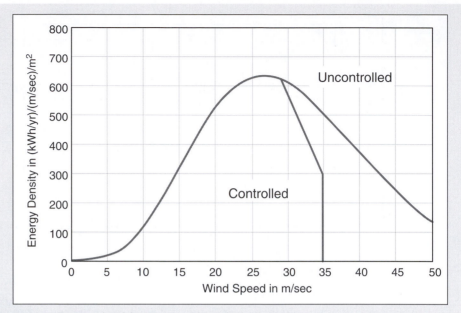

Figure 4.18 Energy densities as a function of wind speed.

The power densities multiplied by the wind speed probability distribution, $h(v, k, c)$, and the number of hours in a year, 8760, yield the energy densities. Figure 4.18 is a representation of the energy densities for the controlled and uncontrolled conditions. This figure is perhaps the most revealing in the solution. The effects of the cut-out speed and the generator input restriction are quite evident in the presentation. Since the areas under the curve represent the total energy extracted, the effects of the control constraints are evident, especially at the higher wind speeds.

The energy densities integrated over all the speeds yield the total energy extracted. For the case of no control, the total energy extracted is 18,110 kWh/yr/m^2, the same result as in Example 4.2. The actual energy extracted corresponds to the case with controls implemented. The actual energy extracted is 11,710 kWh/yr/m^2. The capture ratio is defined as the actual energy extracted divided by the maximum possible energy (no controls) for a given wind speed distribution. For the conditions of this problem, the capture ratio is 0.647; that is, 65 percent of the available energy can be extracted. The only ways to significantly change the capture ratio are to increase the rated input power of the generator and to increase the cut-out speed. For the stated conditions of this example, neither of these strategies is appropriate. Increasing the rated input power would be more expensive and would result in more generator operation at less than the rated input. For example, increasing the rated input of the generator to 10.5 kW/m^2 would result in a capture ratio of 0.682, an increase of only 5.5 percent in actual energy extracted. Increasing the cut-out speed to a value much greater than 35 m/sec would require an enhancement of the structural integrity of the tower, nacelle, and blades.

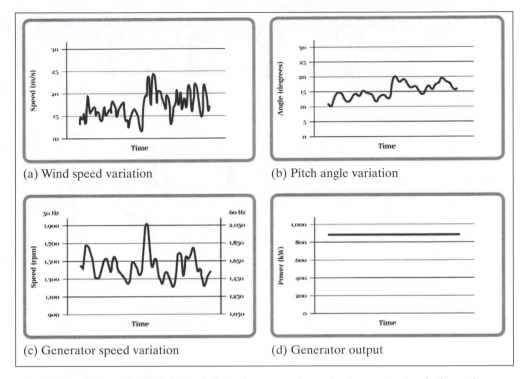

Figure 4.19 Vestas V-52 850-kW wind turbine generator output response to wind speed variations (Vestas website).

Vestas Wind Systems, a leading manufacturer of large (MW range) wind turbines, presents an interesting demonstration of the effectiveness of their control strategy for a Vestas V-52 850-kW wind turbine. Figure 4.19 illustrates the response of the generator output for a V-52 wind turbine subject to variations in wind speed.

The random nature of the wind speed variation is shown in Figure 4.19(a). The purpose of the control protocol is to maintain the generator output at a constant value, 850 kW. To do this the blade pitch and the generator speed are modulated. Figures 4.19(b) and (c) illustrate the required modulations in pitch angle and generator speed. The generator output is tracked in Figure 4.19(d). The success of the control protocol in maintaining a constant generator output is evident.

Thus far, this section has addressed the operation of a single wind turbine. However, wind turbine "farms," employing arrays of turbines, are becoming common. For wind turbines employed in arrays, the recommended space is 2–4 rotor diameters facing the prevailing wind and 8–12 rotor diameters parallel to the wind. For more than a single row of wind turbines in an array, the turbine locations in the succeeding rows are staggered. Figure 4.20 provides a schematic illustration of the recommended spacing of wind turbines on wind farms.

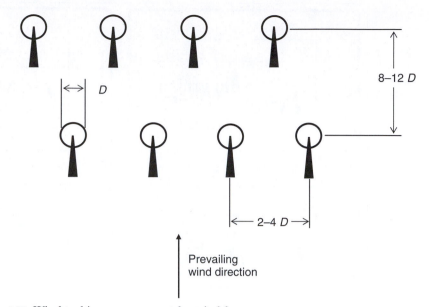

Figure 4.20 Wind turbine arrangement for wind farms.

4.5 COMMERCIAL WIND TURBINE EXAMPLES

This section contains some information and technical data on a sampling of wind turbines that are commercially available. The American Wind Energy Association provides a list of U.S. manufacturers of small wind turbines at www.awea.org. A list of manufacturers worldwide for large wind turbines is available at www.ecobusinesslinks.com. The information in this section was obtained from the various company websites and literature and from the NREL database. Additional technical data and price information can be obtained by contacting the individual companies. Examples included herein are for a GE Energy 1.5-MW wind turbine, a Vestas V52 850-kW wind turbine, and a Bergey 10-kW Excel wind turbine. Almost all of the companies included in the small and large wind turbine manufacturers list have websites.

GE Energy 1.5 MW

GE Energy (www.gepower.com) manufactures large wind turbines with nominal outputs of 1.5 MW, 2.5 MW, 3.0 MW, and 3.6 MW. The family of 1.5 MW devices will be examined here. Figure 4.21 contains pictures of GE Energy 1.5 MW wind turbines. Figure 4.21(a) shows the 1.5-MW turbines in a wind farm arrangement. Figure 4.21(b), a view of the nacelle and blade arrangement with a maintenance person, is included because it conveys an idea of the size of a 1.5-MW wind turbine.

The technical information for the turbine is presented in Table 4.2, and the power curve is shown in Figure 4.22.

Figure 4.21 GE Energy 1.5-MW wind turbine (NREL). (a) Wind farm example. (b) Nacelle with person.

TABLE 4.2 GE Power 1.5-MW specifications

	1.5s	1.5se	1.5sl	1.5sle	1.5xle
Rated capacity (kW)	1500	1500	1500	1500	1500
Cut-in speed (m/sec)	4	4	3.5	3.5	3.5
Cut-out speed (m/sec)	25	25	20	25	20
Rated wind speed (m/sec)	13	13	14	14	12.5
Rotor diameter (m)	70.5	70.5	77	77	82.5
Swept area m^2	3904	3904	4657	4657	5346
Rotor speed (RPM)	12–22.2	12–22.2	11–20.4	11–20.4	10.1–18.7

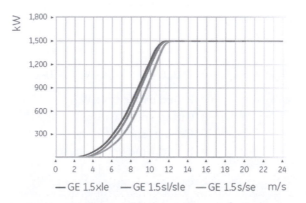

Figure 4.22 Power versus wind speed for a GE 1.5-MW wind turbine (GE Energy).

Figure 4.23 Vestas V52 850-kW wind turbine (Vestas).

Vestas V52 850 kW

The Vestas Wind Systems V52 has a rated output of 850 kW and is a popular wind turbine worldwide. The Vestas website address is www.vestas.com. Vestas reported that in 2005, the company averaged more than 50 installations per week for the year. The effectiveness of the control system for the V52 was discussed in a previous section. Technical information on the V52 850-kW wind turbine is presented in Table 4.3, and a photograph is provided in Figure 4.23. The power output–wind speed performance characteristics are contained in Figure 4.24. The performance characteristics in the figure are parameterized in terms of the sound level.

TABLE 4.3 Vestas V52 850-kW specifications

Rated capacity (kW)	850
Cut-in speed (m/sec)	4
Cut-out speed (m/sec)	25
Rated wind speed (m/sec)	16
Rotor diameter (m)	52
Swept area m^2	2124
Rotor speed (RPM)	14–31.4

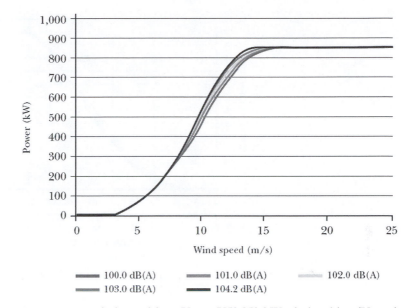

Figure 4.24 Power versus wind speed for a Vestas V52 850-kW wind turbine (Vestas).

Bergey 10-kW Excel

The previous examples were for relatively large wind turbines used for commercial power generation. The Bergey Wind Power Company (www.bergey.com) manufactures small wind turbines suitable for use in residential and small commercial applications.

Technical information on the Bergey 10-kW Excel wind turbine is presented in Table 4.4, and a photograph is shown in Figure 4.25. The power output–wind speed performance characteristics are given in Figure 4.26.

Figure 4.25 Bergey 10-kW Excel wind turbine (Bergey).

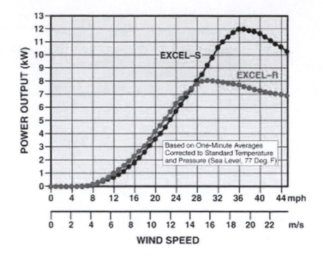

Figure 4.26 Power versus wind speed for a Bergey 10-kW Excel wind turbine (Bergey).

TABLE 4.4 Bergey 10-kW Excel specifications

Rated capacity (kW)	10
Cut-in speed (m/sec)	3.1
Cut-out speed (m/sec)	None (furled at 15.6 m/sec)
Rated wind speed (m/sec)	13.8
Rotor diameter (m)	6.7
Swept area m^2	35.3
Rotor speed (RPM)	14–31.4

Prices for a complete system, including a voltage regulator and a line-commutated inverter, ranged in 2006 from $20,000 to $25,000, depending on options. The Excel is most often installed on a guyed lattice tower, which is available in heights of 18 m (60 ft) to 37 m (120 ft) with prices ranging from $6200 to $9200. Thus, the cost ranges from $2600/kW to $3400/kW, again depending on the options chosen.

The three examples presented here are a relatively small sample of commercially available wind turbines. No recommendation or endorsement of the manufacturer is implied by the inclusion of any of the examples in this section; they are merely presented as samples of commercially available wind turbines.

4.6 CLOSURE

This chapter has explored wind power by developing the operating principles for wind turbines, exploring how the wind speed distribution can be used, addressing how a wind turbine must be controlled, and examining some commercially available wind turbines.

REVIEW QUESTIONS

1. Why is wind speed such an important parameter in determining the power available from the wind?

2. What is the power coefficient?

3. What is the Betz limit?

4. Explain the important variables in the Weibull distribution.

5. Discuss the operating protocol for a wind turbine.

6. Why isn't a wind turbine operated continuously at the maximum power coefficient?

7. Would you recommend constructing a wind turbine at your location? Explain.

EXERCISES

1. A 15-m/sec wind at 101.3 kPa and 20°C enters a two-bladed wind turbine with a diameter of 15 m. Calculate the following:
   (a) The power of the incoming wind
   (b) The theoretical maximum power that could be extracted
   (c) A reasonable attainable turbine power
   (d) The speed in RPM required for part (c)
   (e) The torque for part (c)

2. Compute the power coefficients for a Vestas V52 850-kW (Table 4.3) wind turbine for wind speeds of 10 m/sec and 15 m/sec. What are the advance ratios for these wind speeds? How do the power coefficients compare with the expected values (Figure 4.6)?

3. A General Electric 1.5se wind turbine is used for this exercise. Information on the GE 1.5se is provided in Table 4.2.
   (a) Estimate the power coefficient at the rated speed and at 10 m/sec
   (b) Explain the importance of the result of part (a)
   (c) Estimate the kWh production of a GE 1.5se device in a wind distribution with $c = 10$ m/sec and $k = 2$. Show plots similar to those examined in the chapter

4. A 27-mph wind at 14.65 psia and 70°F enters a wind turbine with a 1000-ft^2 cross-sectional area. Calculate:
   (a) The power of the incoming wind
   (b) The theoretical maximum power that could be extracted
   (c) A reasonable attainable turbine power
   (d) The torque for part (c)

5. Consider two cases: (1) a constant wind velocity twice the mean wind velocity and operating half the time, and (2) a constant wind velocity three times the same mean velocity operating one-third of the time. At all other times the wind velocity is zero. Determine for each case the ratio of the total energy available from the wind to the total wind energy available at the mean velocity continuously.

6. A 15-ft diameter wind turbine operates in a 25 ft/sec wind at 1 atm and 60°F. The turbine is used to pump 60°F water from a 30-ft deep well. How much water (in cubic feet per second) can be pumped if the overall efficiency of the wind-turbine-pump system is 0.25?

7. A wind turbine–generator is designed to attain full-load capacity with a wind velocity of 48 km/h. The rotor diameter is 50 m. If the power coefficient is 0.48 and generator efficiency is 0.85, calculate the rated output for 1 atm and 22°C.

8. The U.S. Department of Energy constructed a Darrieus vertical-axis wind turbine in Sandia, New Mexico. The machine was 60 ft tall and 30 ft in diameter, and swept an area of 1200 ft². Estimate the power this device can produce at a wind speed of 20 mph.

9. An early NASA/DOE wind turbine consisted of a 125-ft diameter, two-bladed, horizontal-axis rotor. Maximum power was achieved at a wind speed of 19 mph. For these conditions estimate:

   (a) The power generated in kW

   (b) The rotor speed in RPM

   (c) The velocity downstream of the rotor (V_o)

10. A wind turbine with a rotor diameter of 40 m has a power coefficient of 0.30 in an 8-m/sec wind. The density is 1.2 kg/m³. The turbine is to be used in a wind farm that is to serve a community of 100,000 (average family size of 4). Each house will require 3 kW. The wind farm will have a turbine spacing of 2.4 rotor diameters perpendicular to the wind and 8 rotor diameters parallel to the wind. The wind farm will have 10 turbines perpendicular to the wind.

    (a) Estimate the power production from one turbine.

    (b) How many turbines will be required in the wind farm for the community?

    (c) Estimate the dimensions of the wind farm.

    (d) How many acres will be required for the wind farm?

    (e) If the average house is on a 0.25-acre lot, how large will the wind farm be in comparison to the community?

    (f) What does this problem imply about wind power feasibility in an urban setting?

11. What is the power coefficient for the Vestas V52 850-kW wind turbine at the rated conditions?

12. What is the power coefficient for the Bergey Excel wind turbine at the rated conditions?

13. The Utopia Wind Turbine Company advertises that its two-bladed, 20-m diameter wind turbine–generator will produce 600 kW in a 15-m/sec wind. The air density is 1.18 kg/m^3. Do you believe its claim? Explain.

14. Repeat Example 4.3 using wind turbine characteristics from a manufacturer's website. If available, use a wind distribution consistent with your location.

REFERENCES

Johnson, G. L. 2001. "Wind Turbine Power, Energy, and Torque." Available at www.eece.ksu.edu/~gjohnson/.

Kreith, F., and West, R. E. 1997. *CRC Handbook of Energy Efficiency*. Boca Raton, FL: CRC Press.

Logan, E. 1993. *Turbomachinery*, 2nd ed. New York: Marcel Dekker.

Patel, M. R. 2005. *Wind and Solar Power Systems*, 2nd ed. New York: Taylor & Francis.

CHAPTER 5

Combustion Turbines

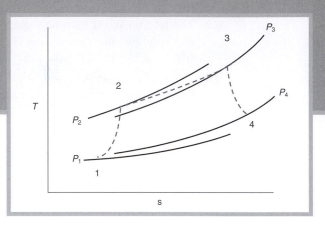

INTRODUCTION

Gas turbines were developed for aeronautical applications prior to and during World War II. Wilson (1984) provides an excellent historical account of the engineering aspects of the development of gas turbine engines. Frank Whittle in the United Kingdom and Hans von Ohain in Germany are credited with essentially simultaneously producing the first useful, flight-worthy gas turbine engines. In the decades following World War II, the gas turbine engine became the preferred propulsion system for most aircraft. At the same time, the inherent simplicity (compared with other prime movers) and operational characteristics of the gas turbine engine resulted in its adaptation for a variety of uses, including power generation and marine propulsion. When used for electric power generation, gas turbines are often called combustion turbines to distinguish them from steam turbines, a longtime mainstay of the electric power industry.

Combustion turbines are used for peaking, base-load, backup/emergency, and grid-independent power generation, and also as power sources for remote or isolated facilities such as offshore drilling platforms. When used for electric power generation, combustion turbines are operated in two modes: (1) electric power generation with no heat recovery and (2) electric power generation with heat recovery. The latter mode is called CHP (combined heat and power) and is the subject of Chapter 11. CHP is often referred to as co-generation—the recovery and use of thermal energy in conjunction with the decentralized generation of electricity. This chapter develops the principles for combustion turbines.

5.2 THE COMBUSTION TURBINE

Figure 5.1 is a schematic of a combustion turbine showing the primary components: the compressor, the combustor, and the turbine. Air is compressed as it passes through the compressor. The work required to drive the compressor is provided by the turbine.

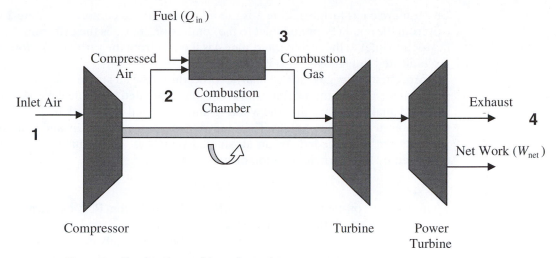

Figure 5.1 Combustion turbine schematic.

Fuel is burned in the combustor, and the temperature of the air and combustion products mixture is increased by the combustion process. As the mixture expands through the turbine, the power extracted by the turbine is used to drive the compressor. In the combustion turbine, the additional energy extracted by the "power" turbine is available as output power from the engine. The combustion turbine output power is typically used to drive a generator for the production of electricity. In many combustion turbines, only a single turbine section is present; part of the power extracted is used to drive the compressor, and the remainder is the output power.

The Brayton cycle is often used as a representation of the gas turbine. The Brayton cycle is composed of isentropic compression and expansion processes and constant-pressure heat addition and rejection processes. P-v (pressure–specific volume) and T-s (temperature-entropy) diagrams are provided in Figure 5.2 for a

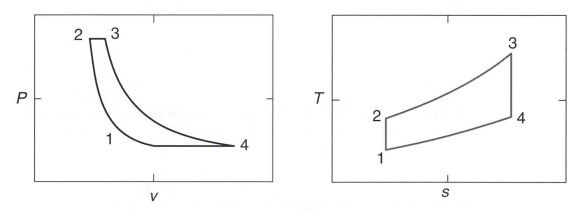

Figure 5.2 P-v and T-s diagrams for a Brayton gas turbine cycle.

Brayton cycle gas turbine. State 1 is the inlet state to the compressor, state 2 is the exit from the compressor and inlet to the combustor, state 3 is the exit from the combustor and inlet to the turbine, and state 4 is the exit from the turbine. The four states are indicated in Figures 5.1 and 5.2.

Most first-order gas turbine analysis procedures use an air-standard analysis. Air standard means that the working fluid is air that is treated as an ideal gas. A cold air-standard analysis means that the working fluid is treated as air with all properties at their room-temperature values. Thermodynamic properties for air and combustion products are thus needed for an air-standard analysis. A thermally perfect gas is described by the thermal equation of state

$$P = \rho RT \tag{5-1}$$

where the gas constant is $R = 287 \text{ J/kg K} = 53.35 \text{ ft lbf/lbm R}$. A calorically perfect gas is a gas for which the specific heat at constant volume, c_v, and the specific heat at constant pressure, c_p, are constant. A gas that is both thermally perfect and calorically perfect is labeled a perfect gas. The ratio of specific heats is defined as $k = c_p/c_v$. For a thermally perfect gas,

$$R = c_p - c_v \quad \text{and} \quad R = \frac{k-1}{k} c_p \tag{5-2}$$

The isentropic relations are expressed as

$$\frac{T_a}{T_b} = \left(\frac{P_a}{P_b}\right)^{(k-1)/k} = \left(\frac{\rho_a}{\rho_b}\right)^{k-1} = \left(\frac{v_b}{v_a}\right)^{k-1} \tag{5-3}$$

For air and combustion products, the accepted values of the thermodynamic constants are as follows:

Air	Combustion Products
$c_p = 1.004 \text{ kJ/kg K} = 0.24 \text{ Btu/lbm R}$	$c_p = 1.148 \text{ kJ/kg K} = 0.2744 \text{ Btu/lbm R}$
$k = 1.40$	$k = 1.33$

In a gas turbine, the combustion products cause the properties at the exit of the combustor to be different from those of air. However, in virtually all gas turbines, the fuel-to-air ratio is less than 0.02, so the actual deviations in combustor exit properties from those of air at the same temperature and pressure are small. A reasonable estimate is that the thermodynamic constants in the heat addition process are the average of the air and combustion products. Thus, for the heat addition (combustion) process:

$$c_p = 1.076 \frac{\text{kJ}}{\text{kg K}} = 0.2572 \frac{\text{Btu}}{\text{lbm R}}$$

$$k = 1.367$$

THE AIR-STANDARD BRAYTON CYCLE

The cold air-standard Brayton cycle is the simplest model of a gas turbine. For the cold air-standard Brayton cycle, the compressor work (per unit mass), W_c, and the expansion work, W_e, are, respectively,

$$W_c = h_1 - h_2$$
$$W_e = h_3 - h_4 \qquad (5\text{-}4)$$

where h represents the enthalpy, which for a thermally perfect gas can be cast as $h = c_p T$. The heat supplied is

$$Q_s = h_3 - h_2 \qquad (5\text{-}5)$$

The net work out of the turbine is the difference between the turbine work and the compressor work. In the Brayton cycle, combustion is replaced by heat addition, so the mass flow rates through all components are the same. The thermal efficiency is defined as the net work divided by the heat supplied. Hence, for the cold air-standard analysis,

$$\eta_t = \frac{\sum W}{Q_s} = \frac{h_1 - h_2 + h_3 - h_4}{h_3 - h_2}$$

$$= \frac{c_p(T_1 - T_2) + c_p(T_3 - T_4)}{c_p(T_3 - T_2)} = \frac{T_1 - T_2 + T_3 - T_4}{T_3 - T_2}$$

$$= \frac{T_3 - T_2 + T_1 - T_4}{T_3 - T_2} = 1 + \frac{T_1 - T_4}{T_3 - T_2} = 1 - \frac{T_1\left(\dfrac{T_4}{T_1} - 1\right)}{T_2\left(\dfrac{T_3}{T_2} - 1\right)} \qquad (5\text{-}6)$$

But since the processes are either isentropic or constant pressure, the following can be verified:

$$\frac{T_4}{T_1} = \frac{T_3}{T_2} \qquad (5\text{-}7)$$

So that the thermal efficiency, η_t, becomes

$$\eta_t = 1 - \frac{T_1}{T_2} = 1 - \frac{1}{\left(\dfrac{T_2}{T_1}\right)} = 1 - \frac{1}{\left(\dfrac{P_2}{P_1}\right)^{(k-1)/k}} \qquad (5\text{-}8)$$

Thus, for a cold air-standard Brayton cycle, the thermal efficiency is a function only of the pressure ratio. A plot of the thermal efficiency as a function of pressure ratio is presented in Figure 5.3. A pressure ratio of 20 represents a reasonable upper limit for typical gas turbines. The figure also demonstrates that even for high pressure

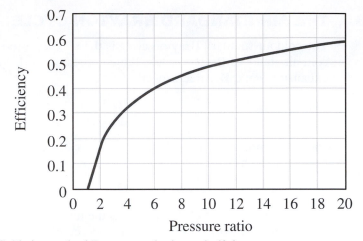

Figure 5.3 Cold air-standard Brayton cycle thermal efficiency.

ratios, the thermal efficiency is not likely to exceed 60 percent. The cold air-standard Brayton cycle thermal efficiency also represents an effective upper bound for gas turbine engine performance expectations.

5.4 | ACTUAL GAS TURBINE CYCLE ANALYSIS

The cold air-standard Brayton cycle is useful for quickly establishing limits and understanding the working principles of a gas turbine engine, but for more realistic analyses, irreversibilities (inefficiencies) must be included in the component models. The usual irreversibilities considered are losses in the compression and expansion processes and pressure drops in the heat addition process. Each will be examined in turn.

Consider the compression process with irreversibilities. Figure 5.4 illustrates a T-s diagram with an isentropic compression from P_1 to P_2 and a compression process, also from P_1 to P_2, with losses. The actual process is from state 1 to state 2, and the isentropic process is from state 1 to state 2s. The isentropic compression efficiency, η_c, is defined as the ratio of the work required in the isentropic process to the work required by the actual process, or

$$\eta_c = \frac{h_1 - h_{2s}}{h_1 - h_2} = \frac{c_p(T_1 - T_{2s})}{c_p(T_1 - T_2)} = \frac{T_1 - T_{2s}}{T_1 - T_2} = \frac{\dfrac{T_{2s}}{T_1} - 1}{\dfrac{T_2}{T_1} - 1} \tag{5-9}$$

But the process from 1 to 2s is isentropic, so that

$$\frac{T_{2s}}{T_1} = \left(\frac{P_2}{P_1}\right)^{(k-1)/k} \tag{5-10}$$

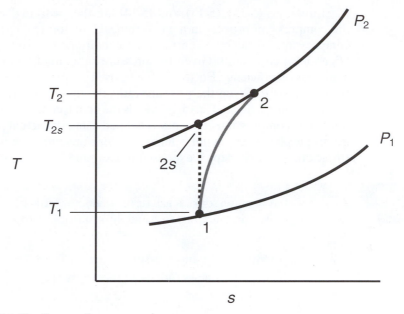

Figure 5.4 *T-s* diagram for compression processes.

The isentropic compression efficiency cast in terms of the compressor pressure ratio and the actual exit temperature is

$$\eta_c = \frac{\left(\dfrac{P_2}{P_1}\right)^{(k-1)/k} - 1}{\dfrac{T_2}{T_1} - 1} \tag{5-11}$$

The work of compression is

$$W_c = c_p(T_1 - T_2) = c_p T_1\left(1 - \frac{T_2}{T_1}\right) \tag{5-12}$$

Substituting Eq. (5-11) into Eq. (5-12) results in

$$W_c = \frac{c_p T_1}{\eta_c}\left[1 - \left(\frac{P_2}{P_1}\right)^{(k-1)/k}\right] \tag{5-13}$$

Equation (5-13) provides some insight into the thermodynamics of the compression process in gas turbines. The higher the inlet temperature T_1, the more work (per unit mass) required to compress a gas through a given pressure ratio. Not surprisingly, the lower the isentropic compression efficiency, the more compressor work required. Solving Eq. (5-11) for the actual compressor exit temperature yields

$$T_2 = T_1\left\{1 + \frac{1}{\eta_c}\left[\left(\frac{P_2}{P_1}\right)^{(k-1)/k} - 1\right]\right\} \tag{5-14}$$

Equations (5-11), (5-13), and (5-14) are the "working" expressions needed for the compression process in a gas turbine. Equation (5-11) provides the isentropic compression efficiency given the inlet conditions (P_1 and T_1) and the actual exit (P_2 and T_2) conditions. Given the pressure ratio, inlet temperature, and isentropic compression efficiency, Eq. (5-13) can be used to compute the work required to drive the compressor. Given the pressure ratio, the inlet temperature, and the isentropic compression efficiency, Eq. (5-14) yields the actual compressor exit temperature. The isentropic compression efficiency is defined using the actual process and a reference isentropic process and is not defined as a work-out and work-in relationship. The isentropic efficiency concept is exercised in Example 5.1.

Example 5.1	Air enters the compressor of a gas turbine engine at 100 kPa (absolute) and 300 K. If the compressor pressure ratio is 8 and the isentropic compression efficiency is 0.87, find the following:

(a) The exit temperature and pressure of the air from the compressor

(b) The work (per unit mass) added to the air by the compressor

(c) The exit temperature of the air if the compression process were isentropic

Solution   The process path on the *T-s* diagram is as indicated in Figure 5.4. From the problem statement,

$$T_1 = 300 \text{ K}$$

$$P_1 = 100 \text{ kPa}$$

$$\frac{P_2}{P_1} = 8$$

Since the working fluid is air, $k = 1.4$ and $c_p = 1.004$ kJ/kg K. The exit temperature can be calculated using Eq. (5-14):

$$T_2 = T_1\left\{1 + \frac{1}{\eta_c}\left[\left(\frac{P_2}{P_1}\right)^{(k-1)/k} - 1\right]\right\} = 300 \text{ K}\left[1 + \frac{1}{0.87}\left(8^{(1.4-1)/1.4} - 1\right)\right] = 580 \text{ K}$$

And the exit pressure is

$$P_2 = \frac{P_2}{P_1}P_1 = 8 \cdot 100 \text{ kPa} = 800 \text{ kPa}$$

The work imparted to the air is

$$W_c = c_p(T_1 - T_2) = 1.004 \frac{\text{kJ}}{\text{kg K}}(300 - 580)\text{ K} = -281 \frac{\text{kJ}}{\text{kg}}$$

The negative sign indicates that work is done on the air. If the compression process were isentropic, then the exit temperature would correspond to that of an isentropic compression, or

$$T_{2s} = T_1\left(\frac{P_2}{P_1}\right)^{(k-1)/k} = 300 \text{ K } (8)^{(1.4-1)/1.4} = 543 \text{ K}$$

The work for the isentropic compression would be

$$W_{\text{isen}} = \eta_c W_c = 0.87 \cdot \left(-281 \frac{\text{kJ}}{\text{kg}} \right) = -244 \frac{\text{kJ}}{\text{kg}}$$

A similar approach is used in defining the isentropic expansion efficiency. The T-s diagram for the expansion process is shown in Figure 5.5. The actual process is from state 3 to state 4, and the isentropic process is from state 3 to state 4s. The isentropic expansion efficiency, η_e, is defined as the ratio of the actual work extracted in the expansion process to the work that would be extracted if the process were isentropic, or

$$\eta_e = \frac{h_3 - h_4}{h_3 - h_{4s}} = \frac{c_p(T_3 - T_4)}{c_p(T_3 - T_{4s})} = \frac{T_3 - T_4}{T_3 - T_{4s}} = \frac{1 - \dfrac{T_4}{T_3}}{1 - \dfrac{T_{4s}}{T_3}} \tag{5-15}$$

But the process from 3 to 4s is isentropic, so

$$\frac{T_{4s}}{T_3} = \left(\frac{P_4}{P_3} \right)^{(k-1)/k} \tag{5-16}$$

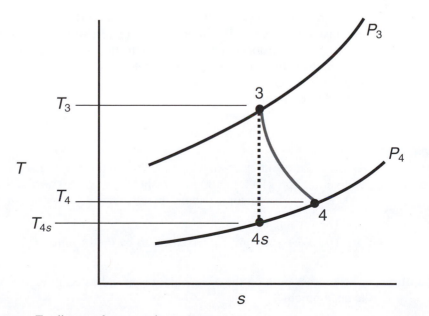

Figure 5.5 T-s diagram for expansion processes.

The isentropic expansion efficiency cast in terms of the turbine pressure ratio and the actual turbine exit temperature is

$$\eta_e = \frac{1 - \dfrac{T_4}{T_3}}{1 - \left(\dfrac{P_4}{P_3}\right)^{(k-1)/k}} \tag{5-17}$$

The work (per unit mass) extracted by the turbine is

$$W_e = c_p(T_3 - T_4) = c_p T_3 \left(1 - \frac{T_4}{T_3}\right) \tag{5-18}$$

Substituting Eq. (5-17) into Eq. (5-18) yields

$$W_e = \eta_e c_p T_3 \left[1 - \left(\frac{P_4}{P_3}\right)^{(k-1)/k}\right] \tag{5-19}$$

Guidance for the expansion process can be found in Eq. (5-19). The higher the turbine inlet temperature, T_3, the more work (per unit mass) can be extracted from a turbine for a given pressure ratio. This dependence on the turbine inlet temperature has been the driving impetus in the quest for materials that can tolerate higher temperatures in the turbine section. And the lower the isentropic expansion efficiency, the less work can be extracted. Solving Eq. (5-19) for the actual turbine exit temperature yields

$$T_4 = T_3 \left\{1 - \eta_e \left[1 - \left(\frac{P_4}{P_3}\right)^{(k-1)/k}\right]\right\} \tag{5-20}$$

Equations (5-17), (5-19), and (5-20) for the turbine (expansion process) are used in a fashion similar to Equations (5-11), (5-13), and (5-14) for the compressor. The isentropic expansion efficiency, similar to the isentropic compression efficiency, is defined using the actual process and a reference isentropic process, and is not defined as a work-in and work-out relationship. The isentropic expansion concept is demonstrated in Example 5.2.

Example 5.2

The inlet and outlet absolute pressures for the turbine section of a gas turbine engine are 550 kPa and 100 kPa, respectively. If the turbine inlet temperature is 1000 K and the isentropic expansion efficiency is 0.91, compute the following:

(a) The turbine exit temperature

(b) The work (per unit mass) extracted by the turbine

Solution The process path on the T-s diagram is as indicated in Figure 5.5. From the problem statement,

$$T_3 = 1000 \text{ K}$$

$$P_3 = 550 \text{ kPa}$$

$$\frac{P_4}{P_3} = \frac{100 \text{ kPa}}{550 \text{ kPa}} = 0.182$$

Since the working fluid is the combustion product, $k = 1.334$ and $c_p = 1.148$ kJ/kg K. The exit temperature can be calculated using Eq. (5-20):

$$T_4 = T_3\left[1 - \eta_e\left[1 - \left(\frac{P_4}{P_3}\right)^{(k-1)/k}\right]\right] = 1000 \text{ K}\left[1 - 0.91\left(1 - 0.182^{(1.33-1)/1.33}\right)\right] = 686 \text{ K}$$

The work extracted by the turbine is

$$W_e = c_p(T_3 - T_4) = 1.148\frac{\text{kJ}}{\text{kg K}}(1000 - 686)\text{ K} = 360\frac{\text{kJ}}{\text{kg}}$$

Some considerations for the combustor are appropriate. For first-order estimates, modeling the combustion process as a heat addition process does not introduce unacceptable errors. For the air-standard Brayton cycle analysis, $P_2 = P_3$; that is, no pressure change occurs across the combustor. In actual gas turbine combustors, P_3 is less than P_2: a pressure drop is experienced in the combustor. The pressure drop is caused by two effects: (1) friction and mixing losses in the combustor and (2) the Rayleigh effect ("simple" heat addition in an open system results in a loss in pressure) (Hodge and Koenig 1995). For well-designed combustors, the drop in pressure is small, typically less than 5 percent of the pressure at the inlet of the combustor.

The heat of reaction per unit mass resulting from the complete combustion of a fuel is defined as the fuel heating value. Heating values are specified either as the lower heating value (LHV) or the higher heating value (HHV). The use of HHV means that the water in the combustion products is in liquid form—that is, the water vapor has been condensed. Use of the LHV value means that the water in the combustion products is in a gaseous state. For most gas turbines, the LHV is used since the temperatures are above the condensing temperature of the water vapor. Table 5.1 provides a compilation of the nominal heating values for typical gas turbine fuels.

TABLE 5.1 Heating values of selected fuels

Fuel	Heating value (H_v)	
	Btu/lbm	MJ/kg
Jet A	18,400	42.8
Jet B/JP4	18,400	42.8
JP5	18,315	42.6
Natural gas	20,250	47.1
Propane	19,820	46.1
Ethanol	11,500	26.7

A simple model of the combustor equates the rate of energy release by the combustion process to the rate of heating of the air, or

$$\dot{m}_{\text{fuel}} H_v = \dot{m}_{\text{air}} c_p''(T_3 - T_2) \tag{5-21}$$

where c_p'' is the specific heat for the heat addition process, so that the fuel-to-air ratio, f/a, becomes

$$\frac{f}{a} = \frac{\dot{m}_{\text{fuel}}}{\dot{m}_{\text{air}}} = \frac{c_p''(T_3 - T_2)}{H_v} \tag{5-22}$$

The fuel-to-air ratio in a gas turbine is small because the amount of fuel that can be added is constrained by the high-temperature-limit mechanical properties of the materials used in the combustor and the turbine. Example 5.3 illustrates the concept.

Example 5.3	A combustion turbine utilizing natural gas as a fuel has a combustor exit temperature of 1500 K and a compressor exit temperature of 690 K. Estimate the fuel-to-air ratio.

Solution   Using the simple model for the combustor,

$$\frac{f}{a} = \frac{\dot{m}_{\text{fuel}}}{\dot{m}_{\text{air}}} = \frac{c_p''(T_3 - T_2)}{H_v} = \frac{1.076 \dfrac{\text{kJ}}{\text{kg K}}(1500\,\text{K} - 690\,\text{K}))}{46.1\,\dfrac{\text{MJ}}{\text{kg}}} = 0.019$$

The numbers in this example are typical for gas turbines. The fuel-to-air ratio is 0.019, which verifies that the fuel flow rate is quite low compared to the air flow rate.

The thermal efficiency of an actual gas turbine, neglecting the effects of the fuel flow rate, is defined as in Eq. (5-5).

$$\eta_t = \frac{\sum W}{Q_s} = \frac{W_{\text{net}}}{Q_s} = \frac{c_p(T_1 - T_2) + c_p'(T_3 - T_4)}{c_p''(T_3 - T_2)} \tag{5-23}$$

where c_p' is the specific heat for the turbine process.

For a combustion turbine that is producing electricity with no heat recovery, the most important metric is the heat rate, the input energy per kWh output. The smaller the heat rate, the better. Thus,

$$\text{Heat rate} = \frac{\text{energy in}}{\text{energy out}} = \frac{Q_s}{W_{\text{net}}} = \frac{c_p''(T_3 - T_2)}{c_p(T_1 - T_2) + c_p'(T_3 - T_4)} = \frac{1}{\eta_t} \tag{5-24}$$

The heat rate from Eq. (5-24) is based on the shaft power available from the operation of the turbine. If the generator (including any mechanical losses) has an efficiency of η_{gen}, then the dimensionless heat rate can be expressed as

$$\text{Heat rate} = \frac{1}{\eta_t \cdot \eta_{\text{gen}}}$$

However, the heat rate, especially for electrical generation, is usually expressed in the units of kJ/kWh or Btu/kWh. A more convenient representation of the heat rate thus becomes

$$\text{Heat rate} = \frac{1}{\eta_t \cdot \eta_{\text{gen}}} \cdot 3412 \frac{\text{Btu}}{\text{kWh}} = \frac{3412}{\eta_t \cdot \eta_{\text{gen}}} \frac{\text{Btu}}{\text{kWh}} = \frac{3600}{\eta_t \cdot \eta_{\text{gen}}} \frac{\text{kJ}}{\text{kWh}} \quad (5\text{-}25)$$

where 1 kWh = 3412 Btu is the conversion factor. In many instances, the generator efficiency, η_{gen}, is taken as unity. Many companies specify the heat rates of their combustion turbines in both Btu/kWh and kJ/kWh. Consider Example 5.4.

| **Example 5.4** | A combustion turbine possesses a thermal efficiency, η_t, of 31 percent and a generator efficiency of 0.98. What is the heat rate? |

Solution   The heat rate can be calculated directly from Eq. (5-25).

$$\text{Heat rate} = \frac{3412}{\eta_t \cdot \eta_{\text{gen}}} \frac{\text{Btu}}{\text{kWh}} = \frac{3412}{0.31 \cdot 0.98} \frac{\text{Btu}}{\text{kWh}} = 11{,}231 \frac{\text{Btu}}{\text{kWh}} = 11{,}850 \frac{\text{kJ}}{\text{kWh}}$$

The heat rate indicates that for every 11,231 Btu of input energy, 1 kWh of electricity is produced by the combustion turbine.

A more comprehensive example is in order.

| **Example 5.5** | A combustion turbine possesses the following characteristics: |

Compressor

   97 kPa (abs) and 30°C, inlet conditions
   Pressure ratio: 5.5
   Isentropic compression efficiency: 0.84

Combustor

   Outlet temperature: 1000°C
   Pressure loss: 3 percent
   Fuel: natural gas

Turbine

   Exit pressure: 100 kPa (abs)
   Isentropic expansion efficiency: 0.88

Generator

   Generator efficiency: 0.98

Determine the overall thermal efficiency and the heat rate, and sketch the T-s diagram. Is the fuel flow rate low compared to the air flow rate?

Solution If the fuel flow rate is low compared to the air flow rate, the thermal efficiency is described by Eq. (5-22). T_1, T_2, T_3, and T_4 are needed for calculating the thermal efficiency. The inlet temperature, T_1, is given as 30°C or 303 K. The compressor exit temperature, T_2, can be calculated as

$$T_2 = T_1 \left\{ 1 + \frac{1}{\eta_c} \left[\left(\frac{P_2}{P_1} \right)^{(k-1)/k} - 1 \right] \right\}$$

$$= 303 \text{ K} \left[1 + \frac{1}{0.84} \left(5.5^{(1.4-1)/1.4} - 1 \right) \right] = 529 \text{ K}$$

The compressor exit pressure is 5.5 · 97 kPa = 533.5 kPa. The combustor exit temperature, T_3, is specified as 1000°C or 1273 K, and the combustor exit pressure is

$$P_3 = (1 - 0.03) \cdot P_2 = 0.97 \cdot 533.5 \text{ kPa} = 517.5 \text{ kPa}$$

The turbine expands the flow from 517.5 kPa to 100 kPa, or

$$\frac{P_4}{P_3} = \frac{100 \text{ kPa}}{517.5 \text{ kPa}} = 0.193$$

The turbine exit temperature, T_4, becomes for these conditions

$$T_4 = T_3 \left\{ 1 - \eta_e \left[1 - \left(\frac{P_4}{P_3} \right)^{(k-1)/k} \right] \right\}$$

$$= 1273 \text{ K} \left[1 - 0.88 \left(1 - 0.193^{(1.33-1)/1.33} \right) \right] = 898 \text{ K}$$

The thermal efficiency can now be computed as

$$\eta_t = \frac{c_p(T_1 - T_2) + c_p'(T_3 - T_4)}{c_p''(T_3 - T_2)}$$

$$= \frac{1.004 \frac{\text{kJ}}{\text{kg K}} (303 \text{ K} - 529 \text{ K}) + 1.148 \frac{\text{kJ}}{\text{kg K}} (1273 \text{ K} - 898 \text{ K})}{1.076 \frac{\text{kJ}}{\text{kg K}} (1273 \text{ K} - 529 \text{ K})} = 0.256$$

The heat rate becomes

$$\text{Heat rate} = \frac{3412}{\eta_t \cdot \eta_{\text{gen}}} \frac{\text{Btu}}{\text{kWh}} = \frac{3412}{0.256 \cdot 0.98} \frac{\text{Btu}}{\text{kWh}} = 13,600 \frac{\text{Btu}}{\text{kWh}} = 14,349 \frac{\text{kJ}}{\text{kWh}}$$

The fuel-to-air ratio is calculated using the simple combustor model:

$$\frac{f}{a} = \frac{\dot{m}_{\text{fuel}}}{\dot{m}_{\text{air}}} = \frac{c_p''(T_3 - T_2)}{H_v} = \frac{1.076 \frac{\text{kJ}}{\text{kg K}} (1273\text{K} - 529\text{K})}{47.1 \frac{\text{MJ}}{\text{kg}}} = 0.017$$

Thus, the fuel flow rate is low compared to the air flow rate.

Taking into account the states (1, 2, 3, and 4) of the combustion turbine, the *T-s* diagram appears as in Figure 5.6.

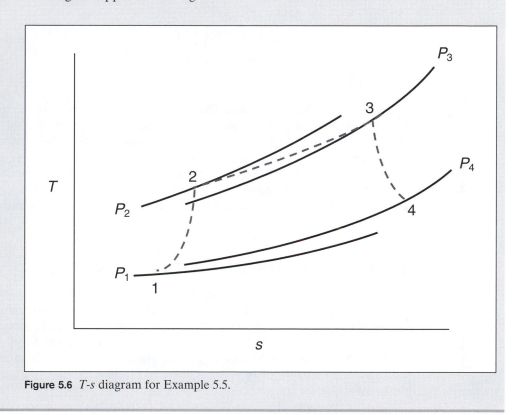

Figure 5.6 *T-s* diagram for Example 5.5.

Most of the development in this chapter to this point has been based on the assertion that the fuel flow rate is low in comparison to the air flow rate. Equation (5-23) for the thermal efficiency of a gas turbine was formulated based on this assertion. The thermal efficiency of a gas turbine including the effects of fuel flow rate can be developed by considering the power required by the compressor and extracted by the turbine, or

$$\dot{W}_c = \dot{m}_{air}c_p(T_1 - T_2)$$

$$\dot{W}_e = (\dot{m}_{air} + \dot{m}_{fuel})c_p'(T_3 - T_4) \tag{5-26}$$

The rate at which heat is supplied to the combustor is

$$\dot{Q}_s = (\dot{m}_{air} + \dot{m}_{fuel})c_p''(T_3 - T_4) \tag{5-27}$$

The thermal efficiency can be cast as

$$\eta_t = \frac{\sum \dot{W}}{\dot{Q}_s} = \frac{\dot{m}_{air}c_p(T_1 - T_2) + (\dot{m}_{air} + \dot{m}_{fuel})c_p'(T_3 - T_4)}{(\dot{m}_{air} + \dot{m}_{fuel})c_p''(T_3 - T_2)}$$

$$= \frac{c_p(T_1 - T_2) + \left(1 + \frac{f}{a}\right)c_p'(T_3 - T_4)}{\left(1 + \frac{f}{a}\right)c_p''(T_3 - T_2)} \tag{5-28}$$

where f/a is the fuel-to-air ratio. Thus, to account for the effect of the fuel mass flow rate, the thermal efficiency expression with the fuel mass flow rate included is very similar to the thermal efficiency expression, except that the factor $(1 + f/a)$ appears in the turbine work and heat addition terms. Cohen et al. (1972) point out that in many gas turbines, 1 or 2 percent of the compressed air is bled from the compressor and that the fuel flow rate is 1 or 2 percent of the air flow rate. Hence, the two cancel out and, as a result, the mass flow rates through the compressor and turbine tend to be equal.

5.5   COMBUSTION TURBINE CYCLE VARIATIONS

Figure 5.1 and the bulk of this chapter have been devoted to aspects of Brayton cycle–like gas turbines. However, as fuel prices have continued to escalate and as the electrical utility grid becomes more stressed, more thermally efficient combustion turbines are becoming increasingly attractive. Although there are many variations of the basic combustion turbine cycle, two adaptations are especially effective: (1) the addition of a recuperator or regenerator, and (2) the addition of an intercooler in the compressor section.

As the example problems demonstrate, turbine exhaust temperatures are well above ambient conditions and offer the potential for heat recovery. A recuperator is a heat exchanger that uses the turbine exhaust gases to preheat the air exiting the compressor before the air enters the combustor. Increasing the temperature of the air entering the combustor reduces the amount of fuel necessary for a given combustor exit temperature. A decrease in $\dot{Q}_s$ the denominator term in the thermal efficiency expression, Eq. (5-28), results in increased thermal efficiency. The schematic of a combustion turbine with a regenerator is provided in Figure 5.7.

The other common variation of the basic combustion turbine cycle is produced with the addition of an intercooler. As Eq. (5-13) indicates, the compressor work required to achieve a given pressure ratio is directly proportional to the inlet temperature. Thus, if the compression process is broken into two or more stages, the heat removed from an upstream stage results in a reduction in the inlet temperature of the downstream stage, reducing the compressor work required for the latter stage. Figure 5.8 illustrates a combustion turbine with an intercooler between the low-pressure and high-pressure compressor sections.

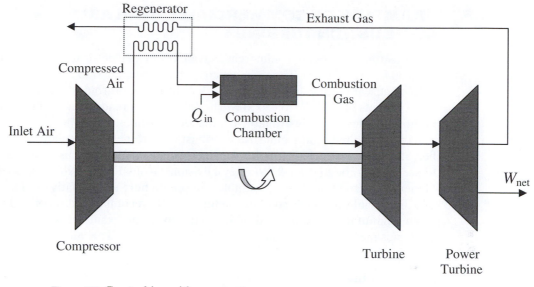

Figure 5.7 Gas turbine with regenerator.

For ground-based gas turbines, such as combustion turbines, the added weight and volume of a regenerator or intercooler are not problems. The inclusion of a regenerator or intercooler decreases the heat rate and increases the overall cycle thermal efficiency.

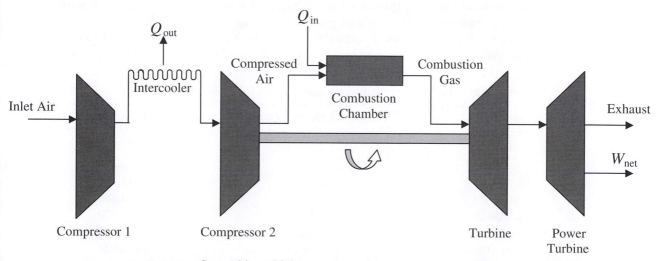

Figure 5.8 Gas turbine with intercooler and two compression stages.

5.6 EXAMPLES OF COMMERCIALLY AVAILABLE COMBUSTION TURBINES

Combustion turbines are commercially available in an electrical power output range from 30 kW to 250 MW. The term "microturbine" is applied to combustion turbines with electrical outputs less than 60–100 kW. Industrial turbines typically have electrical outputs of 0.5 MW to 250 MW. The Internet is a ready source of general information on commercially available combustion turbines.

In the United States, industrial combustion turbines are available from a number of companies, including Solar Turbines, Siemens-Westinghouse, United Technologies Corporation (Pratt & Whitney Division), Rolls-Royce (North America), and General Electric Energy. The examples presented here are a relatively small sample of commercially available combustion turbines. No recommendation or endorsement of any manufacturer is implied by its inclusion in the examples in this section; the examples merely present samples of what is commercially available.

Solar Turbines

Solar Turbines Incorporated manufactures a line of combustion turbines with electrical output in the range 1 to 15 MW. More then 12,000 Solar Turbines units have been installed worldwide. Table 5.2 lists the family of turbines available along with their nominal outputs and ISO heat rates.

For the Solar Turbines family of combustion turbines, the effect of size is interesting, as the heat rate is essentially monotonic with electrical output—the heat rate decreases with increasing output—except for the Mars turbines.

Figure 5.9 is a portion of the Solar Turbines data sheet for the Mars 90 combustion turbine. The nominal ISO performance is for 15°C at sea level with 60 percent relative humidity. The impact of the inlet air temperature on the electrical output and the heating rate is illustrated graphically in Figure 5.9. At the ISO rating conditions, the Mars 90 electrical output is 9.86 MW with a heat rate of 10,830 Btu/kWh. The turbine exhaust flow rate and temperature are also specified at the ISO condi-

TABLE 5.2 Solar Turbines family of combustion turbines

Name	Nominal Output (MW)	Heat Rate (Btu/kWh)
Saturn 20	1.185	13,900
Centaur 40	3.5	12,230
Centaur 50	4.57	11,400
Taurus 60	5.74	10,680
Taurus 70	7.69	9,800
Mars 90	9.86	10,260
Mars 100	11.185	10,040
Titan 130	15.29	9,421

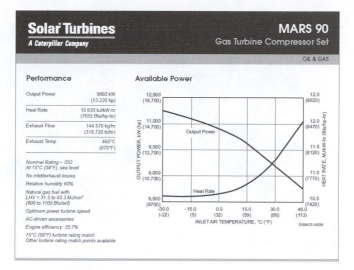

Figure 5.9 Performance data on the Solar Turbines Mars 90 (Solar Turbines).

tions. The effects of inlet air temperature changes from −22°F to 113°F are graphically illustrated. As the inlet temperature increases from 59°F (ISO rating condition) to 113°F, the output decreases from 9.86 MW to approximately 7.2 MW—a decrease of nearly 27 percent. The heat rate also increases substantially in this inlet temperature range. Decreasing output as the ambient temperature increases is a characteristic of gas turbines.

General Electric (GE) Energy

GE Energy is a major provider of a variety of energy systems in the United States. More than 9700 GE Energy gas turbines have been installed worldwide. The company offers two different lines of combustion turbines: (1) heavy-duty and (2) aeroderivative. The heavy-duty gas turbines are available with electrical outputs ranging from 26 MW to 480 MW and were specifically developed for utility and industrial applications. Details are available on the GE Energy website, www.gepower.com.

The GE Energy aeroderivative combustion turbines are based on their very successful line of aircraft engines. The company's aeroderivative line of gas turbines provides an output range of 13 MW to 100 MW as well as the ability to utilize a variety of fuels. In addition to power generation, the GE Energy aeroderivative gas turbines have been used for marine applications and a variety of oil and gas sector applications. Included are the LMS100, the LM6000, the LM2500, the LMS2000, and the LMS 1600. Details are available on each of the gas turbines on the GE Energy website. The LM2500 is a family of gas turbine engines with different electrical output energy ratings, different applications (including marine), and different cycle features. The LM2500 was derived from the CF6 aircraft engine, used on large commercial and military aircraft. The LM2500 and the LMS100 will be examined in more detail.

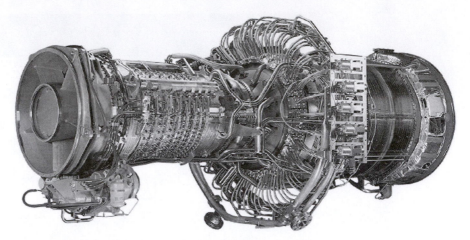

Figure 5.10 LM2500 cutaway drawing (GE Energy).

Figure 5.10 shows a cutaway drawing of the LM2500. Table 5.3 presents a summary of the LM2500 gas turbine performance, and Figure 5.11 illustrates the electrical output and heat rate as a function of the inlet temperature.

The LMS100 is an advanced aeroderivative combustion turbine that utilizes intercooling between the low- and high-pressure compressor stages. A unique feature of the LMS100 is that the intercooler heat exchanger is not adjacent to the engine package but is located away from the engine installation. Discharge air from the low-pressure compressor section is ducted to the intercooler heat exchanger, cooled, and returned to the inlet of the high-pressure compressor section. Such an arrangement provides significant flexibility in the choice of intercooler heat exchanger design. A cutaway drawing of the LMS100 is presented in Figure 5.12. The intercooler inlet and exit scrolls are indicated in the figure. A detailed discussion of the LMS100 is given by Reale (2004) and is available on the GE Energy website.

TABLE 5.3 LM2500 nominal performance summary

Power output	22,000 kW
Heat rate	9,565 Btu/kWh
Exhaust flow	149 lb/sec
Exhaust temperature	990°F
Power turbine speed	3,600 RPM
Number of compressor stages	16
Number of turbine stages	6

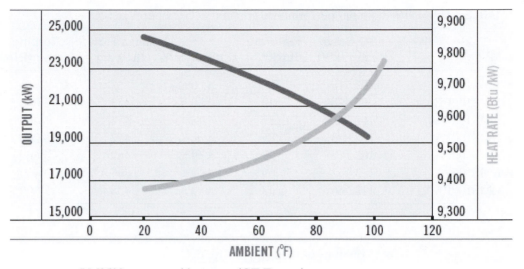

Figure 5.11 LM2500 output and heat rate (GE Energy).

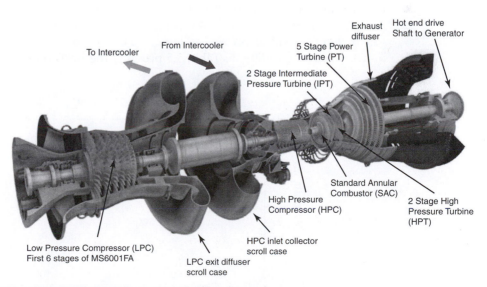

Figure 5.12 LMS100 cutaway drawing (GE Energy).

TABLE 5.4 LMS100 nominal performance summary (GE Energy)

Model	ISO Base Rating (kW)	Heat Rate (Btu/kWh)	Efficiency %	Mass Flow (lb/sec)	Turbine Speed (RPM)	Exhaust Temp (°F)
LMS100PB	97,718	7,592	45.0%	453	3,600	783
LMS100PB	97,878	7,579	45.0%	453	3,000	784
LMS100PA	102,795	7,287	46.8%	472	3,600	747
LMS100PA	102,833	7,285	46.8%	472	3,000	749
LMS100PA	103,112	7,773	43.9%	469	3,600	770
LMS100PA	103,162	7,769	43.9%	469	3,000	767
LMS100PA STIG	112,166	6,845	49.9%	438	3,600	729

Nominal performance characteristics at ISO standard rating conditions for different versions of the LMS100 are delineated in Table 5.4. The LMS100 output as a function of inlet temperature is illustrated in Figure 5.13, and Figure 5.14 shows the heat rate as a function of inlet temperature for the LMS100 STIG (steam-injected gas turbine) version.

Capstone Turbines

Capstone Turbines manufactures microturbines in the 30-kW to 200-kW range and is arguably the best known of the microturbine manufacturers in the United States. Capstone offers a nominal 30-kW unit, the C30, and a nominal 65-kW unit, the C65. Both the C30 and the C65 are available in versions that utilize different fuels and offer a number of options for exhaust stream heat recovery. The basic, natural gas–fueled C30 will be previewed here. The C30 is 76.5 in high, 30 in wide, and 60 in deep and weighs 891 lb. Table 5.5 presents the performance specifications at

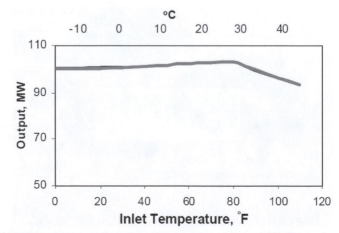

Figure 5.13 LMS100 output as a function of inlet temperature (GE Energy).

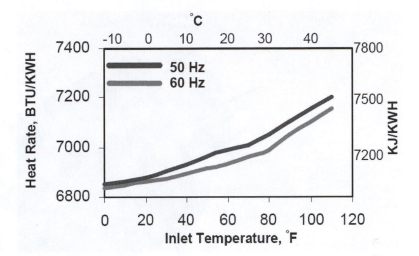

Figure 5.14 LMS100 STIG heat rate as a function of inlet temperature (GE Energy).

ISO conditions. Additional information is available on the website www.microturbine.com.

A cutaway view of the basic C30 is provided in Figure 5.15. The compact C30 layout is much different from layouts of the much-higher-output combustion turbines previously examined in this section. Natural gas–fired microturbines without recuperators produce electricity at efficiencies around 15 percent. Thus, most microturbines are equipped with a recuperator, which results in electrical generation efficiencies from 20 to 30 percent. Recuperated microturbines yield fuel savings of 30 to 40 percent by preheating the incoming combustion air. The C30 recuperator is indicated in the figure, as is the casing for the recuperator. The rotating components are attached to a single shaft and are supported by air bearings. The air bearings permit rotational speeds as high as 96,000 RPM. Because the generator is cooled by the inlet air flow and air bearings are utilized, no oil, lubricants, or coolants are required.

TABLE 5.5 Capstone C30 performance specifications (Capstone Turbines)

Power output	30 kW
Heat rate	13,100 Btu/kWh
Electrical efficiency (LHV)	26 percent
Exhaust temperature	530°F
Mass flow rate	0.68 lb/sec
Exhaust gas energy	310,000 Btu/h

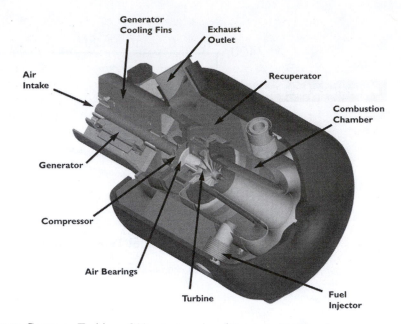

Figure 5.15 Capstone Turbines C30 cutaway view (Capstone Turbines).

The output electrical power and the electrical efficiency as a function of input temperature are illustrated in Figure 5.16. In a fashion similar to that of the large-power-output combustion turbines, the performance of the C30 decreases as the ambient temperature is increased.

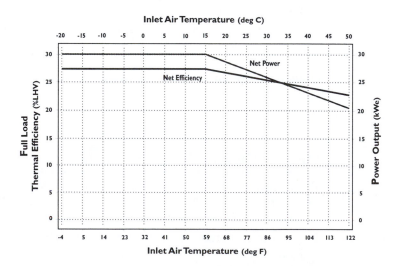

Figure 5.16 Power output and efficiency for the C30 (Capstone Turbines).

Combustion turbines are relatively inexpensive compared with most decentralized power generation systems. The basic cost of a combustion turbine ranges from $300 to $1000 per kW and generally decreases with increasing power output. Combustion turbines cost more than internal combustion (IC) engines with smaller power output but cost less than larger-power-output IC engines. Installation costs and the costs of other required components, plus other owner and miscellaneous costs increase the initial capital outlay by 30 to 50 percent. A natural gas compressor is an example of a component that is required in sequence since natural gas compressors are needed to meet the high gas pressure requirements of combustion turbines. Natural gas compressors increase the cost of a combustion turbine system by 5 to 10 percent. Also, if heat recovery capability is desired, the capital cost increases by $100 to $200 per kW. Fully installed, the typical cost of a combustion turbine with heat recovery is in the range of $1000 to $1200 per kW.

5.7 CLOSURE

The essentials of combustion turbines have been developed in this chapter, and some typical commercially available combustion turbines have been examined. In this chapter the focus has been on combustion turbines used for the generation of electricity only. However, as is evident from the exhaust temperatures given in the examples, the potential for significant thermal energy recovery exists. Chapter 11 examines the concept of CHP (combined heat and power), or the decentralized generation of electricity coupled with the recovery and use of exhaust thermal energy as from a combustion turbine.

REVIEW QUESTIONS

1. Explain the ideal Brayton cycle.
2. Why does it require more work to compress high-temperature air through a given pressure ratio than cooler air?
3. What is the heat rate of a combustion turbine?
4. Which heat rate is better, 10,000 Btu/kWh or 8000 Btu/kWh? Explain.
5. How does the heat rate generally vary with respect to turbine rated output?
6. Why is a high turbine inlet temperature desirable?
7. Why is the turbine exit flow rate often taken as equal to the inlet compressor flow rate?
8. Why does heat addition result in a pressure drop in the combustor?

EXERCISES

1. A gas turbine power plant produces 5000 kW of shaft power from inlet air at 101.3 kPa and 30°C. The compressor has a compression ratio of 5.5 and an isentropic efficiency of 0.84. In the combustion chamber, there is a pressure loss equal to 3 percent of the inlet air pressure, and the outlet temperature is 1000°C. If the turbine has an isentropic efficiency of 0.88 and an exhaust pressure of 100 kPa, determine the air flow rate and power plant thermal efficiency. How much thermal energy would be available for use by a regenerator?

2. A stationary gas turbine has compressor and turbine efficiencies of 0.85 and 0.90, respectively, and a compressor pressure ratio of 20. Determine the work of the compressor and the turbine, the net work, the turbine exit temperature, and the thermal efficiency for 80°F ambient and 1900°F turbine inlet temperatures. What is the heat rate for this device?

3. A gas turbine engine cycle has a compression ratio of 8 to 1. The compressor and turbine inlet temperatures are, respectively, 300 K and 1400 K. The compressor and turbine efficiencies are 0.85 and 0.90, respectively. Calculate the thermal efficiency of the cycle. What is the thermal efficiency of an air-standard Brayton cycle with a compression ratio of 8?

4. The following data are for a gas turbine power plant:

   Shaft output of 5000 kW

   Inlet air at 97 kPa and 30°C

   Compressor pressure ratio of 5.5

   Compressor isentropic efficiency of 0.84

   Combustion chamber outlet temperature of 1000°C

   Combustion chamber pressure loss is 3 percent of the inlet pressure

   Turbine isentropic efficiency of 0.88

   Turbine exhaust pressure of 100 kPa

   Determine the air flow rate and the thermal efficiency.

5. A gas turbine has 86 percent compressor and 89 percent turbine efficiencies, a compressor pressure ratio of 6, a 4 percent loss in pressure in the combustor, and a turbine inlet temperature of 1400°F. Ambient conditions are 60°F and 1 atmosphere. Determine the net work and the thermal efficiency for the system. Assume that the fuel flow rate is negligible compared with that of the air.

6. A stationary gas turbine has compressor and turbine efficiencies of 0.85 and 0.90, respectively, and a compressor pressure ratio of 20. Determine the work of the compressor and the turbine, the net work, the turbine exit temperature, and the thermal efficiency for 20°C ambient and 1200°C turbine inlet temperatures.

7. A gas turbine is designed for a turbine inlet temperature of 1450°F, a compressor pressure ratio of 12, and compressor and turbine efficiencies of 84 percent and 88 percent, respectively. Ambient conditions are 85°F and 14.5 psia.

(a) Draw and label a *T-s* diagram.

(b) Determine the compressor, turbine, and net work for this engine.

(c) Determine the thermal efficiency.

REFERENCES

Bathie, W. W. 1984. *Fundamentals of Gas Turbines*. New York: Wiley.

Cohen, H., Rogers, G. F. C., and Saravanamuttoo, H. I. H. 1972. *Gas Turbine Theory*, 2nd ed. London: Longman.

Harman, R. T. C. 1981. *Gas Turbine Engineering*. New York: Wiley/Halsted Press.

Hodge, B. K., and Koenig, K. 1995. *Compressible Fluid Dynamics with Personal Computer Applications*. Englewood Cliffs, NJ: Prentice-Hall.

Reale, M. J. 2004. "New High Efficiency Simple Cycle Gas Turbine—GE's LMS100." General Electric Company, GER 4222A, available at www.gepower.com.

Wilson, D. G. 1984. *The Design of High-Efficiency Turbomachinery and Gas Turbines*. Cambridge, MA: MIT Press.

WEBSITES

Combustion Turbine Manufacturers

www.gepower.com
www.siemenswestinghouse.com
www.pratt-whitney.com
www.rolls-royce.com
www.vericor.com

Microturbine Manufacturers

www.capstone.com
www.ingersoll-rand.com
www.turbec.com

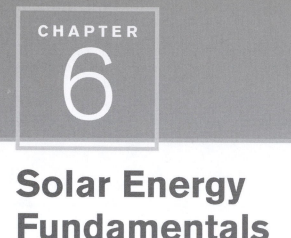

CHAPTER
6

Solar Energy Fundamentals

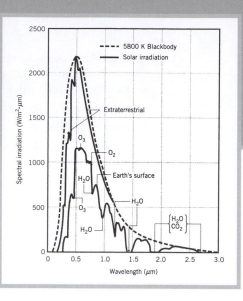

6.1 INTRODUCTION

This and the next three chapters will examine solar energy. The purposes of this chapter are to review radiation heat transfer and to present fundamental solar energy information necessary to understand the applications of the next three chapters.

6.2 RADIATION HEAT TRANSFER REVIEW

An understanding of solar energy must be based on familiarity with the spectral nature of solar radiation from the sun. This section reviews the fundamental radiation heat transfer information needed for studying solar energy engineering processes.

From a solar energy standpoint, two areas of interest are the characteristics of energy from the sun and the response of a surface to that energy. The starting point is to consider the characteristics of radiation heat transfer. Thermal radiation may be viewed as the transport of photons or the propagation of an electromagnetic wave. For the propagation of radiation in a medium, the frequency and wavelength are related as

$$\lambda = \frac{c}{\nu} \tag{6-1}$$

where c is the speed of light in the medium, λ is the wavelength, and ν is the frequency. In a vacuum, $c = 2.998 \times 10^8$ m/sec.

The electromagnetic continuum is divided into different regimes as a function of the wavelength. The regimes and wavelength ranges are delineated in Table 6.1.

Of interest in solar energy engineering is thermal radiation, in the wavelength range $10^{-1} < \lambda < 10^2$ μm. The visible part of the spectrum is in the range $0.4\,\mu m < \lambda < 0.7\,\mu m$ and is bounded by the ultraviolet and the infrared. Parts of the ultraviolet and the infrared regimes are also in the thermal radiation regime.

TABLE 6.1 The electromagnetic continuum and regimes

Regime	Range
Gamma rays	$\lambda < 10^{-4}$ μm
X-rays	$10^{-4} < \lambda < 10^{-2}$ μm
Ultraviolet	$10^{-2} < \lambda < 0.4$ μm
Thermal radiation	$\mathbf{10^{-1} < \lambda < 10^2}$ **μm**
Infrared	$0.7 < \lambda < 10^2$ μm
Microwave	$\lambda > 10^2$ μm

The magnitude of thermal radiation varies with the wavelength and gives rise to the spectral distribution. In addition to the spectral nature of thermal radiation, a key feature is the directional distribution. In order for thermal radiation to be quantified, both the spectral and directional attributes must be examined.

A good starting point is to consider the characteristics of a blackbody:

(1) A blackbody absorbs all incident radiation, regardless of wavelength and direction.

(2) For a prescribed temperature and wavelength, no surface can emit more energy than a blackbody.

(3) The blackbody is a diffuse (independent of direction) emitter.

The spectral emissive power density of a blackbody is given by the Planck distribution (Incropera and deWitt, 1999),

$$E_{\lambda,b}(\lambda,T) = \frac{C_1}{\lambda^5 \cdot \left[\exp\left(\dfrac{C_2}{\lambda \cdot T}\right) - 1\right]} \tag{6-2}$$

where $C_1 = 3.742 \cdot 10^8$ W μm^4/m^2 and $C_2 = 1.439 \cdot 10^4$ μm K. The spectral emissive power density, $E_{\lambda,b}$, is customarily expressed in the units of W/μm m^2 and is interpreted as the emissive power per unit area per wavelength. The subscripts, λ and b, in Eq. (6-2) signify that the power density is given per wavelength (spectral) and is from a blackbody, respectively. The total emissive power of a blackbody is the integral of the blackbody spectral emissive power density over all wavelengths, or

$$E_b(T) = \int_0^\infty E_{\lambda,b}d\lambda = \int_0^\infty \frac{C_1}{\lambda^5 \cdot \left[\exp\left(\dfrac{C_2}{\lambda \cdot T}\right) - 1\right]} d\lambda = \sigma T^4 \tag{6-3}$$

where $\sigma = 5.67 \cdot 10^{-8}$ W/m^2 K^4, the Stefan-Boltzmann constant. The T^4 functional dependence of the blackbody emissive power is one of the strongest functional dependencies in physics and emphasizes the importance of temperature in thermal radiation.

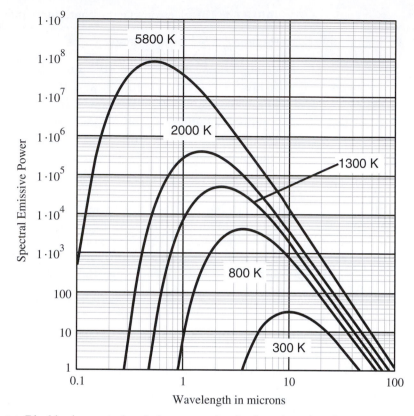

Figure 6.1 Blackbody spectral emissive power density for selected surface temperatures.

Additional examination of Eq. (6-2) is warranted. Figure 6.1, plotted in log-log coordinates, presents a graphical study of the blackbody spectral emissive power density for several different temperatures. $T = 5800$ K is one of the temperatures considered because the spectral distribution of thermal radiation from the sun is close to that of a blackbody at 5800 K. The pronounced effect of temperature on the blackbody spectral emissive power is demonstrated by its range—eight orders of magnitude, from 1 W/μm m^2 to nearly 10^8 W/μm m^2. Also of interest is the location of the maximum spectral emissive power as a function of surface temperature. For example, at $T = 300$ K, close to room temperature, the maximum occurs near $\lambda = 10$ μm, but as the temperature is increased, the maximum shifts to shorter wavelengths; thus, at $T = 5800$ K, the maximum is near 0.5 μm. Not surprisingly, $\lambda = 0.5$ μm is the middle of the visible spectrum for which vision on Earth is optimally adapted. The thermal radiation range given in Table 6.1 is congruent with the results of Figure 6.1, as the $10^{-1} < \lambda < 10^2$ μm range of Figure 6.1 contains virtually all of the emitted energy for these relevant temperatures. The total blackbody emissive power is the "area under the curve," and the blackbody emissive power at $T = 5800$ K is 140,000 times that at $T = 300$ K.

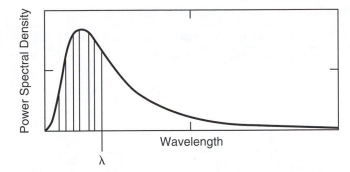

Figure 6.2 Blackbody radiation function illustration.

For many radiation heat transfer investigations, the fraction of energy contained between a wavelength interval is needed. The basis for obtaining such fractions is the blackbody radiation function. Using the Planck distribution, the fraction of thermal energy in the wavelength range 0 to λ is cast as follows:

$$F(0 - \lambda T) = \frac{\int_0^\lambda E_{\lambda,b}\,d\lambda}{\int_0^\infty E_{\lambda,b}\,d\lambda} = \frac{\int_0^\lambda E_{\lambda,b}\,d\lambda}{\sigma T^4} = \int_0^{\lambda T} \frac{E_{\lambda,b}}{\sigma T^5}\,d(\lambda T) \qquad (6\text{-}4)$$

Figure 6.2 illustrates the concept graphically.

Many heat transfer textbooks contain tabulations of the blackbody radiation function; alternatively, Eqs. (6-2) and (6-4) can be integrated to give the blackbody radiation function values. Moreover, Siegel and Howell (2002) have devised a closed-form series that replicates, to within a fraction of a percent, the accepted blackbody radiation function values and also provides a convenient expression for use in many problem solutions. Their series appears as

$$F(\zeta) = \frac{15}{\pi^4} \sum_{n=1}^{20} \frac{e^{-n\zeta}}{n} \left(\frac{6}{n^3} + 6\frac{\zeta}{n^2} + \zeta^3 + 3\frac{\zeta^2}{n} \right) \qquad (6\text{-}5)$$

where $\zeta = C_2/\lambda T$. Table 6.2 contains a tabular listing of Eq. (6-5) values as a function of λT. Table 6.2 and Eq. (6-5) will be used in the solution of example problems requiring the use of the blackbody radiation function. Example 6.1 illustrates the use of the blackbody radiation function.

Example 6.1

For a blackbody at 5800 K, what is the percentage of energy contained in the visible part of the spectrum? Contrast this with the percentage of energy contained in the visible spectrum for a blackbody at 3000 K.

Solution Two different approaches for obtaining the solution will be presented and discussed: (1) the use of Table 6.2 and (2) the use of Eq. (6-5) and Mathcad.

TABLE 6.2 Blackbody radiation function tabulation

λT (μm·K)	F(0 − λT)	λT (μm·K)	F(0 − λT)	λT (μm·K)	F(0 − λT)	λT (μm·K)	F(0 − λT)
200	0	5200	0.6579	10500	0.9236	25000	0.9922
400	$1.8554 \cdot 10^{-12}$	5400	0.6802	11000	0.9318	30000	0.9953
600	$9.263 \cdot 10^{-8}$	5600	0.7009	11500	0.9389	35000	0.997
800	$1.6396 \cdot 10^{-5}$	5800	0.72	12000	0.945	40000	0.9979
1000	0.0003	6000	0.7377	12500	0.9504	45000	0.9985
1200	0.0021	6200	0.754	13000	0.9551	50000	0.9989
1400	0.0078	6400	0.7691	13500	0.9592	55000	0.9992
1600	0.0197	6600	0.7831	14000	0.9628	60000	0.9993
1800	0.0393	6800	0.796	14500	0.9661	65000	0.9995
2000	0.0667	7000	0.808	15000	0.9689	70000	0.9996
2200	0.1008	7200	0.8191	15500	0.9715	75000	0.9997
2400	0.1402	7400	0.8294	16000	0.9738	80000	0.9997
2600	0.183	7600	0.839	16500	0.9758	85000	0.9998
2800	0.2278	7800	0.8479	17000	0.9776	90000	0.9998
3000	0.2731	8000	0.8562	17500	0.9793	95000	0.9998
3200	0.318	8200	0.8639	18000	0.9808	$1 \cdot 10^5$	0.9998
3400	0.3616	8400	0.8711	18500	0.9822		
3600	0.4035	8600	0.8778	19000	0.9834		
3800	0.4433	8800	0.8841	19500	0.9845		
4000	0.4808	9000	0.8899	20000	0.9855		
4200	0.5159	9200	0.8954				
4400	0.5487	9400	0.9006				
4600	0.5792	9600	0.9054				
4800	0.6074	9800	0.9099				
5000	0.6336	10000	0.9141				

The visible spectrum spans the wavelength range $0.4 \, \mu m \leq \lambda \leq 0.7 \, \mu m$. Figure 6.3 shows the range of interest for the 5800 K blackbody. $F(0 - \lambda_2 T)$ provides the fraction of radiant energy from 0 to λ_2, and $F(0 - \lambda_1 T)$ provides the fraction from 0 to λ_1. The fraction of radiant energy contained between λ_1 and λ_2 can be expressed as the difference

$$F(\lambda_2 T - \lambda_1 T) = F(0 - \lambda_2 T) - F(0 - \lambda_1 T) \tag{6-6}$$

The Table 6.2 approach will be illustrated first. For the blackbody at 5800 K,

$$\lambda_1 T = 0.4 \; \mu m \; 5800 \; K = 2320 \; \mu m \, K$$
$$\lambda_2 T = 0.7 \; \mu m \; 5800 \; K = 4060 \; \mu m \, K$$

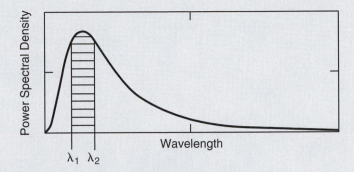

Figure 6.3 Visible spectrum range for a 5800 K blackbody.

Interpolating between the entries in Table 6.2 yields

$$F(2320 \ \mu m \, K) = 0.1244$$
$$F(4060 \ \mu m \, K) = 0.4913$$

so the fractional radiant energy contained between $\lambda_1 = 0.4 \ \mu m$ and $\lambda_2 = 0.7 \ \mu m$ becomes

$$F(\lambda_2 T - \lambda_1 T) = 0.4913 - 0.1244 = 0.3669$$

Thus, for a 5800 K blackbody, 37 percent of the radiant energy is contained in the visible part of the spectrum.

The same procedure is applied for the 3000 K blackbody except that the temperature used is 3000 K, not 5800 K. The results can be summarized as

$$\lambda_1 T = 0.4 \ \mu m \cdot 3000 \ K = 1200 \ \mu m \, K$$
$$\lambda_2 T = 0.7 \ \mu m \cdot 5800 \ K = 2100 \ \mu m \, K$$

$$F(1200 \ \mu m \, K) = 0.0021$$
$$F(2100 \ \mu m \, K) = 0.0838$$

$$F(\lambda_2 T - \lambda_1 T) = 0.0830 - 0.0021 = 0.0817$$

Hence, for the 3000 K blackbody, only 8 percent of the total radiant energy is in the visible range.

Figure 6.4 presents the solution using Eq. (6-5) and Mathcad. The same procedure is followed, with the only major difference being the use of Eq. (6-5) for the blackbody radiation function. The use of the Siegel-Howell expression is very convenient, as it obviates the need for table lookups and interpolations to solve spectral distribution problems. As the remainder of the section will demonstrate, this ability will be important for a number of thermal radiation problems not limited to blackbody formulations. The results in Figure 6.4 are more accurate than using Table 6.2 since no interpolation is required, but the differences between the two approaches are so negligible that for most purposes either will suffice.

$$\sigma := 5.67 \cdot 10^{-8} \cdot \frac{W}{m^2 \cdot K^4} \qquad \mu m := \frac{m}{10^6}$$

The Stefan-Boltzmann constant and μm.

$$C_1 := 3.742 \cdot 10^8 \cdot W \cdot \frac{\mu m^4}{m^2} \qquad C_2 := 1.439 \cdot 10^4 \cdot \mu m \cdot K$$

C_1 and C_2 for the Planck power spectral density expression.

$$F(\zeta) := \frac{15}{\pi^4} \cdot \left[\sum_{n=1}^{20} \frac{e^{-n \cdot \zeta}}{n} \cdot \left(\frac{6}{n^3} + 6 \cdot \frac{\zeta}{n^2} + \zeta^3 + 3 \cdot \frac{\zeta^2}{n} \right) \right] \qquad \zeta = \frac{C_2}{\lambda \cdot T} \quad \text{Definition of } \zeta$$

Blackbody at 5800 K.

$$\lambda T_1 := 0.4 \cdot \mu m \cdot 5800 \cdot K \qquad\qquad F\left(\frac{C_2}{\lambda T_1} \right) = 0.12392$$

$$\lambda T_2 := 0.7 \cdot \mu m \cdot 5800 \cdot K \qquad\qquad F\left(\frac{C_2}{\lambda T_2} \right) = 0.49154$$

$$Fvisible := F\left(\frac{C_2}{\lambda T_2} \right) - F\left(\frac{C_2}{\lambda T_1} \right) \qquad Fvisible = 0.36762$$

Blackbody at 3000 K:

$$\lambda T_1 := 0.4 \cdot \mu m \cdot 3000 \cdot K \qquad\qquad F\left(\frac{C_2}{\lambda T_1} \right) = 2.13115 \times 10^{-3}$$

$$\lambda T_2 := 0.7 \cdot \mu m \cdot 3000 \cdot K \qquad\qquad F\left(\frac{C_2}{\lambda T_2} \right) = 0.083$$

$$Fvisible := F\left(\frac{C_2}{\lambda T_2} \right) - F\left(\frac{C_2}{\lambda T_1} \right) \qquad Fvisible = 0.08087$$

Figure 6.4 Mathcad worksheet for the solution to Example 6.1.

The source of solar energy on the Earth is, obviously, the sun. The sun is located 1.50×10^{11} m from the Earth and has a diameter of 1.39×10^9 m. The solar constant, defined as the flux of solar energy incident on the surface oriented normal to the rays of the sun, has a mean value of 1353 W/m². However, the solar constant is the flux incident at the outer edge of the atmosphere. For terrestrial applications, what is important is the solar flux incident on the surface of the Earth, after the sun's rays have passed through the atmosphere. Figure 6.5 provides a quantitative assessment of the solar flux. The spectral distribution is illustrated for extraterrestrial conditions (incident on the edge of the atmosphere) and on the Earth's surface. The extraterrestrial solar flux has the approximate spectral distribution of blackbody radiation

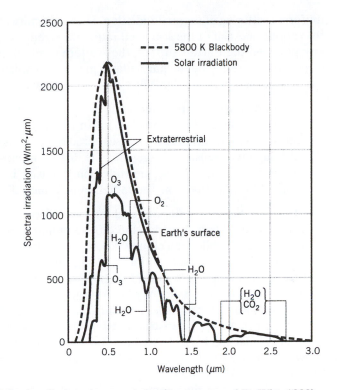

Figure 6.5 Solar irradiation characteristics (Incropera and DeWitt, 1999).

at 5800 K, *but not the total emissive power of a blackbody* — as a comparison of Figures 6.1 and 6.2 will confirm.

As the solar flux passes through the atmosphere, it is attenuated by the presence of oxygen (O_2), ozone (O_3), water vapor (H_2O), and carbon dioxide (CO_2). As a result, the solar flux at the Earth's surface is significantly less than the extraterrestrial solar flux. As the solar flux traverses the atmosphere, both the magnitude and directional distributions are altered. These alterations are due to absorption and scattering. The lower plot in Figure 6.5 demonstrates the absorption of radiant energy by gases in the atmosphere. In the ultraviolet regime, $10^{-2} < \lambda < 0.4$ μm, ozone is responsible for the absorption and results in virtually no solar flux reaching the surface for $\lambda < 0.3$ μm. Part of the concern about depletion of the ozone layer is that without the ozone, much more of the thermal radiation in the ultraviolet regime would reach the surface of the Earth, with long-term deleterious results. In the infrared regime, the absorption is primarily due to water vapor, although CO_2 also plays a role. For all wavelengths, thermal radiation is also absorbed by the particulate contents of the atmosphere.

In addition to absorption, scattering occurs, resulting in the redirection of the sun's rays. Rayleigh scattering, attributable to the gas molecules, provides for virtually uniform scattering of thermal radiation. The uniformity of Rayleigh scattering means that about one-half of the scattered radiation is redirected away from the Earth. At any point on the Earth's surface, the scattered radiation is incident from all directions. Mie scattering by the particulates in the atmosphere does little to reorient the sun's rays. Incropera and DeWitt (1999) have provided a cogent summary:

> That portion of the radiation that has penetrated the atmosphere without having been scattered (or absorbed) is in the direction of the zenith angle and is termed the direct radiation. The scattered radiation is incident from all directions, although its intensity is largest for the directions close to that of the direct radiation. However, because the radiation intensity is often assumed to be independent of the direction, the radiation is termed diffuse.

The total thermal radiation reaching the surface is the sum of the direct and the diffuse amounts. On a clear day, the diffuse radiation may account for only 10 percent of the total, while on a cloudy day, it is essentially 100 percent of the total.

With the characteristics of thermal radiation from the sun understood, consideration of radiation heat transfer fundamentals for a surface is appropriate. Specific topics include (1) what happens when thermal radiation is incident on a surface and (2) the emission characteristics of a surface. Emission will be examined first.

The emissive characteristics of a blackbody have been established. Emissions from a non-blackbody can have directional and spectral characteristics. Consider, as shown in Figure 6.6, a hemisphere surrounding an elemental area that is emitting thermal radiation. The hemisphere about the emitting area dA is a convenient mechanism for visualizing the geometry of thermal radiation. The spectral intensity, $I_{\lambda,e}$, of the emitted radiation is defined as the rate at which radiant energy is emitted at the wavelength λ in the (θ, ϕ) direction per unit area of the emitting surface normal to this direction, per unit solid angle about this direction, and per unit wavelength interval $d\lambda$ about λ. The normal to the emitting surface area in the (θ, ϕ) direction is the area dA_n in Figure 6.6. If the blackbody surface is used as the basis, the spectral directional emissivity, $\varepsilon_{\lambda,\theta}(\lambda, \theta, \phi, T)$, is defined as the ratio of the emitted radiation intensity in a given direction at a given wavelength, divided by the radiation intensity of a blackbody, or

$$\varepsilon_{\lambda,\theta}(\lambda,\theta,\phi,T) = \frac{I_{\lambda,e}(\lambda,\theta,\phi,T)}{I_{\lambda,b}(\lambda,T)} \tag{6-7}$$

Equation (6-7) provides a process for computing the actual emission in reference to that of a blackbody if $\varepsilon_{\lambda,\theta}(\lambda, \theta, \phi, T)$ is known. If the spectral directional intensity is integrated over $0 \leq \phi \leq 2\pi$ and $0 \leq \theta \leq \pi/2$, the spectral hemispherical emissivity, $\varepsilon_\lambda(\lambda, T)$, results and is defined as

$$\varepsilon_\lambda(\lambda,T) = \frac{I_{\lambda,e}(\lambda,T)}{I_{\lambda,b}(\lambda,T)} = \frac{E_\lambda(\lambda,T)}{E_{\lambda,b}(\lambda,T)} \tag{6-8}$$

where $E_\lambda(\lambda, T)$ is the spectral hemispherical emissive power and $\varepsilon_\lambda(\lambda, T)$ is the spectral hemispherical emissivity. Since the angular dependence has been inte-

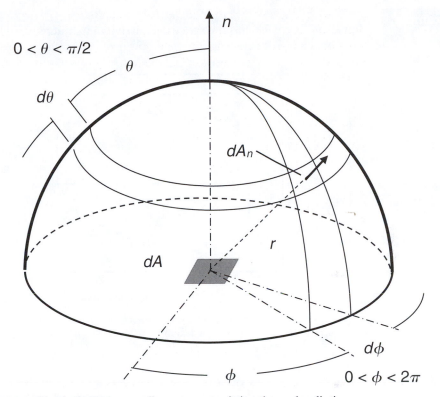

Figure 6.6 Hemisphere surrounding an area emitting thermal radiation.

grated out, the word "hemispherical" rather than "directional" is used. The transition from $I_{\lambda,e}(\lambda, T)$ to $E_\lambda(\lambda, T)$ is seamless since the emission intensity is based on the projected area and the emission is based on the actual area, and they are related as $E_\lambda(\lambda, T) = \pi\, I_{\lambda,e}(\lambda, T)$. Details are provided in heat transfer textbooks such as Incropera and DeWitt (1999). If the spectral hemispherical emissive power is integrated over all wavelengths, the total hemispherical emissive power, $E(T)$, is recovered and appears as

$$\varepsilon(T) = =\frac{E(T)}{E_b(T)} \tag{6-9}$$

where $E(T)$ is the total hemispherical emissive power and $\varepsilon(T)$ is the total hemispherical emissivity, generally referred to as simply the emissivity. The sequence is from spectral directional to spectral hemispherical to total hemispherical. Values for the spectral hemispherical and total hemispherical emissivities for various materials are available from handbooks and textbooks and on a number of websites (www.icess.ucsb/modis/emis, for example).

<table>
<tr><td>**Example 6.2**</td><td>A wall at $T_s = 500$ K has the spectral emissivity illustrated in Figure 6.7. Determine the total emissivity and the emissive power of the surface.</td></tr>
</table>

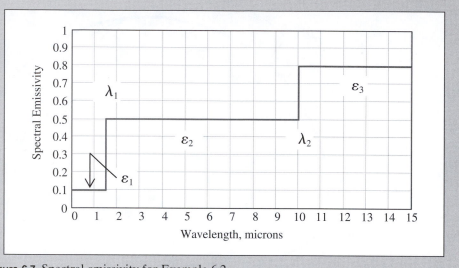

Figure 6.7 Spectral emissivity for Example 6.2.

Solution Once the total emissivity is known, Eq. (6-9) can be used to calculate the emissive power. The emissive power of the surface is composed of the individual contribution for each wavelength range associated with each spectral hemispherical emissivity, or

$$\varepsilon E_b = \varepsilon_1 \int_0^{\lambda_1} E_{\lambda,b} d\lambda + \varepsilon_2 \int_{\lambda_1}^{\lambda_2} E_{\lambda,b} d\lambda + \varepsilon_3 \int_{\lambda_2}^{\infty} E_{\lambda,b} d\lambda \tag{6-10}$$

which can be expressed as

$$\varepsilon = \varepsilon_1 F(0 - \lambda_1 T) + \varepsilon_2 [F(0 - \lambda_2 T) - F(0 - \lambda_1 T)] + \varepsilon_3 [1 - F(0 - \lambda_2 T)] \tag{6-11}$$

Thus, by using the spectral emissive description of the surface behavior, the blackbody radiation function can be used to make computations for non-blackbody surfaces.

   The Mathcad worksheet for this problem is presented in Figure 6.8. The total emissivity is 0.6099, and the total emissive power is 2161 W/m². By comparison, if the surface were a blackbody, the total emissive power would be 3544 W/m².

Evaluation of the total emissivity:

$$\lambda T_1 := 1.5 \cdot \mu m \cdot 500 \cdot K \qquad F\left(\frac{C_2}{\lambda T_1}\right) = 5.93352 \times 10^{-6}$$

$$\lambda T_2 := 10 \cdot \mu m \cdot 500 \cdot K \qquad F\left(\frac{C_2}{\lambda T_2}\right) = 0.63363$$

$$\varepsilon := 0.1 \cdot F\left(\frac{C_2}{\lambda T_1}\right) + 0.5 \cdot \left(F\left(\frac{C_2}{\lambda T_2}\right) - F\left(\frac{C_2}{\lambda T_1}\right)\right) + 0.8 \cdot \left(1 - F\left(\frac{C_2}{\lambda T_2}\right)\right)$$

$$\varepsilon = 0.60991$$

Compute the total emissive power per unit area:

$$T_s := 500 \cdot K \qquad E := \varepsilon \cdot \sigma \cdot T_s^{\,4} \qquad E = 2.16137 \times 10^3 \frac{W}{m^2}$$

Blackbody emissive power:

$$E_b := \sigma \cdot T_s^{\,4} \qquad E_b = 3.54375 \times 10^3 \frac{W}{m^2}$$

Figure 6.8 Mathcad solution for Example 6.2.

Equations (6-7) to (6-9) involve surface emissions of thermal radiation. Radiation incident on a surface is called irradiation and is denoted G, and the spectral irradiation (the irradiation at a given wavelength) is denoted G_λ. Three things happen to thermal radiation incident on a surface: (1) some is absorbed, (2) some is reflected, and (3) some is transmitted. Figure 6.9 schematically illustrates these processes. Consider each of the possibilities in turn. Irradiation, like emitted radiation, has a spectral nature; but unlike emitted radiation, whose spectral characteristics are determined by

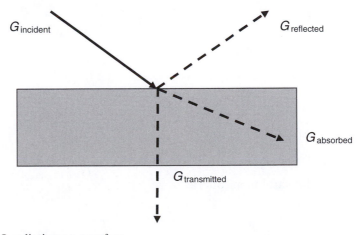

Figure 6.9 Irradiation on a surface.

the surface temperature, the spectral characteristics of irradiation are determined by the source of the irradiation. And just like emitted radiation, irradiation has directional attributes. Surface properties associated with irradiation are described by three terms: (1) *absorptivity*, for the absorbed irradiation; (2) *reflectivity*, for the reflected irradiation, and (3) *transmissivity*, for the transmitted irradiation. The same hierarchy of spectral directional, to spectral hemispherical, to total hemispherical is used. Working definitions are as follows:

Absorptivity

Spectral directional

$$\alpha_{\lambda,\theta}(\lambda,\theta,\phi) = \frac{I_{\lambda,i,\text{absorbed}}(\lambda,\theta,\phi)}{I_{\lambda,i}(\lambda,\theta,\phi)} \tag{6-12}$$

Spectral hemispherical

$$\alpha_\lambda(\lambda) = \frac{G_{\lambda,\text{absorbed}}(\lambda)}{G_\lambda(\lambda)} \tag{6-13}$$

Total hemispherical

$$\alpha = \frac{\int_0^\infty \alpha_\lambda G_\lambda(\lambda) d\lambda}{\int_0^\infty G_\lambda d\lambda} \tag{6-14}$$

where $I_{\lambda,i}$ is the incident irradiation flux and G_λ is spectral irradiation. An analogous sequence is defined for the reflectivity and the transmissivity.

Reflectivity

Spectral directional

$$\rho_{\lambda,\theta}(\lambda,\theta,\phi) = \frac{I_{\lambda,i,\text{reflected}}(\lambda,\theta,\phi)}{I_{\lambda,i}(\lambda,\theta,\phi)} \tag{6-15}$$

Spectral hemispherical

$$\rho_\lambda(\lambda) = \frac{G_{\lambda,\text{reflected}}(\lambda)}{G_\lambda(\lambda)} \tag{6-16}$$

Total hemispherical

$$\rho = \frac{\int_0^\infty \rho_\lambda G_\lambda(\lambda) d\lambda}{\int_0^\infty G_\lambda d\lambda} \tag{6-17}$$

Transmissivity

Spectral directional

$$\tau_{\lambda,\theta}(\lambda,\theta,\phi) = \frac{I_{\lambda,i,\text{transmitted}}(\lambda,\theta,\phi)}{I_{\lambda,i}(\lambda,\theta,\phi)} \tag{6-18}$$

Spectral hemispherical

$$\tau_\lambda(\lambda) = \frac{G_{\lambda,\text{transmitted}}(\lambda)}{G_\lambda(\lambda)} \tag{6-19}$$

Total hemispherical

$$\tau = \frac{\int_0^\infty \tau_\lambda G_\lambda(\lambda) d\lambda}{\int_0^\infty G_\lambda d\lambda} \tag{6-20}$$

For an opaque surface, a surface that does not transmit thermal radiation, the transmissivity terms are all zero.

Conservation of energy and the above definitions demand that

$$1 = \alpha_\lambda + \rho_\lambda + \tau_\lambda$$
$$1 = \alpha + \rho + \tau \tag{6-21}$$

Example 6.3	If the spectral absorptivity is equal to the spectral emissivity in Example 6.2, $\varepsilon_\lambda = \alpha_\lambda$, and if the irradiation, G, of 750 W/m^2 has the spectral characteristics of a blackbody at 2000 K, find the total absorptivity, the total reflectivity, and the irradiation absorbed if the surface is opaque.

Solution   The spectral absorptivity is known from Example 6.2. Since the irradiation has the spectral characteristics of a blackbody at 2000 K, the blackbody emission function for a temperature of 2000 K can be used to describe the spectral distribution of the irradiation. Since the surface is opaque, the transmissivity is zero. The Mathcad solution is presented in Figure 6.10. The solution follows the general approach of

$$\mu m := \frac{m}{10^6} \qquad \text{The definition of } \mu m.$$

$$C_1 := 3.742 \cdot 10^8 \cdot W \cdot \frac{\mu m^4}{m^2} \qquad C_2 := 1.439 \cdot 10^4 \cdot \mu m \cdot K \qquad \begin{array}{l} C_1 \text{ and } C_2 \text{ for the Planck power spectral} \\ \text{density expression.} \end{array}$$

$$F(\zeta) := \frac{15}{\pi^4} \cdot \left[\sum_{n=1}^{20} \frac{e^{-n \cdot \zeta}}{n} \cdot \left(\frac{6}{n^3} + 6 \cdot \frac{\zeta}{n^2} + \zeta^3 + 3 \cdot \frac{\zeta^2}{n} \right) \right] \qquad \zeta = \frac{C_2}{\lambda \cdot T} \quad \text{Definition of } \zeta$$

$$G := 750 \frac{W}{m^2} \qquad \text{Irradiation.}$$

$$\lambda T_1 := 1.5 \cdot \mu m \cdot 2000 \cdot K \qquad F\left(\frac{C_2}{\lambda T_1} \right) = 0.27312 \qquad \begin{array}{l} \text{The radiation function for the} \\ \text{absorptivity must be calculated} \\ \text{using the spectral characteristics} \end{array}$$

$$\lambda T_2 := 10 \cdot \mu m \cdot 2000 \cdot K \qquad F\left(\frac{C_2}{\lambda T_2} \right) = 0.98555 \qquad \text{of the source of the irradiation.}$$

$$\alpha := 0.1 \cdot F\left(\frac{C_2}{\lambda T_1} \right) + 0.5 \cdot \left(F\left(\frac{C_2}{\lambda T_2} \right) - F\left(\frac{C_2}{\lambda T_1} \right) \right) + 0.8 \cdot \left(1 - F\left(\frac{C_2}{\lambda T_2} \right) \right)$$

$$\alpha = 0.39509$$

$$\rho := 1 - \alpha \qquad \rho = 0.60491 \qquad \text{Reflectivity}$$

$$Gabsorbed := \alpha \cdot G \qquad Gabsorbed = 296.31455 \frac{W}{m^2}$$

Irradiation absorbed by surface

Figure 6.10 Mathcad solution for Example 6.3.

Example 6.2. Several things are of interest in this problem. Because of the spectral characteristics of the surface, $\alpha \neq \varepsilon$, which is an assumption sometimes made for problems such as this. The irradiation is not that of a blackbody, but it has the spectral characteristics of a blackbody at 2000 K.

In the problem statement for Example 6.3, $\varepsilon_\lambda = \alpha_\lambda$; under what conditions is this a viable assertion? When is $\alpha = \varepsilon$ acceptable? Most heat transfer textbooks address these issues. The usual question to ask is, when is $\varepsilon_{\lambda\theta} = \alpha_{\lambda\theta}$? This is always true since $\varepsilon_{\lambda\theta}$ and $\alpha_{\lambda\theta}$ are properties of the surface and are independent of the emission or irradiation. The spectral hemispherical emissivity and absorptivity are equal under two conditions: (1) when the irradiation is diffuse or (2) when the surface is diffuse. For many problems of interest, one of these conditions is often approximated. Moving up the hierarchy, when is $\alpha = \varepsilon$? The total absorptivity is equal to the total emissivity under two circumstances: (1) when the irradiation corresponds to that of blackbody or (2) when the surface is gray. A gray surface is a surface for which $\varepsilon_\lambda = \alpha_\lambda$ for all λ (in the ranges of interest). Many spectrally selective surfaces, including the one in Examples 6.2 and 6.3, are *not* gray surfaces, and $\alpha \neq \varepsilon$.

Table 6.3 presents the emissivity and solar absorptivity for selected surfaces. The ratio α/ε of the absorptivity for solar irradiation and the emissivity is a useful solar engineering parameter as small values reject heat and large values absorb solar energy.

The foregoing review of radiation heat transfer is sufficient for the needs of this book. The next section will explore the motion of the sun with respect to the Earth.

TABLE 6.3 Solar absorptivity, emissivity, and their ratio for selected surfaces

Surface	Solar Absorptivity, α	Emissivity, ε	Ratio, α/ε
Aluminium			
Evaporated film	0.09	0.03	3.00
Hard anodized	0.03	0.80	0.04
White paint	0.21	0.96	0.22
Black paint	0.97	0.97	1.00
Black chrome	0.95	0.15	6.47
Red brick	0.63	0.93	0.68
Snow	0.28	0.97	0.29
Teflon	0.12	0.85	0.14
Vegetation (corn)	0.76	0.97	0.78

6.3 | SUN PATH DESCRIPTION AND CALCULATION

An understanding of the position of the sun relative to the Earth is important for solar engineering studies. Perhaps the most effective tool for explaining and understanding the position of the sun is the sun path—the path of the sun in relation to a location on the surface of the Earth. A number of textbooks, reference books, and websites contain discussions, descriptions, tables, and illustrations of sun paths for various locations. However, few describe how to calculate sun paths or contain software elements to aid in their computation. The purpose of this section is to provide a procedure for calculating and understanding sun path lines.

Consider the Earth-sun geometric relationship. As illustrated in Figure 6.11, the Earth rotates at an angle of 23.45° with respect to the ecliptic orbital plane. This rotation angle is responsible for the seasons (and for much of the difficulty in computing sun paths). Although the Earth moves around the sun, the simplest way to understand the motion of the Earth is to adopt a Ptolemaic view—to consider the Earth as stationary and the sun as being in motion. In the Ptolemaic view, the sun's relation to a point on the surface of the Earth is described by the solar altitude angle, α, and the solar azimuth angle, a_s. These angles are shown in Figure 6.12. The altitude angle is the angle between a line collinear with the sun and the horizontal plane. The azimuth angle is the angle between a due-south line and the projection of the

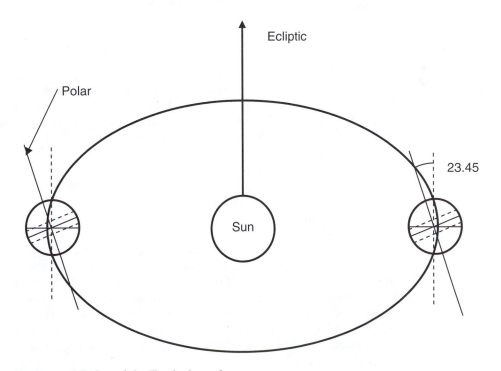

Figure 6.11 Motion of the Earth about the sun.

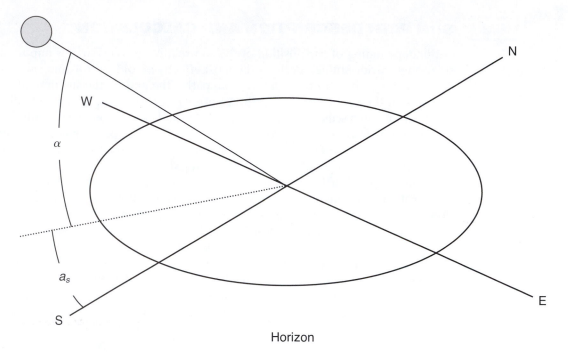

Figure 6.12 Solar azimuth angle and solar altitude angle definitions.

site-to-sun line on the horizontal plane. The sun path for June 21 for Mississippi State University (MSU), 33.455° north latitude, is illustrated in Figure 6.13. In addition to the relation between the azimuth and altitude angles, the solar time is shown in the figure. At MSU, June 21 has almost 14 hours of daylight, and at solar noon the sun is nearly vertical ($\alpha \approx 80°$). The sun rises almost 30° north of east ($a_s \approx -120°$) and sets almost 30° north of west ($a_s \approx +120°$). An interesting aspect of this sun path line is the rapidity of movement of the sun between 11:00 a.m. and 1:00 p.m. During this time, the sun traverses a total azimuth angle of nearly 120°—almost half of the total azimuth angle traversal of 240°. The motion of the sun is also symmetrical about solar noon, with the difference between morning and afternoon being just the sign of the azimuth angle.

However, as Goswami et al. (2000) point out, the altitude and azimuth angles are not fundamental but are functions of the location (the latitude), the time (the solar hour angle), and the solar declination. (As an aside, longitudes and latitudes for locations in the United States can be found at www.geonames.usgs.gov and those for locations worldwide at www.astro.com.) The solar declination, δ_s, varies between −23.45° and +23.45° and has the same numerical value as the latitude at which the sun is directly overhead at solar noon on a given day. The solar declination can be approximated as

$$\delta_s = 23.45° \sin\left[\frac{360(284 + n)}{365}\right] \tag{6-22}$$

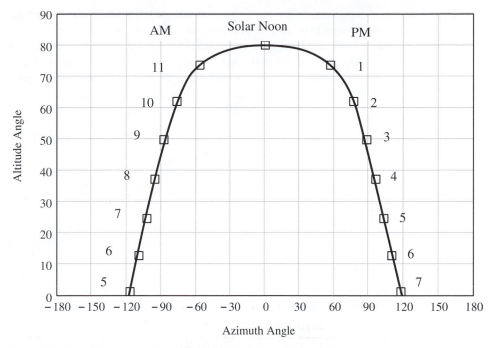

Figure 6.13 June 21 sun path line for Mississippi State University.

where n is the Julian day. The Julian day is the number of the day as measured sequentially from January 1 (Julian day 1) to December 31 (Julian day 365). Strictly interpreted, the number of the day from January 1 is the ordinal day, but in many solar engineering applications the Julian day is taken as the ordinal day. The solar declination as a function of the Julian day (or ordinal day) is illustrated in Figure 6.14.

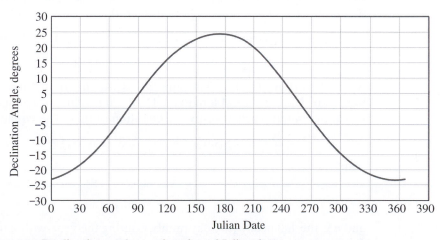

Figure 6.14 Declination angle as a function of Julian date.

The solar hour angle, h_s, is based on the 24 hours required for the sun to move 360° around the Earth. The motion is 15° per hour with $-180° \leq h_s \leq +180°$. The solar hour angle at solar noon (the time when the azimuth angle is zero and the sun altitude angle is the maximum for a given day) is defined as zero, with values east of south (morning) being negative and values west of south (afternoon) being positive.

The solar altitude angle, α, is related to the declination (δ_s), the latitude (L), and the solar hour (h_s) angles as

$$\sin(\alpha) = \sin(L)\sin(\delta_s) + \cos(L)\cos(\delta_s)\cos(h_s) \qquad (6\text{-}23)$$

The solar azimuth angle, a_s, is given as

$$\sin(a_s) = \frac{\cos(\delta_s) \cdot \sin(h_s)}{\cos(\alpha)} \qquad (6\text{-}24)$$

For a given day, the generation of a sun path line using Eqs. (6-23) and (6-24) is straightforward for a latitude greater than the solar declination. Care must be taken for solar azimuth angles greater than +90° for the afternoon or less than −90° for the morning. Since the principal angle range for the arcsin function is +90° to −90°, for azimuth angles greater/less than ±90°, logic must be used to ensure that a value greater/less than 90° is obtained. One way to accomplish this is to determine the hour angle, hlimit, that corresponds to $a_s = \pm90°$ (the sun position due west/east). Then for all hour angles greater/less than ±hlimit, the azimuth angle must be greater/less than ±90°. Goswami et al. (2000) provide the following expression for hlimit:

$$\cos(\text{hlimit}) = \pm\frac{\tan(\delta_s)}{\tan(L)} \qquad (6\text{-}25)$$

When the hour angle is greater/less than ±hlimit, the azimuth angle is evaluated as $\pm\{\pi - \arcsin[\sin(a_s)]\}$, properly preserving $|a_s| > 90°$ for hour angles greater/less than ±hlimit.

The relation between standard time and local time must be incorporated if the sun path times are to be cast as local or "clock" times. The relationship between solar time and local (clock) time is presented in the following expressions:

$$\text{Solartime} = \text{ST} + 4(\text{SL} - \text{LL}) \text{ (minutes)} + E \text{ (minutes)} \qquad (6\text{-}26)$$

$$\text{ST} + 1 \text{ hour} = \text{DST} \qquad (6\text{-}27)$$

where Solartime is the solar time, ST is the standard time, SL is the standard longitude, LL is the local longitude, DST is daylight savings time, and E is the correction in minutes provided by the equation of time. The equation of time is

$$E = 9.87 \cdot \sin(2 \cdot B_n) - 7.53 \cdot \cos(B_n) - 1.5 \cdot \sin(B_n) \qquad (6\text{-}28)$$

with

$$B_n = \frac{360 \cdot (n - 81) \cdot \pi}{364 \cdot 180} \qquad (6\text{-}29)$$

and n is the Julian day (the day number during the year, with January 1 being day 1). Table 6.3 provides the standard longitudes for United States time zones.

TABLE 6.3 Standard longitudes for United States time zones

75° for Eastern
90° for Central
105° for Mountain
120° for Pacific

The equation of time is plotted as a function of the Julian day in Figure 6.15. Using Eqs. (6-26)–(6-28) or Eqs. (6-26)–(6-27) and Figure 6.15, the solar time can be related to the standard time for any longitude and Julian day. In this context, standard time is the local time, or the time indicated by a clock.

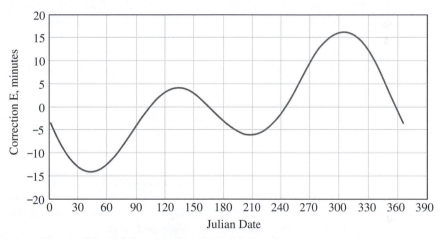

Figure 6.15 The equation of time as a function of Julian day.

Example 6.4	What is the standard time corresponding to 1:00 p.m. solar time on August 20 at Mississippi State University?

Solution   Mississippi State University is located in the Central time zone at latitude 33.455° (north) and longitude 88.789° (west). The Julian day or ordinal day for August 20 is 232. Daylight savings time is observed. Inserting Eq. (6-27) into Eq. (6-26) and solving for the standard time yields

$$\text{DST} = \text{Solartime} + 1\text{ h} - 4(\text{SL} - \text{LL})\text{ min} - E\text{ min} \qquad (6\text{-}30)$$

The equation of time correction can be read from Figure 6.15 or calculated for day 232 using Eq. (6-28), with a result of −2.946 minutes. Inserting the values into Eq. (6-30), the local daylight savings time becomes

$$\text{DST} = 1:00\text{ p.m.} + 1\text{ h} - 4(90 - 88.789)\text{ min} - (-2.95\text{ min})$$
$$= 1:58.10\text{ p.m.}$$

6.4 SUN PATH DEVELOPMENT USING MATHCAD

A number of web-based aids discuss all or some aspects of sun path computations. The website www.uni.edu/darrow/frames/geosol.html, contains a compilation of solar and geographical website addresses that provide information for many aspects of sun paths. The United States Naval Observatory (USNO) website, www.usno.navy.mil, provides a wide range of data and computation capabilities. USNO capabilities include not only site-specific solar and lunar daily/yearly characteristics, but also sun path generation in the form of tabular listings for United States locations or for specified latitudes and longitudes worldwide. This is a particularly useful website, but since all calculations are hidden, it is not a good instructional website for studying sun paths. The National Oceanic and Atmospheric Administration (NOAA) website, www.srrb.noaa.gov/highlights/sunrise/azel.html, has an online sun position calculator as well as some explanation of the approach. Additionally, www.susdesign.com/ sunangle/ offers an online sun position calculator, but with no explanation as to the method(s) used.

Commercial software to compute solar positions and sun paths is also available. For example, the Florida Solar Energy Center (Cocoa, FL) markets SunPath™, a software element for sun path calculation. SUNPATH™, available from Film-tools (Burbank, CA), is designed for determining lighting issues associated with filming and photography but can also be used to generate conventional sun paths. An out-of-print book by Petherbridge (1966) presents sun paths and overlays for heat gain calculations.

With the Julian day and the latitude known, Eqs. (6-23)–(6-29) are sufficient to construct the sun path line for the corresponding day and location. Figure 6.13, the June 21 sun path line for MSU, was generated from these equations using the Mathcad software element as described next.

Figure 6.16 shows the complete Mathcad worksheet needed to compute and plot the sun path for June 21 at MSU (Figure 6.13). Only the date and latitude need be changed to generate and plot the sun path line for another day or location. Since the worksheet represents the kernel needed to construct sun path lines, an examination of the procedure is warranted.

The Julian day is used to compute the declination using Eq. (6-22). The latitude is entered, and the hour angle, hss, is specified in the range from solar noon (0°) to solar midnight (180°). Equation (6-23) is used to calculate the solar altitude angle for every hour angle for the specified day and latitude. Equation (6-24) provides the corresponding solar azimuth angles. Logic is provided to determine solar altitude angles greater than 90°. "hlimit" is the hour angle for which the azimuth angle is equal to ±90° and is determined using Eq. (6-25). The sun path can then be plotted, or values of the altitude and azimuth angles printed, as a function of hour angle. The initial computational results cover solar noon to solar midnight. However, since only the sign of the azimuth angles differs for morning, the complete day's sun path can be generated simply by plotting $-a_s$ for the morning hours.

The solar times corresponding to the azimuth and altitude angles are then extracted for every solar hour (hss = 0, 15, 30, . . . , 180) from the azimuth and alti-

The generation of the sunpath line for a given day as a function of latitude.

$n := 1 .. 365$ The days of the year.

Declination Angle

The declination angle is the angle between the sun's rays and the zenith (overhead) direction at solar noon on the equator. The declination is dependent on the Earth's position in its orbit around the sun.

$$\delta_n := 23.45 \sin\left[360 \cdot \frac{(n + 284) \cdot \pi}{365 \cdot 180} \right]$$

Declination for specific day (use Julian date). For June 21, the Julian date is 172.

$\delta D := \delta_{172}$ $\delta D = 23.45$ 21 June   Declination angle in degrees for use in sunpath generation.

Input the latitude (in degrees):

$L := 33.455$ Location of Mississippi State University

Establish range variables for days and hours.   Degrees to radian conversion:

$hss := 0 .. 180$ $hsp_{hss} := hss$ Hours $dr := \dfrac{\pi}{180}$

Calculation of sunpath angles following Goswami et al.

$$\sin\alpha_{hss} := \sin(L \cdot dr) \cdot \sin(\delta D \cdot dr) + \cos(L \cdot dr) \cdot \cos(\delta D \cdot dr) \cdot \cos\left(hsp_{hss} \cdot dr\right) \quad \text{Altitude angle}$$

$\alpha_{hss} := \operatorname{asin}(\sin\alpha_{hss})$ $ang_{hss} := \dfrac{\alpha_{hss}}{dr}$ Altitude angle in degrees

$sinas_{hss} := \cos(\delta D \cdot dr) \cdot \dfrac{\sin\left(hsp_{hss} \cdot dr\right)}{\cos\left(\alpha_{hss}\right)}$ Azimuth angle

Test for azimuth angle > 90 degrees.

Since the principal values of the arcsin are defined for -90 degrees < angle < 90 degrees, logic is needed for any azimuth angle greater than 90 degrees.

$$hlimit := \left| \begin{array}{l} \left(\operatorname{acos}\left(\dfrac{\tan(\delta D \cdot dr)}{\tan(L \cdot dr)} \right) \right) \cdot \dfrac{1}{dr} \quad \text{if } L > \delta D \\ 0 \quad \text{otherwise} \end{array} \right.$$

$hlimit = 48.968$ Hour angle at 90-degree azimuth for given day.

Definition of arcsin function to include azimuth angles > 90 degrees.

$$as_{hss} := \left| \begin{array}{l} \left(\pi - \operatorname{asin}\left(sinas_{hss}\right) \right) \quad \text{if } hsp_{hss} > hlimit \\ \operatorname{asin}\left(sinas_{hss}\right) \quad \text{otherwise} \end{array} \right.$$

Change all angles from radians to degrees $\alpha_{hss} := \dfrac{\alpha_{hss}}{dr}$ $as_{hss} := \dfrac{as_{hss}}{dr}$

Plot the sunpath taking advantage of the symmetry of the morning and afternoon segments.

(continued on next page)

Figure 6.16 Sun path Mathcad worksheet for a single day.

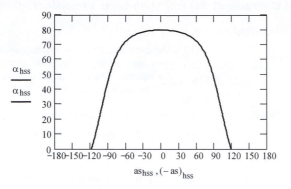

Establish the lines of constant solar time (hour angle) on the sunpath. Each solar hour corresponds to 15 degrees of hour angle. Thus, 1 PM solar time is 15 degress from solar noon.

$\text{solar}_1 := \alpha_{15}$	$\text{azi}_1 := \text{as}_{15}$	$\text{solar}_2 := \alpha_{30}$	$\text{azi}_2 := \text{as}_{30}$
$\text{solar}_3 := \alpha_{45}$	$\text{azi}_3 := \text{as}_{45}$	$\text{solar}_4 := \alpha_{60}$	$\text{azi}_4 := \text{as}_{60}$
$\text{solar}_5 := \alpha_{75}$	$\text{azi}_5 := \text{as}_{75}$	$\text{solar}_6 := \alpha_{90}$	$\text{azi}_6 := \text{as}_{90}$
$\text{solar}_7 := \alpha_{105}$	$\text{azi}_7 := \text{as}_{105}$	$\text{solar}_8 := \alpha_{120}$	$\text{azi}_8 := \text{as}_{120}$
$\text{solar}_9 := \alpha_{135}$	$\text{azi}_9 := \text{as}_{135}$	$\text{solar}_{10} := \alpha_{150}$	$\text{azi}_{10} := \text{as}_{150}$
$\text{solar}_{11} := \alpha_{165}$	$\text{azi}_{11} := \text{as}_{165}$	$\text{solar}_{12} := \alpha_{180}$	$\text{azi}_{12} := \text{as}_{180}$

$\text{solar}_0 := \alpha_0$ $\text{azi}_0 := \text{as}_0$ Solar noon

Add the solar time to the sunpath plot taking advantage of the symmetry..

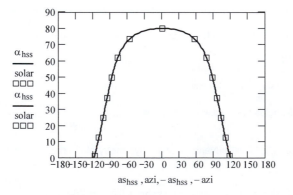

Figure 6.16 *(continued)*

tude angles. The solar times are added to the sun path plot to complete the presentation in Figure 6.13.

Although the sun path for one day is useful for that particular day, a complete understanding of the sun's yearly path at a given location is also needed. Figure 6.17, shows the sun path lines and solar times for seven days spanning a year for MSU. Figure 6.18 illustrates the Mathcad worksheet that is used to generate the sun path for different days of the year. For this example, the 21st of each month was chosen. Because of symmetry, the sun paths for May 21 and July 21, April 21 and August 21, March 21 and September 21, February 21 and October 21, and January 21 and November 21 are the same. Only June 21 and December 21 are lacking symmetry

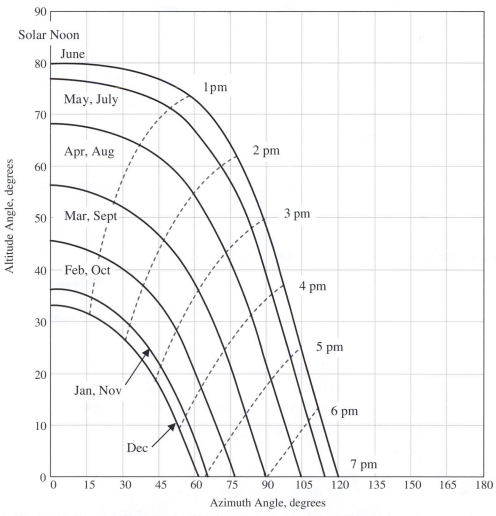

Figure 6.17 Sun path lines for the 21st day of every month for MSU.

The generation of the sunpath chart as a function of latitude.

$n := 1..365$ The days of the year.

Declination Angle

$$\delta_n := 23.45 \cdot \sin\left[360 \cdot \frac{(n+284)\cdot\pi}{365\cdot180}\right]$$

The declination angle is the angle between the sun's rays and the zenith (overhead) direction at solar noon on the equator. The declination is dependent on the Earth's position in its orbit around the sun.

Declination for specific days

$\delta D_0 := \delta_{355}$ 21 Dec $\delta D_1 := \delta_{21}$ 21 Jan $\delta D_2 := \delta_{52}$ 21 Feb $\delta D_3 := \delta_{80}$ 21 March

$\delta D_4 := \delta_{111}$ 21 April $\delta D_5 := \delta_{141}$ 21 May $\delta D_6 := \delta_{172}$ 21 June

$$\delta D = \begin{pmatrix} -23.45 \\ -20.138 \\ -11.226 \\ -0.404 \\ 11.579 \\ 20.138 \\ 23.45 \end{pmatrix}$$

Declination angle in degrees for use in sunpath generation.

Input the latitude (in degrees):

$L := 33.455$ Latitude of Mississippi State University

Establish range variables for days and hours.

$i := 0..6$ Days of interest   Hours

$hss := 0..180$ $hsp_{hss} := hss$

Degrees to radian conversion:

$$dr := \frac{\pi}{180}$$

Calculation of sunpath angles following Goswami, Kreith, and Kreider.

$$\sin\alpha_{hss,i} := \sin(L \cdot dr) \cdot \sin(\delta D_i \cdot dr) + \cos(L \cdot dr) \cdot \cos(\delta D_i \cdot dr) \cdot \cos(hsp_{hss} \cdot dr)$$ Altitude angle

$$\alpha_{hss,i} := asin(\sin\alpha_{hss,i})$$ $$ang_{hss} := \frac{\alpha_{hss,6}}{dr}$$ Altitude angle in degrees

$$\sin as_{hss,i} := \cos(\delta D_i \cdot dr) \cdot \frac{\sin(hsp_{hss} \cdot dr)}{\cos(\alpha_{hss,i})}$$ Azimuth angle

Test for azimuth angle > 90 degrees.

Since the principal values of the arcsin are defined for -90 degrees < angle < 90 degrees, logic is needed for any azimuth angle greater than 90 degrees.

$$hlimit := \frac{1}{dr} acos\left(\frac{\tan(\delta D_i \cdot dr)}{\tan(L \cdot dr)}\right)$$

$$hlimit^T = (131.032 \quad 123.709 \quad 107.481 \quad 90.611 \quad 71.936 \quad 56.291 \quad 48.968)$$

(continued on next page)

Figure 6.18 Mathcad worksheet for the sun path for MSU.

Hour angles at a 90-degree azimuth for a given latitude.

Definition of arcsin function to include azimuth angles > 90 degrees.

$$as_{hss,i} := \begin{vmatrix} (\pi - asin(sinas_{hss,i})) & \text{if } hsp_{hss} > hlimit \\ asin(sinas_{hss,i}) & \text{otherwise} \end{vmatrix}$$

Change all angles from radians to degrees $\alpha_{hss,i} := \dfrac{\alpha_{hss,i}}{dr}$ $as_{hss,i} := \dfrac{as_{hss,i}}{dr}$

Redefine $\alpha_{hss,i}$ and $as_{hss,i}$ as Y_i and X_i, respectively. $X_i := as^{\langle i \rangle}$ $Y_i := \alpha^{\langle i \rangle}$

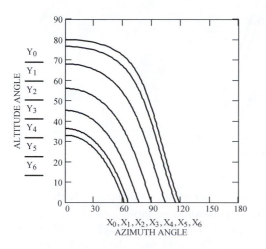

Establish the lines of constant solar time (hour angle) on the sunpath. Each solar hour corresponds to 15 degrees of hour angle. Thus, 1 PM solar time is 15 degrees from solar noon.

$solar1_i := \alpha_{15,i}$	$azi1_i := as_{15,i}$	$solar2_i := \alpha_{30,i}$	$azi2_i := as_{30,i}$
$solar3_i := \alpha_{45,i}$	$azi3_i := as_{45,i}$	$solar4_i := \alpha_{60,i}$	$azi4_i := as_{60,i}$
$solar5_i := \alpha_{75,i}$	$azi5_i := as_{75,i}$	$solar6_i := \alpha_{90,i}$	$azi6_i := as_{90,i}$
$solar7_i := \alpha_{105,i}$	$azi7_i := as_{105,i}$	$solar8_i := \alpha_{120,i}$	$azi8_i := as_{120,i}$
$solar9_i := \alpha_{135,i}$	$azi9_i := as_{135,i}$	$solar10_i := \alpha_{150,i}$	$azi10_i := as_{150,i}$
$solar11_i := \alpha_{165,i}$	$azi11_i := as_{165,i}$	$solar12_i := \alpha_{180,i}$	$azi12_i := as_{180,i}$

(continued on next page)

Figure 6.18 *(continued)*

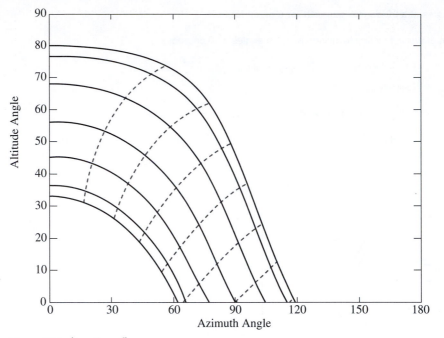

Figure 6.18 (*continued*)

months. On the multi-day sun paths, to avoid cluttered presentations, only the after-noon portions are presented since symmetry provides the mornings. The seven sun path lines representing the 21st day of each month provide a relatively complete pic-ture of the variations in the sun's path (Ptolemaic view) over the year for a given location. The variations in length of day, in solar azimuth sunset (and sunrise) angles, and in solar altitude angle at noon are evident on inspection of the figure. Additionally, the azimuth angles swept per hour can be determined.

An examination of Figure 6.18 is appropriate. The solar declinations are com-puted for each of the seven days, and the altitude and azimuth angles are generated for each solar hour angle for each day. The same logic as used in Figure 6.16 for a single day is used in Figure 6.18 for each day. Thus, the altitude and azimuth angles are arrays rather than vectors (in the nomenclature of Mathcad).

Obviously, sun path lines could be presented for a wide range of latitudes. In particular, a sun path at a higher latitude is worth examining. Consider Figure 6.19, which is for a latitude above the Arctic Circle. Such latitudes are characterized by the phenomenon "the sun never sets" during some summer months. The figure, for latitude 80° (north), illustrates that the sun does not set from April through August. Even in the summer, the maximum solar altitude angle is only 33.45°, which is rela-tively low on the horizon. However, if the sun never sets during some summer months, that means the sun never rises during some winter months. For the latitude of 80° (north), the sun barely rises (has a positive solar altitude angle) during March and September, and, as the sun path lines indicate, the sun does not rise during October through February. All in all, Figure 6.19 portrays the sun's path at this high

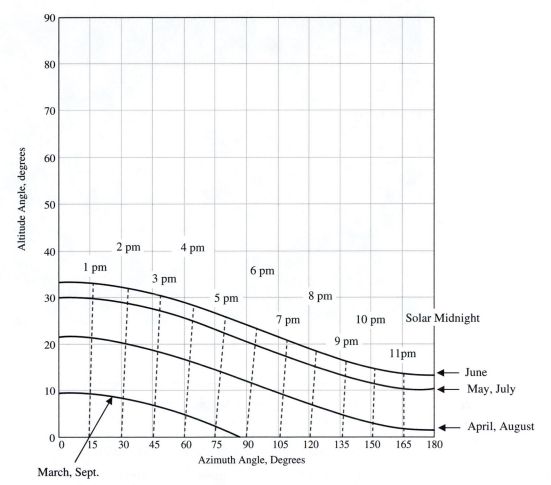

Figure 6.19 Sun path lines for latitude above the Arctic Circle.

latitude as skirting the horizon in the summer and sinking below the horizon during the winter.

Additional information and details on Mathcad procedures described herein can be found in Hodge (2003).

6.5   THE SOLAR ENERGY DATABASE

For any quantitative considerations of solar energy applications, hourly, monthly, and yearly information about the climate, including irradiation (the solar "insolation"), is needed. The United States Department of Energy's National Renewable Energy Laboratory (NREL) at Golden, Colorado, provides an astonishingly wide range of climatic and energy engineering information, much of which is available on its website, www.rredc.nrel.gov.

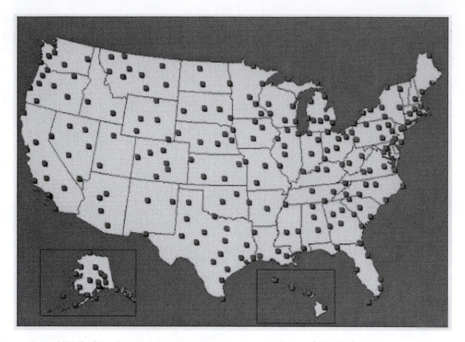

Figure 6.20 NWS sites for which NSRDB data are available (NCDC).

The National Solar Radiation Data Base (NSRDB) contains 30 years (1961–1990) of solar radiation and supplementary meteorological data from 237 National Weather Service (NWS) sites in the United States, plus sites in Guam and Puerto Rico. Figure 6.20 shows the sites for which NSRDB data are available.

The following data products are available on the NREL website:

1. Daily statistics files (monthly averages of daily totals)
2. Hourly data files
3. *Solar Radiation Data Manual for Buildings* [30-year (1961–1990) average of solar radiation and illuminance for each month]
4. *Solar Radiation Data Manual for Flat-Plate and Concentrating Collectors*
   a. Averages of solar radiation for the 360 months from 1961 to 1990
   b. Thirty-year (1961–1990) average of solar radiation for each month.
5. Typical Meteorological Year 2 (TMY2) files
6. United States solar radiation resource maps

The Typical Meteorological Year 2 (TMY2) data sets are derived from the 1961–1990 National Solar Radiation Data Base. Because they are based on the NSRDB, these data sets are referred to as TMY2 to distinguish them from the TMY data sets, which were based on older data. The TMY and TMY2 data sets cannot be used interchangeably because of differences in time (solar versus local), formats, elements, and units. The TMY2 data sets contain hourly values for solar radiation and meteorological elements for a one-year period. Their intended use is for computer simulations of solar energy conversion systems and building systems to facilitate performance comparisons of different system types, configurations, and locations in the United States and its territories. Since TMY2 data sets represent typical rather than extreme conditions, they are not suited for designing systems to meet the worst-case conditions occurring at a given location. The TMY2 data sets and manual were produced by the NREL Analytic Studies Division under the Resource Assessment Program, which is funded and monitored by the U.S. Department of Energy Office of Solar Energy Conversion. The data contained in the NREL website will be used in the following chapters.

The NREL website also provides a useful assessment of the potential for solar energy in the United States. Figure 6.21 shows the average daily direct solar energy incident on a perpendicular surface tracking the sun path for the United States. The desert southwest consistently receives 6.5 to 8.5 kWh/m$^2 \cdot$day, while the southeast,

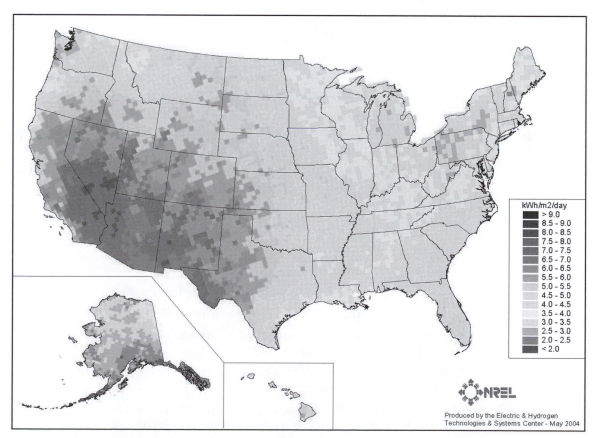

Figure 6.21 Solar potential for the United States (NREL).

because of cloud cover and humidity effects, receives about 4.0 kWh/m$^2 \cdot$ day. Much of the northeast solar energy reception is 3.0 kWh/m$^2 \cdot$ day or less. Even though the southwest is the most favored region, solar energy systems are viable in other parts of the United States.

6.6 CLOSURE

This chapter has provided a review of radiation heat transfer, examined sun path lines, and identified sources of solar energy engineering data. With these topics as a basis, solar energy engineering principles can now be considered. The next three chapters discuss, respectively, active solar application, passive solar applications, and photovoltaic solar concepts.

REVIEW QUESTIONS

1. What is the visible spectrum?

2. Sketch and label the Planck distribution corresponding to a 5800 K blackbody.

3. Explain the differences between the spectral directional, the spectral hemispherical, and the total hemispherical surface properties.

4. Under what circumstances is $\alpha_\lambda = \varepsilon_\lambda$? What is required for $\alpha = \varepsilon$?

5. At 35° N latitude and 80° W longitude, the actual time on June 21 (Julian date of 172) is 1 p.m. Central Daylight Savings Time. Find the solar time.

6. What is the solar time for a clock time (the time read on a clock or a watch) of 2:00 p.m. on September 27 in a location specified by 35:25:30 N and 82:30:00 W?

7. The sun paths in Figure R6.7, are for 33.5° N latitude, and the location is at 88.5° W longitude. Answer the following questions:

   (a) What is the altitude angle at solar noon on June 21?

   (b) How many hours of daylight are there in the shortest day?

   (c) How many hours of daylight are there in the longest day?

   (d) At solar noon on June 21, how long a shadow would a 100-ft flagpole cast?

   (e) If the latitude were increased to 50° N latitude, would the hours of daylight for December 21 be more or less?

8. At 35° N latitude and 88° W longitude, the solar time on January 30 is 11 a.m. Find the actual (clock) time.

9. At 55° N latitude and 98° W longitude, the solar time on the 180th day of the year is 11 a.m. Find the local time (Central Daylight Savings Time).

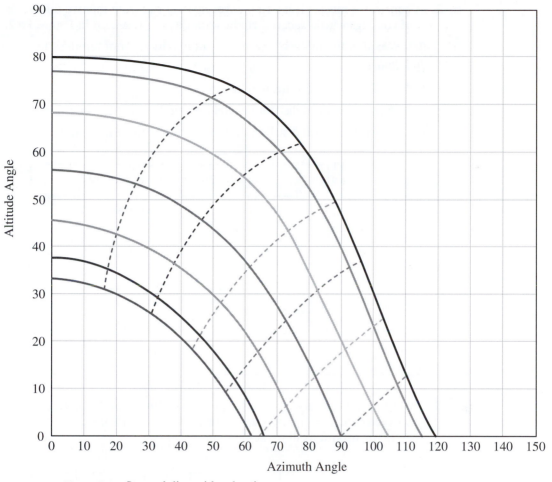

Figure R6.7 Sunpath line with solar times

EXERCISES

1. A radiator on a solar-powered satellite must dissipate the heat being generated with the satellite by radiating the heat into space. The radiator surface has a solar absorptivity of 0.5 and an emissivity of 0.95. What is the surface temperature when the required dissipation is 1500 W/m² for each of the following two conditions?

   (a) The radiator is facing the sun, and the solar irradiation is 1353 W/m².

   (b) The radiator is shielded from the sun, and the solar irradiation is negligible.

2. A contractor must select a roof covering material from the two diffuse $(\varepsilon_\lambda = \alpha_\lambda)$ roof coatings whose spectral characteristics are presented in Figure P6.2.

(a) Which of the materials would result in a lower roof temperature?

(b) Which is preferred for summer use?

(c) Which is preferred for winter use?

(d) Sketch a spectral distribution that would be ideal for summer.

(e) Sketch a spectral distribution that would be ideal for winter.

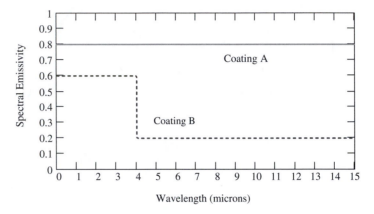

Figure P6.2

3. Two special coatings are available for use on an absorber plate for a flat-plate solar collector. The coatings are diffuse $(\varepsilon_\lambda = \alpha_\lambda)$ and are characterized by the spectral distributions illustrated in Figure P6.3.

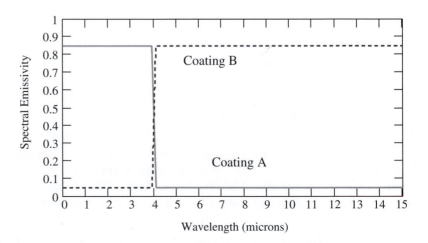

Figure P6.3

(a) If the irradiation incident on the plate is $G = 1000$ W/m^2, what is the radiant energy absorbed per m^2 for each surface?

(b) Which coating would you select for the absorber plate? Explain.

4. The spectral absorptivity, α_λ, and the spectral reflectivity, ρ_λ, for a spectrally selective diffuse surface are as shown in Figure P6.4.

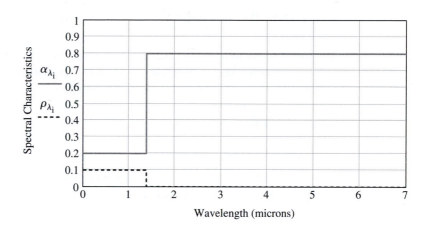

Figure P6.4

(a) Sketch the spectral transmissivity, τ_λ.

(b) If solar irradiation with $G = 750$ W/m^2 and the spectral distribution of a blackbody at 5800 K is incident on the surface, determine the fractions of the irradiation that are transmitted, reflected, and absorbed by the surface.

(c) If the temperature of the surface is 350 K, determine the emissivity, ε.

(d) Determine the net heat flux by radiation at the surface of the material.

5. An opaque solar collector surface is 3 m by 1 m and is maintained at 425 K. The surface is exposed to solar irradiation with $G = 800$ W/m^2. The surface is diffuse, and its spectral absorptivity is

$$\alpha_\lambda = 0 \qquad 0 < \lambda < 0.5 \ \mu m$$
$$= 0.8 \qquad 0.5 \ \mu m < \lambda < 1.0 \ \mu m$$
$$= 0.5 \qquad 1.0 \ \mu m < \lambda < 2.0 \ \mu m$$
$$= 0.3 \qquad \lambda > 2.0 \ \mu m$$

Determine the absorbed radiation, the emissive power, and the net radiation heat transfer from the surface.

6. An opaque surface, 2 m by 2 m, is maintained at 400 K and is exposed to solar irradiation with $G = 1200$ W/m^2. The surface is diffuse and its spectral absorptivity is

$$\alpha_\lambda = 0 \qquad 0 < \lambda < 0.5 \ \mu m$$
$$= 0.8 \qquad 0.5 \ \mu m < \lambda < 1.0 \ \mu m$$
$$= 0.0 \qquad 1.0 \ \mu m < \lambda < 2.0 \ \mu m$$
$$= 0.9 \qquad \lambda > 2.0 \ \mu m$$

Determine the absorbed radiation, the emissive power, and the net radiation heat transfer.

7. Develop a figure showing the sun path lines for the latitude of your hometown. On a separate figure show the sun path line for June 21 and indicate the *actual time* (not the solar time) on the June 21 sun path line. How many hours are there from sunrise to sunset?

8. Develop a figure showing the sun path line for the latitude of your hometown on the day of your birth. Indicate the *actual time* (not the solar time) on the sun path line. How many hours are there from sunrise to sunset?

9. Develop a figure showing the sun path lines for the latitude of Washington, DC. Washington's Reagan-National Airport is located at 38° 51′ N, 77° 2′ W.

REFERENCES

Goswami, D. Y., Kreith, F., and Kreider, J. F. 2000. *Principles of Solar Engineering*, 2nd ed. Philadelphia: Taylor & Francis.

Hodge, B. K. 2003. "Sunpath Lines Using Mathcad," *Computers in Education Journal* 8(4), 78–92.

Incropera, F. P., and DeWitt, D. P. 1999. *Introduction to Heat Transfer*, 3rd ed. New York: Wiley.

Petherbridge, P. 1966. *Sunpath Diagrams and Overlays for Solar Heat Gain Calculations*. London: Building Research Station.

Siegel, R., and Howell, J. R. 2002. *Thermal Radiation Heat Transfer*, 4th ed. New York: Taylor & Francis.

Active Solar Thermal Applications

INTRODUCTION

This is the first of three applications chapters dealing with solar energy. Although not precisely defined, solar energy applications are generally classed as active, passive, or photovoltaic, which are the topics of interest for this and the next two chapters. Active solar energy applications are usually concerned with harvesting thermal energy through the use of solar collectors that employ "active" mechanical components (such as pumps) to collect and transport heat. Passive systems, typically associated with the built environment, collect and transport heat by non-mechanical means. Photovoltaics utilize solar energy to directly produce electricity. As the title indicates, this chapter deals with active solar thermal systems.

Solar collectors make up the ubiquitous images that represent solar energy engineering. The two most common configurations are the flat-plate collector and the parabolic trough. Figure 7.1(a) shows a typical flat-plate collector, and Figure 7.1(b) illustrates a typical parabolic-trough collector. Flat-plate collectors are widely used in residences and in commercial and industrial buildings for hot water and space conditioning (heating). The example in Figure 7.1(a) is mounted at a fixed angle on the south-facing roof of a residence. The parabolic-trough collector illustrated in Figure 7.1(b) possesses the ability to track the altitude angle of the sun. The parabolic-trough collector in this example focuses the sun's rays on a single pipe and provides higher water temperatures than a flat-plate collector. Parabolic-trough collectors are more expensive than flat-plate collectors and, because of the tracking mechanisms, require more maintenance. For these reasons, parabolic-trough collectors are more frequently used by institutional, industrial, and commercial facilities than in residences. Additional configurations for solar collectors include the Fresnel reflector, tubular, paraboloids, and compound parabolic.

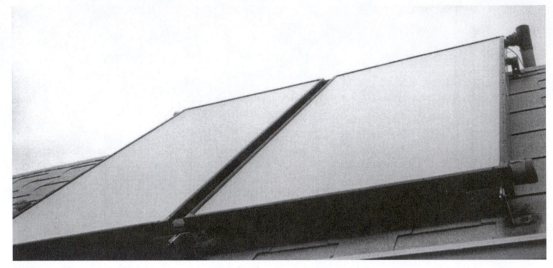

(a)

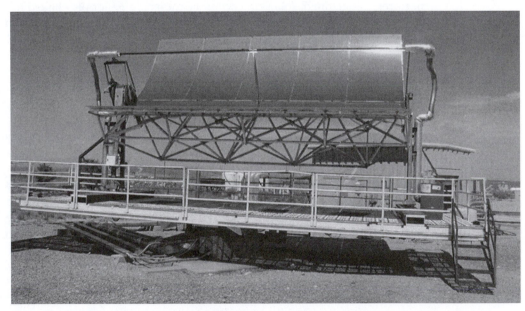

(b)

Figure 7.1 Examples of flat-plate and parabolic-trough solar collectors.
(a) Flat-plate (NREL 09572). (b) Parabolic trough (NREL 14087).

Figure 7.2 Heliostat (NREL 00036).

The heliostat with a central receiver, shown in Figure 7.2, is most impressive and produces temperatures much higher than are possible with the conventional flat-plate or parabolic-trough collector. As illustrated in the figure, heliostats have many reflecting mirrors, with each mirror capable of being oriented so as to reflect the sun's rays onto a central receiver. The mirror field surrounding the central receiver is circular, so that energy can be directed to the receiver throughout the day. Very high temperatures are thus obtainable with the central receiver.

Solar thermal collectors can be either tracking or non-tracking. Tracking collectors, as their name implies, track the diurnal motion of the sun (see Chapter 6) about a single- or dual-axis system. The azimuth and altitude angles of non-tracking collectors are fixed, and a common arrangement is a zero azimuth angle and an altitude angle set equal to the latitude of the collector location. Figure 7.3 illustrates arrangements for non-tracking, one-axis tracking, and two-axis tracking flat-plate collectors. Figure 7.4 shows the arrangement for tracking parabolic-trough collectors.

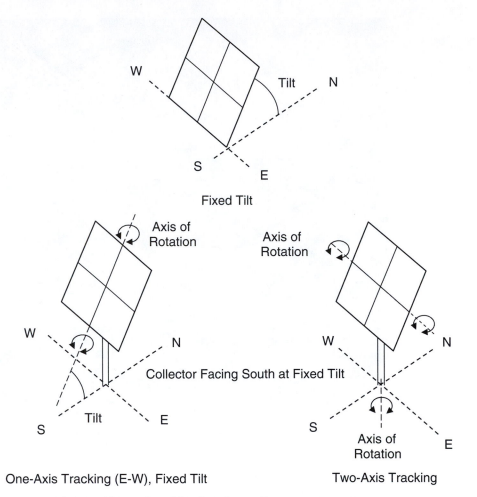

Figure 7.3 Non-tracking and tracking flat-plate collector arrangements.

The focus of this chapter is on flat-plate collectors since they are probably the most frequently used collector. Generally, flat-plate thermal solar collectors are used to heat water for domestic and/or for space-heating purposes. A simple flat-plate solar collector system is schematically illustrated in Figure 7.5. The system consists of two flow loops—one from the solar collector to the storage system and one from the storage system to the load. Water circulated through the collector system is heated by the collector and routed to a storage tank. Hot water for the load is drawn from the storage tank and returned at reduced temperature to the tank. The auxiliary heater is used when the solar energy system is providing insufficient hot water for the load. But the system's reason for being is the solar collector. The next section develops and examines the fundamental principles of operation of a flat-plate solar collector.

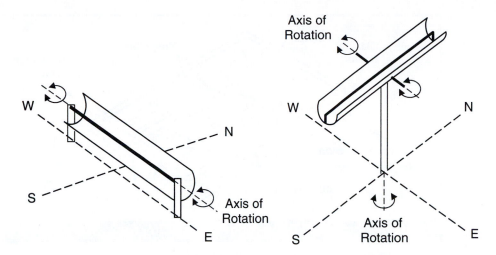

One-Axis Tracking (N-S) Concentrating Collector   Two-Axis Tracking Concentrating Collector

Figure 7.4 Tracking trough collector arrangements.

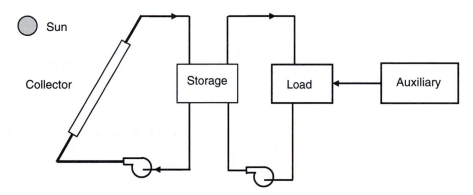

Figure 7.5 Flat-plate solar collector system schematic.

7.2 FLAT-PLATE COLLECTOR FUNDAMENTALS

A cutaway view of a flat-plate solar collector is presented in Figure 7.6. The collector components are contained in an enclosure that provides structural support and protection. Glazings are transparent cover sheets (typically made of glass with high solar transmissivity) that provide protection and pass most of the solar irradiation. Flat-plate solar collectors without glazing are called unglazed. The absorber plate is made of a material that has high solar absorptivity and a low emissivity. Water flowing through the tubes is heated by energy from the absorber plate. To minimize heat loss, the collector bottom, which is in contact with the absorber plate, is insulated.

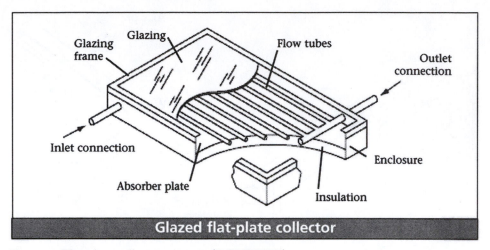

Figure 7.6 Flat-plate collector cutaway (DOE/EERE).

The energy flow for a flat-plate collector is delineated in Figure 7.7. The useful energy, q_{useful}, available is

$$q_{abs} - q_{loss} = q_{useful} \qquad (7\text{-}1)$$

where q_{abs} is the solar irradiation absorbed by the collector and q_{loss} is the sum of the conduction and radiation losses from the collector to the environment. The irradiation absorbed by the absorber plate (α, absorptivity) must first pass through the glass cover plates (τ, transmissivity) so that

$$q_{abs} = I_T A_c \alpha \tau \qquad (7\text{-}2)$$

where A_c is the collector surface area and I_T is the irradiation intensity. The losses due to conduction and radiation, expressed in terms of an overall conductance, U_L, and temperature difference, $T_{ave} - T_a$, can be expressed as

$$q_{losses} = A_c U_L (T_{ave} - T_a) \qquad (7\text{-}3)$$

The useful energy then becomes

$$q_{useful} = A_c [I_T \tau \alpha - U_L (T_{ave} - T_a)] = \dot{m} c_p (T_{out} - T_{in}) \qquad (7\text{-}4)$$

which must also be equal to the increase in the working fluid temperature. This expression provides some interesting insight into the limiting performance of flat-plate collectors. Consider the case for minimizing losses, $T_{ave} = T_a$. But if $T_{ave} = T_a$, the collector and the circulating fluid are at the ambient temperature, and the temperature of the working fluid is so low that the energy is not useful. As the collector temperature increases, the losses increase. The meaningful upper limiting case is for T_{ave} to be so high that $q_{loss} = q_{abs}$, in which case $q_{useful} = 0$. Energy at a high temperature is available, but only at very low flow of the working fluid.

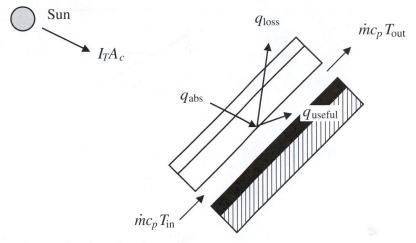

Figure 7.7 Energy flow for a flat-plate collector.

The collector efficiency is defined as the useful energy extracted divided by the total incident irradiation, or

$$\eta_c = \frac{q_{useful}}{I_T A_c} \tag{7-5}$$

which, when q_{useful} is substituted in and the expression simplified, becomes

$$\eta_c = \tau\alpha - U_L \frac{(T_{ave} - T_a)}{I_T} \tag{7-6}$$

The collector efficiency, η_c, is thus a linear function of the temperature difference divided by the irradiation, I_T, and has a slope of $-U_L$. Table 7.1 contains typical values of $\tau\alpha$ and U_L for flat-plate solar collectors (Mitchell 1983).

The behavior of typical collectors using the values from Table 7.1 is illustrated in Figure 7.8. This simple analysis conveys most of the important aspects of flat-plate solar collector performance and is the basis of the procedure used by solar collector manufacturers to describe the performance of their products.

TABLE 7.1 Typical flat-plate solar collector properties

	Black		Selective	
Number of Covers	$\tau\alpha$	U_L (Btu/h ft² °F)	$\tau\alpha$	U_L (Btu/h ft² °F)
0	0.95	6.0	0.90	5.0
1	0.90	1.0	0.85	0.5
2	0.85	0.6	0.80	0.3

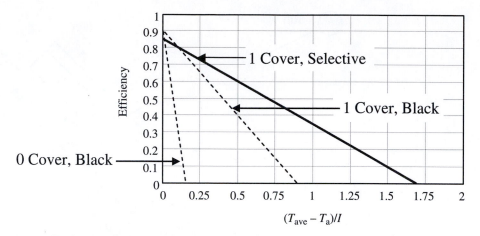

Figure 7.8 Performance characteristics of typical flat-plate solar collectors.

Even though the $\tau\alpha$ product decreases as the cover configuration changes from unglazed to one cover to two covers, the decreases in the U_L values are even more pronounced, with the net result that for a given temperature difference, the efficiencies increase as the glazing progresses from unglazed to two covers.

The simple analysis expresses the collector efficiency in terms of the average collector temperature, T_{ave}. A more useful expression that involves T_{in}, the inlet fluid temperature, can be cast as

$$\eta = F_R\left(\tau\alpha - U_L\frac{T_{in} - T_a}{I_T}\right) = F_R\,\tau\alpha - F_R U_L\frac{T_{in} - T_a}{I_T} \qquad (7\text{-}7)$$

where F_R is the collector heat removal factor. Equation (7-7) is sometimes called the Hottel-Whillier-Bliss equation and is considered to be the most important expression associated with flat-plate solar collectors. This approach incorporates the heat exchange process between the absorber plate and the collector heat exchanger. Collector data are usually represented in terms of the intercept, $F_R\,\tau\alpha$, and the slope, $F_R U_L$. Goswami et al. (2000) and Duffie and Beckman (2006) provide excellent discussions of the details of the heat removal factor.

For more accurate analyses of flat-plate solar collectors, an incident angle modifier, $K_{\tau\alpha}$, is often added to Eq. (7-7):

$$\eta = F_R\left(K_{\tau\alpha}\,(\tau\alpha)_n - U_L\frac{T_{in} - T_a}{I_T}\right) = F_R\,K_{\tau\alpha}\,(\tau\alpha)_n - F_R U_L\frac{T_{in} - T_a}{I_T} \qquad (7\text{-}8)$$

In the Hottel-Whillier-Bliss equation, the product, $\tau\alpha$, will change with the angle of incidence of the solar irradiation. But many flat-plate collectors are fixed, and the angle of incidence will change throughout the day. The incidence angle modifier accounts for that variation by expressing the value of $\tau\alpha$ as

$$\tau\alpha = K_{\tau\alpha}(\tau\alpha)_n \qquad (7\text{-}9)$$

$K_{\tau\alpha}$ is usually cast in the form

$$K_{\tau\alpha} = 1 - b\left(\frac{1}{\cos(i)} - 1\right)$$ (7-10)

where i is the incidence angle. Consider the following example.

Example 7.1

Figure 7.9 provides the results of a performance test for a single-glazed flat-plate collector. The transmissivity, τ, of the glass is 0.90, and the absorptivity, α, of the surface is 0.92. For the collector, find:

(a) The collector heat removal factor, F_R

(b) The overall conductance, U_L in Btu/ft^2 °F.

(c) The rate at which the collector can deliver useful energy when the irradiation incident on the collector per unit area is 200 Btu/ft^2 h, the ambient temperature is 30° F, and the inlet water temperature is 60° F.

(d) The collector temperature when the flow rate is zero ($\eta = 0$).

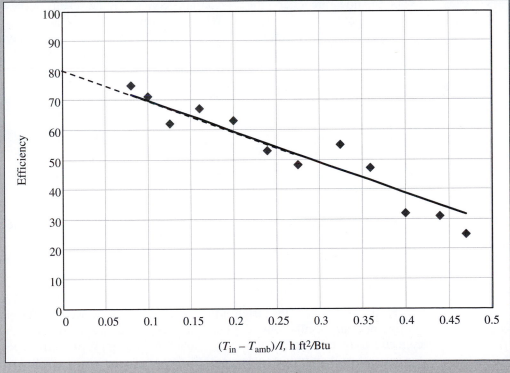

Figure 7.9 Flat-plate solar collector performance data.

Solution   From Figure 7.9, the intercept (extrapolated) is approximately 0.8. Then

$$0.8 = F_R \tau \alpha = F_R (0.9)(0.92) \quad F_R = 0.97$$

From the collector efficiency expression, the slope is $F_R U_L = \Delta \eta / \Delta X$:

$$-0.97 \, U_L = (0.5 - 0.8) \text{ Btu/h ft}^2 \, {}^\circ\text{F} / (0.3 - 0.0) = -1 \text{ Btu/h ft}^2 \, {}^\circ\text{F}$$

from which the overall conductance, U_L, is found as

$$U_L = 1.03 \text{ Btu/h ft}^2 \, {}^\circ\text{F}$$

The collector efficiency expression is $\eta = 0.8 - 1.0 \, (T_{\text{in}} - T_a)/I_T$. The efficiency becomes

$$\eta = 0.8 - 1.0 \text{ Btu/ft}^2 \text{ h } {}^\circ\text{F} \cdot (60 - 30)/200 \text{ Btu/ft}^2 \text{ h} = 0.65$$

The definition of collector efficiency is $\eta = q_{\text{useful}}/I_T A_c$, from which $q_{\text{useful}}/A_c = \eta I_T$:

$$q_{\text{useful}}/A_c = 0.65 \cdot 200 \text{ Btu/ft}^2 \text{ h} = 130 \text{ Btu/ft}^2 \text{ h}$$

When the flow rate is zero, the "stagnation" condition that results in the maximum collector temperature is attained. When the flow rate is zero, no useful energy is removed from the collector by the flowing fluid, and the steady-state temperature of the collector increases until all the incident solar energy is dissipated by convection and radiation. For this condition, $\eta = 0$ and

$$0 = F_R \left(\tau \alpha - U_L \frac{T_{\max} - T_a}{I_T} \right) \quad \text{or} \quad \tau \alpha = U_L \frac{T_{\max} - T_a}{I_T}$$

$$(0.9)(0.92) = 1.03 \frac{\text{Btu}}{\text{ft}^2 \text{h} \, {}^\circ\text{F}} \frac{T_{\max} - 30^\circ \text{ F}}{200 \dfrac{\text{Btu}}{\text{ft}^2 \text{h}}}$$

$$T_{\max} = 191^\circ \text{ F}$$

The stagnation temperature, $T_{\max}$, is 191° F. Thus, if the flow were stopped, the collector would eventually be heated to 191° F.

Example 7.1 provides insight into the information needed for flat-plate collector engineering calculations. Slope and intercept data are needed for any flat-plate collector to be considered, and solar insolation data, appropriate for specific collector azimuth and altitude angle orientation, are necessary to assess the performance of a collector. Sources for and examples of flat-plate solar collector data and solar data are presented in the next section.

7.3 **SOLAR COLLECTOR AND WEATHER DATA**

Third-party certification data on specific flat-plate solar collectors and solar water heating systems are available from the Solar Rating and Certification Corporation (SRCC). The SRCC website, which is well organized and user friendly, is www.solarrating.org. The SRCC website provides, in PDF format, the following online publications:

1. *Summary of SRCC Certified Solar Collector and Water Heating System Ratings*, which lists the performance ratings for solar collectors and systems.

2. *Directory of SRCC Certified Solar Collector Ratings,* which presents construction and rating information on certified solar collectors.

3. *Directory of SRCC Certified Solar Water Heating System Ratings,* which presents schematics and ratings for certified solar water heating systems.

4. *Annual Performance of OG-300 Certified Systems,* which provides estimated annual performance data on SRCC-rated certified systems for various locations.

The SRCC also maintains an up-to-date list of names (and contact information) of companies that participate in the SRCC certification programs, as well as convenient summary sheets (company name, model number, collector area, absorber coating, slope, intercept, and average daily energy collected for standard conditions) for certified solar collectors. Some examples of SRCC data presentations are in order.

Figure 7.10 is a data sheet from the *Directory of SRCC Certified Solar Collector Ratings* for a solar collector. These data were obtained in accordance with ASHRAE Standard 93, Methods of Testing to Determine the Performance of Solar Collectors. Administrative information is presented at the top of the data sheet. The "Collector Thermal Performance Rating" section consists of performance metrics in SI and inch-pound (IP) units for three irradiance conditions (clear, mildly cloudy, and cloudy) and five (A–E) temperature categories. The five categories represent different uses that have one-to-one correspondences with the inlet fluid and ambient air temperature differences. Table 7.2 contains information defining categories A–E. "Collector Specifications,"

TABLE 7.2 Flat-plate collector end-use descriptions

Category	Temperature Difference		Application
A	−5°C	−9°F	Swimming pool heating
B	5°C	9°F	Liquid collectors (heat pump) space heating, air systems
C	20°C	36°F	Service hot water, space heating, air systems
D	50°C	90°F	Service hot water, space heating, liquid systems, air conditioning
E	80°C	144°F	Space heating, liquid systems, air conditioning, process heating

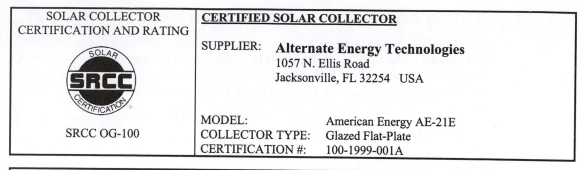

SOLAR COLLECTOR CERTIFICATION AND RATING	**CERTIFIED SOLAR COLLECTOR**
SRCC OG-100	SUPPLIER: **Alternate Energy Technologies** 1057 N. Ellis Road Jacksonville, FL 32254 USA MODEL: American Energy AE-21E COLLECTOR TYPE: Glazed Flat-Plate CERTIFICATION #: 100-1999-001A

COLLECTOR THERMAL PERFORMANCE RATING

Megajoules Per Panel Per Day				Thousands of Btu Per Panel Per Day			
CATEGORY (Ti-Ta)	CLEAR DAY 23 MJ/m²·d	MILDLY CLOUDY DAY 17 MJ/m²·d	CLOUDY DAY 11 MJ/m²·d	CATEGORY (Ti-Ta)	CLEAR DAY 2000 Btu/ft²·d	MILDLY CLOUDY DAY 1500 Btu/ft²·d	CLOUDY DAY 1000 Btu/ft²·d
A (-5°C)	28	21	15	A (-9°F)	27	20	14
B (5°C)	25	18	11	B (9°F)	24	17	11
C (20°C)	20	13	6	C (36°F)	19	12	6
D (50°C)	9	3		D (90°F)	8	3	
E (80°C)				E (144°F)			

A-Pool Heating (Warm Climate) B-Pool Heating (Cool Climate) C-Water Heating (Warm Climate) D-Water Heating (Cool Climate) E-Air Conditioning

Original Certification Date: June 15, 1999

COLLECTOR SPECIFICATIONS

Gross Area:	1.926	m²	20.73	ft²	Net Aperture Area:	1.776	m²	19.12 ft²
Dry Weight:	40.8	kg	90	lb	Fluid Capacity:	3.0	l	0.8 gal
Test Pressure:	1103	kPa	160	psig				

COLLECTOR MATERIALS

Frame:	Anodized Aluminum
Cover (Outer):	Low Iron Tempered Glass
Cover (Inner):	None
Absorber Material:	Tube - Copper / Plate - Copper
Absorber Coating:	Moderately Selective Black Paint
Insulation (Side):	Polyisocyanurate
Insulation (Back):	Polyisocyanurate

PRESSURE DROP

Flow		ΔP	
ml/s	gpm	Pa	in H₂O
20	0.32	55	0.22
50	0.79	306	1.23
80	1.27	745	2.99

TECHNICAL INFORMATION

Efficiency Equation [NOTE: Based on gross area and (P) = Ti-Ta]

				Y Intercept	**Slope**	
S I Units:	$\eta = 0.638$	-4.2645 (P)/I	-0.0297 (P)²/I	0.66	-6.37	W/m²·°C
I P Units:	$\eta = 0.638$	-0.7515 (P)/I	-0.0029 (P)²/I	0.66	-1.123	Btu/hr·ft²·°F

Incident Angle Modifier [(S) = 1/cos θ - 1, 0°≤ θ ≤60°]

						Model Tested:	AE-21E	
$K_{\alpha\tau} =$	1.0	+0.0248 (S)		-0.0861 (S)²		**Test Fluid:**	Water	
$K_{\alpha\tau} =$	1.0	-0.05 (S)		(Linear Fit)		**Test Flow Rate:**	39 ml/s	0.61 gpm

REMARKS:

Figure 7.10 SRCC sample certification information (SRCC).

"Collector Materials," and "Pressure Drop" information is delineated under these headings near the middle of the page. The lower part of the page is devoted to "Technical Information." Slope and intercept as well as quadratic efficiency equations are listed in both SI and IP units, and incident-angle modifier expressions are given. Thus, sufficient information is provided in the certification-rating sheet to make engineering calculations.

As with most energy system components, information on flat-plate solar collectors is also available from manufacturers. For example, the Alternate Energy Technologies website is accessible for information on the AE-21E collector. Figure 7.11 is the Alternate Energy Technologies marketing sheet for the AE-21 solar collector. Presentations such as the ones reproduced in Figures 7.10 and 7.11 are typical for solar collector manufacturers.

Solar energy engineering weather and solar irradiance database information was reviewed in Section 6.5 of the last chapter. In addition to data provided by NREL, various textbook authors have provided solar data required for performance estimates of solar energy systems. For example, Goswami et al. (2000) present hourly solar irradiance information useful for hour-by-hour solar collector calculations such as those used in Example 7.1. Their tabulations are for the 21st day of each month and are provided for 24°, 32°, 40°, 48°, 56°, and 64° north latitude. Total solar insolation in Btu/h ft^2 is provided for south-facing surfaces inclined at the latitude angle as well as at the latitude angle $-10°$, $+10°$, $+20°$, and $+90°$. Insolation values for horizontal

Glazing:
Low-Iron Tempered Glass exclusively using our new "High-T" tempered glass with a total solar energy transmission of 90%.

Manufactured by Thermafin™ Mfg., a 100% copper absorber plate, high frequency forge welded to create a crystalline homogenous connection between the fin and riser tube.

Collector Frame and Battens:
Type 6063-T6 extruded aluminum frame and battens with architectural bronze powder-coat finish that facilitates long life and strength.

Absorber Coating:
Exclusively by Thermafin™ Manufacturing, a Selective "Crystal Clear" Coating
α ≈ 0.96   ε ≈ 0.08

Insulation:
Polyisocyanurate foam board insulation. Foil-faced, glass fiber-reinforced, rigid board Thermax sheathing (1-1/4" in the bed / 3/4" in the sidewalls).

Gasket, Grommets:
A UV durable EPDM, U-channel gasket with molded corners which prohibits water penetration and assures long life. Extruded Silicone Grommet, 1-1/8" Bore

Mounting Hardware:
The variable "Quick Lock" mounting hardware reduces mounting time and makes it simple for anyone to install. The Quick Lock System allows the highest flexibility in mounting and is tested to wind load conditions of 180 mph. Mounting possibilities include: Pitched-roof, Flat-roof, Ground, Balcony, and Facade mounting.

Backsheet:
Type 3105-H14, 0.019" stucco embossed aluminum sheet (bronze) pop-riveted to aluminum frame.

Corner Bracket:
Architechtural aluminum angles inside with aircraft-grade pin grip rivets to insure high stability.

Design Life: 30 Years
Warranty: 10 Year Full
Working Pressure: 165 PSI
Flow Rate: 0.5 to 1.8 GPM
(recommended)

Fasteners:
5056 Aluminum Rivets secure the backsheet. Batten screws are 18-8 SS, 10-24 x 3/8", Hex head screws black oxide coated.

SPECIFICATIONS						
Collector	AE-21	AE-24	AE-26	AE-28	AE-32	AE-40
Length (in)	85.187	97.187	77.187	85.187	97.187	121.187
Width (in)	35.187	35.187	47.187	47.187	47.187	47.187
Height (in)	3.137	3.137	3.137	3.137	3.137	3.137
Gross Area (ft²)	20.8	23.7	25.3	27.9	31.8	39.7
Transparent Area (ft²)	19.2	21.9	23.6	26.1	29.9	37.4
Dry Weight (lb)	74	84	90	99	113	153

Figure 7.11 Alternate Energy Technologies AE-21 information (Alternate Energy Technolgies).

surfaces and for surfaces normal to the sun complete the tabulation. The Goswami et al. (2000) solar insolation tables are provided as Appendix 7A to this chapter.

For many solar engineering calculations, monthly insolation (irradiance) values for various orientations and tracking protocols are needed. The NREL website, www.nrel.gov, provides monthly values based on 30-year averages (1961–1990) for the 239 locations in the National Solar Radiation Data Base (NSRDB). Figure 6.20 shows the sites for which NSRDB data are available. A number of solar energy databases are available on the website under Solar Radiation Resource Information. The NSRDB contains the following:

Daily statistics files (DSFs)

Hourly data files (HDFs)

Solar Radiation Data Manual for Buildings

- 30-year averages (1961–1990) of solar radiation and illuminance for each month

Solar Radiation Data Manual for Flat-Plate and Concentrating Collectors

- Averages of solar radiation for each of the 360 months during the period 1961–1990
- 30-year (1961–1990) average of monthly solar radiation for each month

Typical Meteorological Year 2: 39 sites

In addition, the website contains other broadband solar radiation data and resource maps in atlas form for buildings and for flat-plate and concentrating solar collectors. A number of solar computer codes, algorithms, models, and calculators are also provided and described on the website. The information contained in the website, as well as the references by SRCC and NREL at the end of this chapter, are logical next resorts for engineering details and procedures beyond the scope of this chapter.

Solar Radiation Data Manual for Flat-Plate and Concentrating Collectors (1994) provides the necessary data for the next level of solar energy system analyses and performance estimates. The data for Meridian, MS, from the manual are reproduced as Figure 7.12. An examination of the contents of this typical file is appropriate. The first series of entries contain location information and specify the site as primary or secondary—secondary in the case depicted in the figure. The remaining data are divided into solar radiation–specific and climatic condition–specific sections. The solar radiation–specific data are presented as the incident solar insolation energy per m^2 per day (kWh/m^2 day) for four solar collector configurations. The solar collector configurations referenced in the data are as follows:

1. Flat-plate collectors facing south at a fixed tilt
   a. 0° tilt
   b. Latitude −15° tilt
   c. Latitude tilt
   d. Latitude +15° tilt
   e. 90° tilt

```
"City:  ","MERIDIAN          " "State: ","MS"          "WBAN No:", 13865
"Lat(N): ", 32.33        "Long(W): ", 88.75
"Elev(m): ",   94        "Pres(mb):",  1007    "Stn Type:","Secondary"
```

```
"SOLAR RADIATION FOR FLAT-PLATE COLLECTORS FACING SOUTH AT A FIXED-TILT (kWh/m2/day), Percentage
Uncertainty = 9"
"Tilt(deg)","        ","Jan","Feb","Mar","Apr","May","Jun","Jul","Aug","Sep","Oct","Nov","Dec","Year"
"0         ","Average", 2.6, 3.4, 4.4, 5.4, 5.9, 6.2, 5.9, 5.6, 4.8, 4.1, 2.9, 2.4, 4.5
"          ","Minimum", 2.2, 3.0, 3.7, 4.8, 5.4, 5.3, 5.2, 4.8, 4.1, 3.2, 2.3, 2.0, 4.2
"          ","Maximum", 3.0, 3.9, 5.1, 6.4, 6.8, 6.9, 6.5, 6.3, 5.5, 4.8, 3.4, 2.8, 4.7
"Lat - 15 ","Average", 3.2, 4.1, 4.8, 5.7, 5.9, 6.0, 5.8, 5.7, 5.2, 4.8, 3.7, 3.1, 4.8
"          ","Minimum", 2.6, 3.4, 4.1, 4.9, 5.4, 5.2, 5.1, 4.9, 4.4, 3.6, 2.6, 2.5, 4.6
"          ","Maximum", 4.0, 4.8, 5.8, 6.8, 6.8, 6.7, 6.4, 6.4, 6.0, 5.8, 4.4, 3.8, 5.1
"Lat      ","Average", 3.6, 4.4, 5.0, 5.6, 5.6, 5.6, 5.4, 5.5, 5.2, 5.2, 4.1, 3.5, 4.9
"          ","Minimum", 2.8, 3.6, 4.1, 4.8, 5.1, 4.9, 4.8, 4.7, 4.4, 3.7, 2.7, 2.7, 4.7
"          ","Maximum", 4.6, 5.3, 6.1, 6.7, 6.4, 6.2, 6.0, 6.2, 6.1, 6.3, 5.0, 4.4, 5.2
"Lat + 15 ","Average", 3.8, 4.5, 4.9, 5.2, 5.0, 4.9, 4.8, 5.0, 5.0, 5.2, 4.3, 3.8, 4.7
"          ","Minimum", 2.9, 3.6, 4.0, 4.5, 4.5, 4.3, 4.2, 4.3, 4.2, 3.7, 2.8, 2.8, 4.5
"          ","Maximum", 5.0, 5.5, 6.0, 6.2, 5.7, 5.4, 5.3, 5.7, 5.9, 6.4, 5.3, 4.8, 5.0
"90        ","Average", 3.3, 3.5, 3.3, 2.8, 2.2, 2.0, 2.1, 2.6, 3.1, 3.9, 3.6, 3.3, 3.0
"          ","Minimum", 2.4, 2.8, 2.7, 2.5, 2.1, 1.9, 2.0, 2.3, 2.6, 2.6, 2.2, 2.3, 2.8
"          ","Maximum", 4.4, 4.4, 4.3, 3.4, 2.4, 2.1, 2.2, 2.8, 3.7, 4.8, 4.5, 4.3, 3.2
"SOLAR RADIATION FOR 1-AXIS TRACKING FLAT-PLATE COLLECTORS WITH A NORTH-SOUTH AXIS (kWh/m2/day), Percentage
Uncertainty = 9"
"Axis Tilt","        ","Jan","Feb","Mar","Apr","May","Jun","Jul","Aug","Sep","Oct","Nov","Dec","Year"
"0         ","Average", 3.5, 4.6, 5.7, 7.0, 7.5, 7.7, 7.2, 7.0, 6.1, 5.5, 4.0, 3.2, 5.7
"          ","Minimum", 2.7, 3.6, 4.6, 5.8, 6.5, 6.3, 6.2, 5.6, 4.9, 3.8, 2.6, 2.5, 5.4
"          ","Maximum", 4.4, 5.6, 7.1, 8.9, 8.9, 9.0, 8.4, 8.1, 7.3, 6.9, 4.8, 4.0, 6.2
"Lat - 15 ","Average", 4.0, 5.1, 6.1, 7.2, 7.5, 7.6, 7.2, 7.1, 6.5, 6.1, 4.5, 3.8, 6.0
"          ","Minimum", 3.0, 4.0, 4.9, 6.5, 6.5, 6.2, 6.1, 5.7, 5.1, 4.1, 2.8, 2.8, 5.6
"          ","Maximum", 5.1, 6.3, 7.7, 9.1, 8.9, 8.9, 8.4, 8.2, 7.8, 7.6, 5.6, 4.7, 6.5
"Lat      ","Average", 4.3, 5.3, 6.2, 7.1, 7.2, 7.3, 6.9, 7.0, 6.5, 6.4, 4.9, 4.1, 6.1
"          ","Minimum", 3.1, 4.1, 4.9, 5.9, 6.3, 5.9, 5.9, 5.6, 5.1, 4.3, 3.0, 3.0, 5.7
"          ","Maximum", 5.6, 6.7, 7.8, 9.1, 8.6, 8.5, 8.1, 8.1, 7.8, 8.0, 6.1, 5.2, 6.6
"Lat + 15 ","Average", 4.4, 5.4, 6.1, 6.8, 6.8, 6.8, 6.5, 6.6, 6.3, 6.4, 5.0, 4.3, 6.0
"          ","Minimum", 3.2, 4.2, 4.8, 5.6, 5.9, 5.5, 5.5, 5.3, 4.9, 4.2, 3.0, 3.0, 5.6
"          ","Maximum", 5.9, 6.8, 7.8, 8.7, 8.1, 7.9, 7.6, 7.7, 7.7, 8.1, 6.3, 5.5, 6.5
"SOLAR RADIATION FOR 2-AXIS TRACKING FLAT-PLATE COLLECTORS (kWh/m2/day), Percentage Uncertainty = 9"
"Tracker  ","        ","Jan","Feb","Mar","Apr","May","Jun","Jul","Aug","Sep","Oct","Nov","Dec","Year"
"2-Axis    ","Average", 4.5, 5.4, 6.2, 7.2, 7.6, 7.8, 7.3, 7.1, 6.5, 6.4, 5.1, 4.4, 6.3
"          ","Minimum", 3.2, 4.2, 4.9, 6.0, 6.6, 6.3, 6.2, 5.7, 5.1, 4.3, 3.0, 3.1, 5.9
"          ","Maximum", 6.0, 6.9, 7.9, 9.2, 9.0, 9.1, 8.5, 8.3, 7.9, 8.1, 6.3, 5.6, 6.8
"DIRECT BEAM SOLAR RADIATION FOR CONCENTRATING COLLECTORS (kWh/m2/day), Percentage Uncertainty = 8"
"Tracker  ","        ","Jan","Feb","Mar","Apr","May","Jun","Jul","Aug","Sep","Oct","Nov","Dec","Year"
"1-X, E-W ","Average", 2.5, 2.8, 2.9, 3.3, 3.3, 3.4, 3.1, 3.0, 2.9, 3.4, 2.8, 2.5, 3.0
"Hor Axis ","Minimum", 1.3, 1.8, 1.9, 2.4, 2.5, 2.3, 2.1, 1.9, 1.9, 1.7, 1.2, 1.4, 2.7
"          ","Maximum", 3.7, 4.1, 4.3, 4.9, 4.5, 4.4, 4.0, 3.9, 3.9, 4.7, 3.9, 3.5, 3.4
"1-X, N-S ","Average", 2.1, 2.9, 3.5, 4.4, 4.4, 4.4, 4.0, 3.9, 3.6, 3.6, 2.5, 2.0, 3.4
"Hor Axis ","Minimum", 1.1, 1.7, 2.3, 3.1, 3.3, 2.9, 2.8, 2.5, 2.3, 1.8, 1.0, 1.0, 3.0
"          ","Maximum", 3.2, 4.1, 5.1, 6.5, 6.0, 5.8, 5.3, 5.1, 4.8, 5.1, 3.4, 2.8, 4.0
"1-X, N-S ","Average", 2.8, 3.5, 3.9, 4.5, 4.2, 4.1, 3.8, 3.9, 3.9, 4.3, 3.2, 2.7, 3.7
"Tilt=Lat ","Minimum", 1.5, 2.2, 2.6, 3.2, 3.1, 2.7, 2.6, 2.5, 2.5, 2.1, 1.3, 1.4, 3.3
"          ","Maximum", 4.2, 5.1, 5.7, 6.7, 5.8, 5.4, 5.0, 5.1, 5.2, 6.0, 4.5, 3.8, 4.3
"2-X       ","Average", 2.9, 3.6, 3.9, 4.5, 4.5, 4.4, 4.0, 4.0, 3.9, 4.3, 3.4, 2.9, 3.9
"          ","Minimum", 1.6, 2.2, 2.6, 3.3, 3.3, 2.9, 2.8, 2.6, 2.5, 2.2, 1.4, 1.6, 3.5
"          ","Maximum", 4.5, 5.2, 5.7, 6.8, 6.1, 5.9, 5.4, 5.2, 5.3, 6.1, 4.7, 4.2, 4.5
"AVERAGE CLIMATIC CONDITIONS"
"Element  ","        ","Jan","Feb","Mar","Apr","May","Jun","Jul","Aug","Sep","Oct","Nov","Dec","Year"
"Temp.     ","(deg C)",  7.2,  9.4, 13.7, 17.8, 21.8, 25.6, 27.2, 27.0, 24.1, 17.8, 13.1,  9.1, 17.8
"Daily Min","(deg C)",  0.8,  2.6,  6.4, 10.5, 14.9, 18.9, 21.1, 20.7, 17.7, 10.3,  5.8,  2.6, 11.0
"Daily Max","(deg C)", 13.6, 16.2, 20.9, 25.2, 28.7, 32.2, 33.4, 33.2, 30.5, 25.4, 20.3, 15.6, 24.6
"Record Lo","(deg C)",-17.8,-13.3, -9.4, -2.2,  3.3,  5.6, 12.8, 11.7,  1.1, -4.4, -8.9,-16.7,-17.8
"Record Hi","(deg C)", 28.3, 29.4, 32.2, 35.0, 37.2, 40.0, 41.7, 40.0, 40.6, 36.1, 30.6, 27.8, 41.7
"HDD,Base=","18.3C  ",  349,  253,  162,   55,    8,    0,    0,    0,    3,   68,  167,  292, 1358
"CDD,Base=","18.3C  ",    4,    3,   17,   40,  117,  218,  276,  269,  177,   53,    9,    6, 1188
"Rel Hum  ","percent",   74,   71,   70,   71,   73,   74,   77,   77,   76,   75,   75,   75,   74
"Wind Spd.","(m/s) ",   3.2,  3.4,  3.5,  3.2,  2.6,  2.3,  2.1,  2.1,  2.4,  2.3,  2.8,  3.2,  2.7
```

Figure 7.12 Average monthly solar data for Meridian, MS.

2. One-axis tracking flat-plate collectors with north-south axis
   a. 0° tilt
   b. Latitude −15° tilt
   c. Latitude tilt
   d. Latitude +15° tilt
3. Two-axis tracking flat-plate collectors
4. Direct-beam radiation for concentrating collectors
   a. One-axis, east-west horizontal axis
   b. One-axis, north-south horizontal axis
   c. One-axis, north-south tilt at latitude angle
   d. Two-axis

Monthly average insolation values for each solar collector configuration are given for each month and for specified minimum, average, and maximum irradiation conditions. For example, for June and for a fixed-tilt solar collector tilted at the latitude angle, the average insolation on the collector would be 5.4 kWh/m² day. The four solar collector configurations and the orientation options within each configuration provide the insolation data needed for most applications.

The last segment of the data presents climatic conditions on a monthly basis. The average, daily minimum, and daily maximum temperatures are listed, as are the record lows and record highs for each month. The heating degree-days (HDD, base of 18.3° C) and the cooling degree-days (CDD, base of 18.3° C) are provided along with the average relative humidity and the average wind speed. The HDD and CDD are indicators of the relative need for heating and cooling, respectively, and will be defined and used in the next chapter on passive solar energy.

| Example 7.2 | Evaluate the performance of an Alternate Energy Technologies AE-21E solar collector on January 21 if the collector is located at 32° north latitude and tilted at an angle of 32°. The water inlet temperature is 58° F. The ambient temperature varies from 25° F at 8 a.m. to 45° F at solar noon to 35° F at 4 p.m. Metrics of interest include the hourly efficiency, the hourly solar energy collected, and the percentage of incident solar energy collected. |

| Solution | Information about the AE-21E is available from the SRCC website and is reproduced as Figure 7.10. Example 7.1 evaluated the instantaneous performance of a solar collector given the ambient and water inlet temperatures and the incident solar irradiation flux, and it serves as the basis of the solution for this example. The incident solar energy on January 21 at 32° N with a collector tilt angle of 32° is available on an hourly basis from Appendix 7A of this chapter. The inlet water temperature is taken as constant since no information about its variation is indi- |

cated. The ambient temperature is provided at 8 a.m., solar noon, and 4 p.m. From the ambient temperature information, estimates of the ambient temperatures at 9, 10, and 11 a.m. as well as 1, 2, and 3 p.m. can be made. With hourly ambient temperature, water inlet temperature, and solar irradiation available, the collector performance for every hour of incident sunlight can be evaluated. The daily performance can be generated by summing the hourly performance information. The Mathcad worksheet for the solution is reproduced as Figure 7.13. Information from the AE-21E SRCC data sheet is entered. Consistent with SRCC recommendations, the solar collector area is the gross area, not the aperture area. The hourly input solar irradiation (from Appendix 7A for 32° latitude and 32° inclination from the horizontal for January 21), water inlet temperature, and ambient temperatures are entered. The solar collector efficiency for each hour is calculated using Eq. (7-7). The vector symbol over the efficiency definition is a Mathcad operator indicating that a computation is to be carried out for each of the nine defined "i" values—the range variable. The hourly energy harvested is computed utilizing the hourly value of the collector efficiency. The last calculations are for the daily energy harvested (expressed in kWh) and the fraction of incident solar energy harvested. For the scenario described in the problem statement, the AE-21E collector will harvest 6.755 kWh during the day, or 55 percent of the incident solar energy that day.

Enter information about the flat plate solar collector (obtained for SRCC web site).

$$\text{Area} := 20.73 \cdot \text{ft}^2 \qquad \text{AreaTot} := 1 \cdot \text{Area} \qquad \text{AreaTot} = 20.73\,\text{ft}^2 \qquad F := R$$

$$\text{Intercept} := 0.66 \qquad \text{Slope} := -1.123 \cdot \frac{\text{BTU}}{\text{hr} \cdot \text{ft}^2 \cdot F}$$

Setup vectors for I (solar irradiation), T_{in}, T_{amb}. Obtained from weather data base.

$$\text{ORIGIN} \equiv 1 \qquad i := 1 \ldots 9$$

$$I := \begin{pmatrix} 106 \\ 193 \\ 256 \\ 295 \\ 308 \\ 295 \\ 256 \\ 193 \\ 106 \end{pmatrix} \cdot \frac{\text{BTU}}{\text{ft}^2 \cdot \text{hr}} \qquad T_{in} := \begin{pmatrix} 58 \\ 58 \\ 58 \\ 58 \\ 58 \\ 58 \\ 58 \\ 58 \\ 58 \end{pmatrix} \cdot F \qquad T_{amb} := \begin{pmatrix} 25 \\ 30 \\ 35 \\ 40 \\ 45 \\ 43 \\ 41 \\ 38 \\ 35 \end{pmatrix} \cdot F$$

Figure 7.13 Mathcad solution for Example 7.2.

Calculate the solar collector efficiency.

$$\eta := \left[0.66 - 1.123 \cdot \frac{BTU}{hr \cdot ft^2 \cdot F} \cdot \overrightarrow{\frac{(T_{in} - T_{amb})}{I}} \right]$$

$$\eta = \begin{pmatrix} 0.31 \\ 0.497 \\ 0.559 \\ 0.591 \\ 0.613 \\ 0.603 \\ 0.585 \\ 0.544 \\ 0.416 \end{pmatrix}$$

Calculate the useful energy harvested.

$$Q_{useful} := \overrightarrow{(AreaTot \cdot I \cdot \eta)}$$

$$Q_{useful} = \begin{pmatrix} 682.038 \\ 1.989 \times 10^3 \\ 2.967 \times 10^3 \\ 3.617 \times 10^3 \\ 3.911 \times 10^3 \\ 3.687 \times 10^3 \\ 3.107 \times 10^3 \\ 2.175 \times 10^3 \\ 914.836 \end{pmatrix} \frac{BTU}{hr}$$

The daily harvested energy is the sum of the hourly harvested energy.

$$Q_{Total} := \sum_{i=1}^{9} Q_{useful_i} \cdot 1 \cdot hr \qquad kWh := 1000 \cdot watt \cdot hr \qquad Q_{Total} = 6.755 \, kWh$$

Find the percent of incident solar energy captured.

$$percent := \frac{\displaystyle\sum_{i=1}^{9} Q_{useful_i} \cdot 100}{\displaystyle\sum_{i=1}^{9} AreaTot \cdot I_i} \qquad percent = 55.374$$

Figure 7.13 *(continued)*

Calculations such as those presented in Example 7.2 indicate how to estimate solar collector performance for one day; however, for many engineering applications, the performance over a year is needed. Many simulation programs exist that can accomplish yearly performance estimates on an hour-by-hour basis using TMY2 weather data. Duffie and Beckman (2006) provide an excellent discussion of such programs. However, a viable alternative with acceptable accuracy for preliminary design and analysis purposes is the *f*-chart method, which is covered in the next section.

7.4   THE *f*-CHART METHOD

The *f*-chart method is frequently used for estimating the annual thermal performance of an active solar heating system for a building. Duffie and Beckman (2006) as well as Goswami et al. (2000) provide much more detail on the development and application of the *f*-chart method than is possible here. The *f*-chart method provides a procedure for estimating the fraction, *f*, of the total monthly heating load (space plus domestic hot water) that can be supplied by an active solar energy system. The procedure is based on a correlation for *f* as a function of two system parameters. The correlation was derived from the results of a large number of hour-by-hour solar energy system simulations. The two parameters are the ratio of collector losses to heating loads, X, and the ratio of absorbed solar irradiation to heating loads, Y. *f*-charts have been developed for three system configurations: (1) space and water heating using water as the medium, (2) space and water heating using air as the medium, and (3) process hot water. Only the first configuration will be presented here. Figure 7.14 is a schematic of the "standard" system configuration using a liquid heat transfer medium. The system consists of a solar collector, three heat exchangers, a main storage tank, a preheat tank, the load, and ancillary components. Energy harvested by the collector is used to heat water for the main storage tank via the collector-storage heat exchanger. Water from the storage tank is used in the load heat exchanger to provide heated air to the house. Potable water is preheated before entering the service hot-water tank. Auxiliary systems are provided for both the house load and the service hot-water tank in case the solar energy system is unable to meet the required loads.

Detailed simulations of the system illustrated in Figure 7.14 were used to develop the correlation between the two dimensionless variables (X and Y) and *f*, the monthly fraction of the load carried by solar energy. The first of the two dimensionless parameters is defined as

$$X = \frac{A_c F'_R U_L \Delta t \left(100 - \overline{T}_a\right)}{L} \tag{7-11}$$

where X = the ratio of collector losses to load, A_c is the collector area, F'_R is the corrected collector heat removal factor (usually taken as $0.97 F_R$), U_L is the collector loss coefficient, $\overline{T}_a$ is the monthly average ambient temperature, Δt is the number of hours in the month, and L is the monthly total heating load (space and water). The

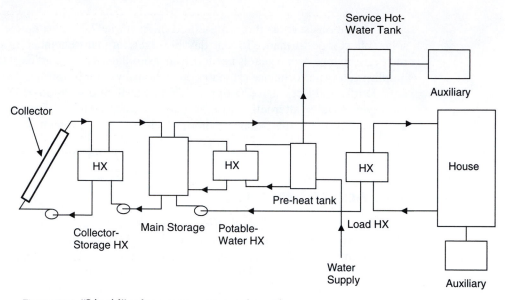

Figure 7.14 "Liquid" solar energy system schematic.

expression is dimensionless. The second dimensionless parameter, Y, is the ratio of the absorbed solar irradiation to the total heating load and is cast as

$$Y = \frac{A_c F_R' \overline{\tau\alpha} H_T N}{L} \tag{7-12}$$

where $\overline{\tau\alpha}$ is the monthly average $\tau\alpha$ product for the collector (usually taken as $0.96\tau\alpha$), H_T is the monthly average daily radiation incident on the collector surface, and N is the number of days in the month.

The monthly solar fraction, f, is obtained for the following correlation for "liquid" solar heating systems:

$$f = 1.029Y - 0.065X - 0.245Y^2 + 0.0018X^2 + 0.0215Y^3 \tag{7-13}$$

Figure 7.15 is a parametric representation of this equation.

A similar relationship has been derived for air heating systems. The monthly solar fraction correlation for air heating systems is

$$f = 1.0409Y - 0.065X - 0.159Y^2 + 0.00187X^2 - 0.0095Y^3 \tag{7-14}$$

Figure 7.16 presents the f-chart relationship for air heating systems.

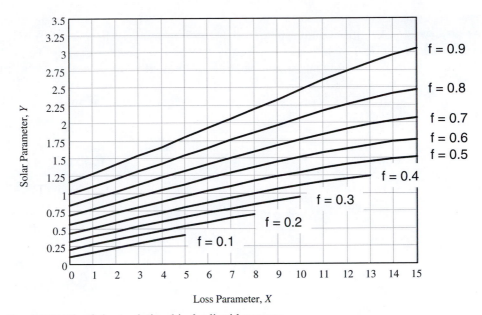

Figure 7.15 The *f*-chart relationship for liquid systems.

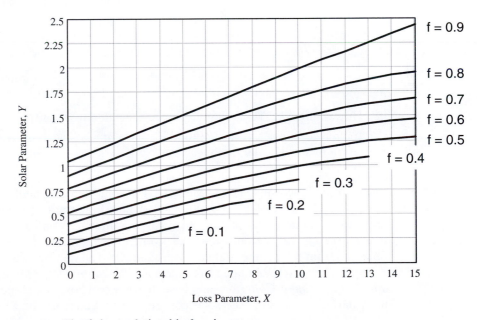

Figure 7.16 The *f*-chart relationship for air systems.

<table>
<tr><td>**Example 7.3**</td><td>An office building in Meridian, MS, is to employ Alternate Energy Technologies AE-21E flat-plate solar collectors for space and water heating. The collectors will be mounted facing south. The estimated monthly load is provided in Table 7. 3.</td></tr>
</table>

TABLE 7.3 Monthly heating energy requirements for an office building

Month	Load (1000 MJ)
January	140
February	120
March	80
April	50
May	28
June	20
July	20
August	20
September	20
October	40
November	80
December	110

The solar energy system is to provide all the water heating requirements for the summer months and a yearly solar heating fraction of 60 percent.

(a) How many AE-21E solar collectors are needed to meet the specifications?

(b) If natural gas costs $10 per million Btu, how much would be saved by the installation of such a solar energy system?

Solution The solution is accomplished in Mathcad. The solution worksheet is represented in Figure 7.17. For ease of use, MJ is defined as megaJoules, and the days of the month are entered for use in calculating the *f*-chart parameters, *X* and *Y*. The average daily irradiation for each month and the average temperature for each month are obtained from the weather data for Meridian, MS. The weather data are taken from Figure 7.12 and were obtained from the *Solar Radiation Data Manual for Flat-Plate and Concentrating Collectors* database at www.nrel.gov. The slope and intercept information for the AE-21E solar collector was taken from the SRCC data sheet that is reproduced as Figure 7.10. Determining the number of AE-21E collectors needed is an iterative process. The required number of collectors is guessed and the analysis completed to ascertain the performance of the guessed number of collectors. When the required performance is achieved, the iterative procedure has

f-chart example for solar heating in Meridian, MS

Set up all input parameters on a monthly basis using the range variable i:

$\text{ORIGIN} \equiv 1 \qquad i := 1 \ .. \ 12$

$MJ := 10^6 \cdot J$

Define month in terms of days.

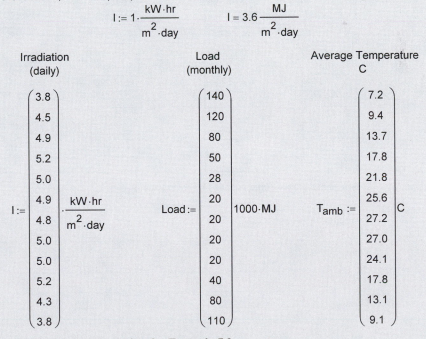

$$month := \begin{pmatrix} 31 \\ 28 \\ 31 \\ 30 \\ 31 \\ 30 \\ 31 \\ 31 \\ 30 \\ 31 \\ 30 \\ 31 \end{pmatrix} \cdot day \qquad \begin{pmatrix} jan \\ feb \\ mar \\ apr \\ may \\ jun \\ jul \\ aug \\ sep \\ oct \\ nov \\ dec \end{pmatrix}$$

The monthly-average conditions from NREL are used to obtain solar and weather data for the location, Meridian, MS, in this case. The load must come from building information.

$$I := 1 \cdot \frac{kW \cdot hr}{m^2 \cdot day} \qquad I = 3.6 \, \frac{MJ}{m^2 \cdot day}$$

Irradiation (daily)	Load (monthly)	Average Temperature C

$$I := \begin{pmatrix} 3.8 \\ 4.5 \\ 4.9 \\ 5.2 \\ 5.0 \\ 4.9 \\ 4.8 \\ 5.0 \\ 5.0 \\ 5.2 \\ 4.3 \\ 3.8 \end{pmatrix} \cdot \frac{kW \cdot hr}{m^2 \cdot day} \qquad Load := \begin{pmatrix} 140 \\ 120 \\ 80 \\ 50 \\ 28 \\ 20 \\ 20 \\ 20 \\ 20 \\ 40 \\ 80 \\ 110 \end{pmatrix} 1000 \cdot MJ \qquad T_{amb} := \begin{pmatrix} 7.2 \\ 9.4 \\ 13.7 \\ 17.8 \\ 21.8 \\ 25.6 \\ 27.2 \\ 27.0 \\ 24.1 \\ 17.8 \\ 13.1 \\ 9.1 \end{pmatrix} C$$

Figure 7.17 Mathcad solution for Example 7.3.

Insert solar collector characteristics:

Number := 171

$$FR\tau\alpha := 0.66 \qquad FRUL := 6.37\frac{W}{m^2 \cdot C} \qquad A := 1.926 \cdot m^2 \cdot Number \qquad A = 329.346\, m^2$$

Implement the f-chart analysis using the definitions of X and Y.

Compute X and Y:

$$X := \overrightarrow{\left[0.97FRUL \cdot \left(100 \cdot C - T_{amb}\right) \cdot \frac{A \cdot month}{Load}\right]} \qquad Y := \overrightarrow{\left(A \cdot 0.96 \cdot 0.97FR\tau\alpha \cdot I \cdot \frac{month}{Load}\right)}$$

Calculate f (the solar fraction per month) based on the f-chart correlation:

$$fall := \overrightarrow{\left(1.029 \cdot Y - 0.065 \cdot X - 0.245 \cdot Y^2 + 0.0018 \cdot X^2 + 0.0215 \cdot Y^3\right)}$$

$$f_i := \begin{vmatrix} fall_i & \text{if} & fall_i < 1 \\ 1 & \text{otherwise} \end{vmatrix} \qquad \begin{array}{l}\text{If solar will more than satisfy the load,} \\ \text{set f = 1.}\end{array}$$

X =

	1
1	3.613
2	3.717
3	5.88
4	8.672
5	15.223
6	19.622
7	19.84
8	19.894
9	20.018
10	11.201
11	5.73
12	4.504

Y =

	1
1	0.613
2	0.765
3	1.384
4	2.274
5	4.034
6	5.356
7	5.421
8	5.647
9	5.465
10	2.937
11	1.175
12	0.78

fall =

	1
1	0.332
2	0.437
3	0.692
4	0.897
5	1.003
6	1.204
7	1.223
8	1.289
9	1.236
10	0.951
11	0.592
12	0.408

f =

	1
1	0.332
2	0.437
3	0.692
4	0.897
5	1
6	1
7	1
8	1
9	1
10	0.951
11	0.592
12	0.408

Determine the monthly load supplied by solar system.

$$Load_{solar} := \overrightarrow{(f \cdot Load)}$$

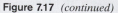

$Load_{solar}{}^T =$

	1	2	3	4	5	6	
1	$4.654 \cdot 10^{10}$	$5.241 \cdot 10^{10}$	$5.534 \cdot 10^{10}$	$4.487 \cdot 10^{10}$	$2.8 \cdot 10^{10}$	$2 \cdot 10^{10}$	J

$Load_{solar}{}^T =$

	7	8	9	10	11	12	
1	$2 \cdot 10^{10}$	$2 \cdot 10^{10}$	$2 \cdot 10^{10}$	$3.805 \cdot 10^{10}$	$4.739 \cdot 10^{10}$	$4.485 \cdot 10^{10}$	J

Figure 7.17 *(continued)*

Estimate the yearly solar fraction, the total solar energy harvested, and the savings.

$$\text{percent} := \frac{\displaystyle\sum_{i=1}^{12} \left(\text{Load}_{\text{solar}_i} \cdot 100\right)}{\displaystyle\sum_{i=1}^{12} \text{Load}_i} \qquad\qquad \text{percent} = 60.09$$

$$\text{TotalSolar} := \sum_{i=1}^{12} \left(\text{Load}_{\text{solar}_i}\right) \qquad\qquad \text{TotalSolar} = 4.375 \times 10^{11} \, \text{J}$$

$$\text{TotalSolar} = 4.146 \times 10^{8} \, \text{BTU}$$

$$\text{Savings} := \frac{\text{TotalSolar}}{10^{6} \cdot \text{BTU}} \cdot 10 \qquad\qquad \text{Savings} = 4.146 \times 10^{3}$$

$$\text{TotalLoad} := \sum_{i=1}^{12} \text{Load}_i \qquad\qquad \text{TotalLoad} = 6.9 \times 10^{8} \, \text{BTU}$$

Figure 7.17 *(continued)*

attained convergence. The parameters X and Y are computed for each month, and the value of the solar fraction, designated *fall*, for each month is obtained from the *f*-chart correlation in the form of Eq. (7-13). Both X and Y must be positive, so in the calculation of X the value of the slope of the collector performance curve is taken as positive. Since the required solar fraction for a given month cannot exceed unity, the value of the solar fraction, *fall*, for months that exceed unity is set to unity by the piecewise continuous "if" statement. Values *of* X, Y, *fall*, and f (the solar fraction) are tabulated in the worksheet. The solar load for each month is computed. The yearly solar fraction, the total solar energy harvested, and the cost savings are then estimated. When the guessed number of required collectors yields the required solar fraction, 60 percent in the problem statement, and solar fractions of unity during the summer, the solution procedure has converged. In this example, 171 AE-21E collectors are needed to meet the specifications.

Example 7.3 illustrates how the *f*-chart procedure can be used to assess the performance of a flat-plate solar collector system. The *f*-chart procedure is based on correlating the outputs from a number of TRYNS systems simulations and as such is an approximate procedure. However, the *f*-chart procedure is easy to apply and yields solutions that are much better than just an order of magnitude. Indeed, with all the uncertainties in many flat-plate solar collector engineering calculations, an argument can be made that the f-chart approach provides sufficient accuracy for most initial assessments of such systems.

7.5 OTHER SOLAR THERMAL SYSTEMS

As is apparent from Figures 7.1(b) and 7.2, solar thermal systems include more than just flat-plate collectors. Figure 7.1(b) shows a parabolic-trough collector, and Figure 7.2 shows a heliostat. Both devices collect solar energy, but they also concentrate solar energy—that is, they enable higher temperatures to be attained. Since higher output temperatures can be attained by concentrating collectors, the usefulness of the energy collected increases and activities requiring higher working temperatures become possible. This section reviews some of the salient features of concentrating collectors.

However, a preliminary review of the fundamentals of concentrating collectors is appropriate. Concentration of incident radiation is achieved by directing (by reflection or refraction) the radiation on an aperture area, A_a, to a smaller absorber or receiver area, A_r. A geometric concentration factor, CR, is defined as the ratio of the aperture to receiver area, or

$$\text{CR} = \frac{A_a}{A_r} \tag{7-15}$$

The energy delivered to the receiver is the incident solar flux, I_c, at the aperture times the product of the aperture area and the optical efficiency of the aperture-to-receiver process, η_{ar}, or

$$q_{\text{del}} = \eta_{ar} I_c A_a \tag{7-16}$$

As with the flat-plate collector, the convective losses are expressed as

$$q_{\text{losses}} = U_c A_r (T_{\text{ave}} - T_a) \tag{7-17}$$

The useful energy then becomes

$$q_{\text{useful}} = \eta_{ar} I_c A_a - U_c A_r (T_{\text{ave}} - T_a) \tag{7-18}$$

The collector efficiency is defined as the useful incident radiation flux divided by the incident irradiation:

$$\eta_c = \frac{q_{\text{useful}}}{I_c A_a} = \eta_{ar} - \frac{U_c(T_{\text{ave}} - T_a)A_r}{I_c A_a} \tag{7-19}$$

Recognizing that the geometric concentration factor, CR, is A_a/A_r, Eq. (7-19) becomes

$$\eta_c = \frac{q_{\text{useful}}}{I_c A_a} = \eta_{ar} - \frac{U_c(T_{\text{ave}} - T_a)}{I_c \, \text{CR}} \tag{7-20}$$

For concentrating collectors, CR > 1. The loss term in Eq.(7-20) is thus smaller for concentrating collectors than for flat-plate collectors (CR ~ 1), and as a result the efficiency for concentrating collectors is higher. For further details see Goswami et al. (2000), who present a particularly lucid evaluation of concentrating solar collectors.

Technical and engineering information for flat-plate solar collectors can be obtained from manufacturers' websites and from the SRCC. Such information for concentrating solar collectors and heliostats is more difficult to find. Most websites for companies manufacturing concentrating solar collectors contain little technical information.

One interesting question in relation to concentrating collectors is the limit of the receiver temperature. Very high temperatures can be obtained in concentrating collectors such as heliostats (Figure 7-2). The ultimate limit of the receiver temperature is limited by the Second Law of Thermodynamics to be less than the effective temperature of the sun. Indeed, Goswami et al. (2000) point out that this conclusion is equivalent to the Clausius statement of the Second Law.

7.6   CLOSURE

Solar thermal systems are becoming increasingly popular as the cost of energy continues to rise. The Energy Information Administration monitors the sale of solar collectors. Figure 7.18 presents data on the yearly installed square footage of solar collections from 1996 through 2005. Athough flat through the last years of the 1990s, solar collector sales after the millennium dramatically increased. This behavior tracks the relatively constant energy prices (see Figures 1.11 and 1.15, for example) of the 1990s up to the escalating energy prices of the last few years.

The Solar Energy Industries Association, SEIA, with the web address www.seia.org, provides an alphabetical listing of manufacturers of solar energy–related items, including solar thermal collectors. The SEIA website is a good source of addresses and contact information, listing companies manufacturing solar collectors of all types.

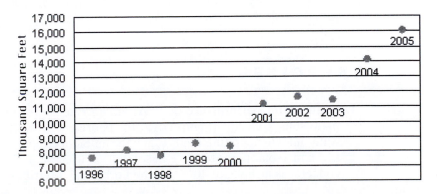

Figure 7.18 Solar collector sales (expressed in ft^2 from 1996 through 2005 [EIA]).

REVIEW QUESTIONS

1. Distinguish between active and passive solar applications.

2. What is a heliostat?

3. How do tracking and non-tracking solar collectors differ?

4. Describe the components of a simple solar thermal system.

5. Why is operating a flat-plate solar collector with $T_{ave} = T_a$ not an effective strategy?

6. What happens in a flat-plate solar collector if $q_{loss} = q_{abs}$?

7. Define the efficiency of a flat-plate solar collector.

8. Describe each of the variables in $\eta = F_R \tau \alpha - F_R U_L (T_{in} - T_a)/I_T$.

9. What does SRCC do?

10. What does the *Solar Radiation Data Manual for Flat-Plate and Concentrating Collectors* (1994) provide?

11. For February in Meridian, MS, how much more daily solar irradiation is available for a two-axis tracking collector than for a fixed collector (tilted at the latitude angle)?

12. What are HDD and CDD? Why are they useful?

13. What do the *f*-chart parameters X and Y represent? What does the *f*–chart procedure yield?

14. What advantage does a concentrating solar collector have over a conventional solar collector?

15. Can a concentrating collector system ever attain a temperature greater than the effective temperature of the sun? Explain.

16. In the last few years, what has happened to the square footage of solar collector installation?

EXERCISES

1. Evaluate the performance of an Alternate Energy Technologies AE-21E flat-plate solar collector (Figure 7.11) for February 21 at 40° north latitude. The water inlet temperature is 60° F, and the ambient temperature varies from 25° F at 7 a.m. (solar time) to 45° F at solar noon to 35° F at 5 p.m. (solar time). The collector is inclined at 40° from the horizontal. Metrics of interest include the

hourly efficiency, the hourly solar energy collected, and the percentage of solar energy captured.

2. Evaluate the performance of a Heliodyne Gobi 408 flat-plate solar collector (performance characteristics available from the SRCC website) for January 21 at 32° north latitude. The water inlet temperature is 58° F, and the ambient temperature varies from 25° F at 7 a.m. (solar time) to 45° F at solar noon to 35° F at 5 p.m. (solar time). The collector is inclined at 32° from the horizontal. Metrics of interest include the hourly efficiency, the hourly solar energy collected, and the percentage of solar energy captured.

3. An office building in Meridian, MS, is to employ Alternate Energy Technologies AE-21E flat-plate solar collectors (Figure 7.11) for space and water heating. The collectors will be mounted facing south and tilted at the same angle as the degrees of latitude. The estimated monthly load is provided in the following table.

Month	Load (1000 MJ)
January	108
February	100
March	80
April	60
May	20
June	15
July	15
August	15
September	15
October	25
November	60
December	95

The solar heating system is to provide all the water heating requirements for the summer months and a yearly solar heating fraction of 50 percent.

(a) How many AE-21E solar collectors are needed to meet the specifications?

(b) If natural gas costs $13 per million Btu, how much is saved by the solar heating system?

4. Repeat Exercise 2 for a New England, a midwestern, and a Pacific Northwest location. Contrast and discuss the results for the different locations.

5. Repeat Exercise 2 using several different flat-plate solar collectors from the SCCR results reported on the SRCC website. Discuss the effect on slope and intercept for the conditions of Exercise 2.

6. An office building in your hometown is to employ Heliodyne Gobi 408 flat-plate solar collectors for space and water heating. The estimated monthly load is provided in the following table.

Month	Load (1000 MJ)
January	130
February	120
March	100
April	70
May	25
June	20
July	20
August	20
September	20
October	25
November	70
December	105

The solar energy system is to provide all the water heating requirements for the summer months and a yearly solar heating fraction of 50 percent.

(a) How many Gobi 408 solar collectors are needed to meet the specifications?

(b) If natural gas costs $9 per million Btu, how much is saved by the active solar system?

REFERENCES

Duffie, J. A., and Beckman, W. A. 2006. *Solar Engineering of Thermal Processes*, 3rd ed. New York: Wiley.

Goswami, Y., Kreith, F., and Kreider, J. F. 2000. *Principles of Solar Engineering*, 2nd ed. New York: Taylor and Francis.

Mitchell, J. W. 1983. *Energy Engineering*. New York: Wiley.

Solar Rating and Certification Corp. 2007. *Directory of SRCC Certified Solar Collector Ratings*. Cocoa, FL: SRCC.

U.S. Dept. of Energy, National Renewable Energy Laboratory. 1994. *Solar Radiation Data Manual for Flat-Plate and Concentrating Collectors*. Golden, CO: NREL.

APPENDIX A
(Reproduced from Goswami et al., 2000)

Solar Position and Insolation Values for 24° North Latitude

Date	Solar time AM	Solar time PM	Solar position Alt	Solar position Azm	Normal[c]	Horiz.	14	24	34	44	90
							South facing surface angle with horiz.				
Jan 21	7	5	4.8	65.6	71	10	17	21	25	28	31
	8	4	16.9	58.3	239	83	110	126	137	145	127
	9	3	27.9	48.8	288	151	188	207	221	228	176
	10	2	37.2	36.1	308	204	246	268	282	287	207
	11	1	43.6	19.6	317	237	283	306	319	324	226
	12		46.0	0.0	320	249	296	319	332	336	232
	Surface daily totals				2766	1622	1984	2174	2300	2360	1766
Feb 21	7	5	9.3	74.6	158	35	44	49	53	56	46
	8	4	22.3	67.2	263	116	135	145	150	151	102
	9	3	34.4	57.6	298	187	213	225	230	228	141
	10	2	45.1	44.2	314	241	273	286	291	287	168
	11	1	53.0	25.0	321	276	310	324	328	323	185
	12		56.0	0.0	324	288	323	337	341	335	191
	Surface daily totals				3036	1998	2276	2396	2436	2424	1476
Mar 21	7	5	13.7	83.3	194	60	63	64	62	59	27
	8	4	27.2	76.8	267	141	150	152	149	142	64
	9	3	40.2	67.9	295	212	226	229	225	214	95
	10	2	52.3	54.8	309	266	285	288	283	270	120
	11	1	61.9	33.4	315	300	322	326	320	305	135
	12		66.0	0.0	317	312	334	339	333	317	140
	Surface daily totals				3078	2270	2428	2456	2412	2298	1022
Apr 21	6	6	4.7	100.6	40	7	5	4	4	3	2
	7	5	18.3	94.9	203	83	77	70	62	51	10
	8	4	32.0	89.0	256	160	157	149	137	122	16
	9	3	45.6	81.9	280	227	227	220	206	186	46
	10	2	59.0	71.8	292	278	282	275	259	237	61
	11	1	71.1	51.6	298	310	316	309	293	269	74
	12		77.6	0.0	299	321	328	321	305	280	79
	Surface daily totals				3036	2454	2458	2374	2228	2016	488
May 21	6	6	8.0	108.4	86	22	15	10	9	9	5
	7	5	21.2	103.2	203	98	85	73	59	44	12
	8	4	34.6	98.5	248	171	159	145	127	106	15
	9	3	48.3	93.6	269	233	224	210	190	165	16
	10	2	62.0	87.7	280	281	275	261	239	211	22
	11	1	75.5	76.9	286	311	307	293	270	240	34
	12		86.0	0.0	288	322	317	304	281	250	37
	Surface daily totals				3032	2556	2447	2286	2072	1800	246
Jun 21	6	6	9.3	111.6	97	29	20	12	12	11	7
	7	5	22.3	106.8	201	103	87	73	58	41	13
	8	4	35.5	102.6	242	173	158	142	122	99	16
	9	3	49.0	98.7	263	234	221	204	182	155	18
	10	2	62.6	95.0	274	280	269	253	229	199	18
	11	1	76.3	90.8	279	309	300	283	259	227	19
	12		89.4	0.0	281	319	310	294	269	236	22
	Surface daily totals				2994	2574	2422	2230	1992	1700	204

Solar Position and Insolation Values for 24° North Latitude

Date	Solar time AM	Solar time PM	Solar position Alt	Solar position Azm	Normal[c]	Horiz.	14	24	34	44	90
								South facing surface angle with horiz.			
Jul 21	6	6	8.2	109.0	81	23	16	11	10	9	6
	7	5	21.4	103.8	195	98	85	73	59	44	13
	8	4	34.8	99.2	239	169	157	143	125	104	16
	9	3	48.4	94.5	261	231	221	207	187	161	18
	10	2	62.1	89.0	272	278	270	256	235	206	21
	11	1	75.7	79.2	278	307	302	287	265	235	32
	12		86.6	0.0	280	317	312	298	275	245	36
	Surface daily totals				2932	2526	2412	2250	2036	1766	246
Aug 21	6	6	5.0	101.3	35	7	5	4	4	4	2
	7	5	18.5	95.6	186	82	76	69	60	50	11
	8	4	32.2	89.7	241	158	154	146	134	118	16
	9	3	45.9	82.9	265	223	222	214	200	181	39
	10	2	59.3	73.0	278	273	275	268	252	230	58
	11	1	71.6	53.2	284	304	309	301	285	261	71
	12		78.3	0.0	286	315	320	313	296	272	75
	Surface daily totals				2864	2408	2402	2316	2168	1958	470
Sep 21	7	5	13.7	83.8	173	57	60	60	59	56	26
	8	4	27.2	76.8	248	136	144	146	143	136	62
	9	3	40.2	67.9	278	205	218	221	217	206	93
	10	2	52.3	54.8	292	258	275	278	273	261	116
	11	1	61.9	33.4	299	291	311	315	309	295	131
	12		66.0	0.0	301	302	323	327	321	306	136
	Surface daily totals				2878	2194	2342	2366	2322	2212	992
Oct 21	7	5	9.1	74.1	138	32	40	45	48	50	42
	8	4	22.0	66.7	247	111	129	139	144	145	99
	9	3	34.1	57.1	284	180	206	217	223	221	138
	10	2	44.7	43.8	301	234	265	277	282	279	165
	11	1	52.5	24.7	309	268	301	315	319	314	182
	12		55.5	0.0	311	279	314	328	332	327	188
	Surface daily totals				2868	1928	2198	2314	2364	2346	1442
Nov 21	7	5	4.9	65.8	67	10	16	20	24	27	29
	8	4	17.0	58.4	232	82	108	123	135	142	124
	9	3	28.0	48.9	282	150	186	205	217	224	172
	10	2	37.3	36.3	303	203	244	265	278	283	204
	11	1	43.8	19.7	312	236	280	302	316	320	222
	12		46.2	0.0	315	247	293	315	328	332	228
	Surface daily totals				2706	1610	1962	2146	2268	2324	1730
Dec 21	7	5	3.2	62.6	30	3	7	9	11	12	14
	8	4	14.9	55.3	225	71	99	116	129	139	130
	9	3	25.5	46.0	281	137	176	198	214	223	184
	10	2	34.3	33.7	304	189	234	258	275	283	217
	11	1	40.4	18.2	314	221	270	295	312	320	236
	12		42.6	0.0	317	232	282	308	325	332	243
	Surface daily totals				2624	1474	1852	2058	2204	2286	1808

BTUH/sq. ft. total insolation on surface[b]

Solar Position and Insolation Values for 32° North Latitude

Date	Solar time AM	Solar time PM	Solar position Alt	Solar position Azm	Normal[c]	Horiz.	22	32	42	52	90
								South facing surface angle with horiz.			
Jan 21	7	5	1.4	65.2	1	0	0	0	0	1	1
	8	4	12.5	56.5	203	56	93	106	116	123	115
	9	3	22.5	46.0	269	118	175	193	206	212	181
	10	2	30.6	33.1	295	167	235	256	269	274	221
	11	1	36.1	17.5	306	198	273	295	308	312	245
	12		38.0	0.0	310	209	285	308	321	324	253
	Surface daily totals				2458	1288	1839	2008	2118	2166	1779
Feb 21	7	5	7.1	73.5	121	22	34	37	40	42	38
	8	4	19.0	64.4	247	95	127	136	140	141	108
	9	3	29.9	53.4	288	161	206	217	222	220	158
	10	2	39.1	39.4	306	212	266	278	283	279	193
	11	1	45.6	21.4	315	244	304	317	321	315	214
	12		48.0	0.0	317	255	316	330	334	328	222
	Surface daily totals				2872	1724	2188	2300	2345	2322	1644
Mar 21	7	5	12.7	81.9	185	54	60	60	59	56	32
	8	4	25.1	73.0	260	129	146	147	144	137	78
	9	3	36.8	62.1	290	194	222	224	220	209	119
	10	2	47.3	47.5	304	245	280	283	278	265	150
	11	1	55.0	26.8	311	277	317	321	315	300	170
	12		58.0	0.0	313	287	329	333	327	312	177
	Surface daily totals				3012	2084	2378	2403	2358	2246	1276
Apr 21	6	6	6.1	99.9	66	14	9	6	6	5	3
	7	5	18.8	92.2	206	86	78	71	62	51	10
	8	4	31.5	84.0	255	158	156	148	136	120	35
	9	3	43.9	74.2	278	220	225	217	203	183	68
	10	2	55.7	60.3	290	267	279	272	256	234	95
	11	1	65.4	37.5	295	297	313	306	290	265	112
	12		69.6	0.0	297	307	325	318	301	276	118
	Surface daily totals				3076	2390	2444	2356	2206	1994	764
May 21	6	6	10.4	107.2	119	36	21	13	13	12	7
	7	5	22.8	100.1	211	107	88	75	60	44	13
	8	4	35.4	92.9	250	175	159	145	127	105	15
	9	3	48.1	84.7	269	233	223	209	188	163	33
	10	2	60.6	73.3	280	277	273	259	237	208	56
	11	1	72.0	51.9	285	305	305	290	268	237	72
	12		78.0	0.0	286	315	315	301	278	247	77
	Surface daily totals				3112	2582	2454	2284	2064	1788	469
Jun 21	6	6	12.2	110.2	131	45	26	16	15	14	9
	7	5	24.3	103.4	210	115	91	76	59	41	14
	8	4	36.9	96.8	245	180	159	143	122	99	16
	9	3	49.6	89.4	264	236	221	204	181	153	19
	10	2	62.2	79.7	274	279	268	251	227	197	41
	11	1	74.2	60.9	279	306	299	282	257	224	56
	12		81.5	0.0	280	315	309	292	267	234	60
	Surface daily totals				3084	2634	2436	2234	1990	1690	370

Solar Position and Insolation Values for 32° North Latitude

Date	Solar time AM	PM	Solar position Alt	Azm	Normal[c]	Horiz.	22	32	42	52	90
							South facing surface angle with horiz.				
Jul 21	6	6	10.7	107.7	113	37	22	14	13	12	8
	7	5	23.1	100.6	203	107	87	75	60	44	14
	8	4	35.7	93.6	241	174	158	143	125	104	16
	9	3	48.4	85.5	261	231	220	205	185	159	31
	10	2	60.9	74.3	271	274	269	254	232	204	54
	11	1	72.4	53.3	277	302	300	285	262	232	69
	12		78.6	0.0	279	311	310	296	273	242	74
	Surface daily totals				3012	2558	2422	2250	2030	1754	458
Aug 21	6	6	6.5	100.5	59	14	9	7	6	6	4
	7	5	19.1	92.8	190	85	77	69	60	50	12
	8	4	31.8	84.7	240	156	152	144	132	116	33
	9	3	44.3	75.0	263	216	220	212	197	178	65
	10	2	56.1	61.3	276	262	272	264	249	226	91
	11	1	66.0	38.4	282	292	305	298	281	257	107
	12		70.3	0.0	284	302	317	309	292	268	113
	Surface daily totals				2902	2352	2388	2296	2144	1934	736
Sep 21	7	5	12.7	81.9	163	51	56	56	55	52	30
	8	4	25.1	73.0	240	124	140	141	138	131	75
	9	3	36.8	62.1	272	188	213	215	211	201	114
	10	2	47.3	47.5	287	237	270	273	268	255	145
	11	1	55.0	26.8	294	268	306	309	303	289	164
	12		58.0	0.0	296	278	318	321	315	300	171
	Surface daily totals				2808	2014	2288	2308	2264	2154	1226
Oct 21	7	5	6.8	73.1	99	19	29	32	34	36	32
	8	4	18.7	64.0	229	90	120	128	133	134	104
	9	3	29.5	53.0	273	155	198	208	213	212	153
	10	2	38.7	39.1	293	204	257	269	273	270	188
	11	1	45.1	21.1	302	236	294	307	311	306	209
	12		47.5	0.0	304	247	306	320	324	318	217
	Surface daily totals				2696	1654	2100	2208	2252	2232	1588
Nov 21	7	5	1.5	65.4	2	0	0	0	1	1	1
	8	4	12.7	56.6	196	55	91	104	113	119	111
	9	3	22.6	46.1	263	118	173	190	202	208	176
	10	2	30.8	33.2	289	166	233	252	265	270	217
	11	1	36.2	17.6	301	197	270	291	303	307	241
	12		38.2	0.0	304	207	282	304	316	320	249
	Surface daily totals				2406	1280	1816	1980	2084	2130	1742
Dec 21	8	4	10.3	53.8	176	41	77	90	101	108	107
	9	3	19.8	43.6	257	102	161	180	195	204	183
	10	2	27.6	31.2	288	150	221	244	259	267	226
	11	1	32.7	16.4	301	180	258	282	298	305	251
	12		34.6	0.0	304	190	271	295	311	318	259
	Surface daily totals				2348	1136	1704	1888	2016	2086	1794

Solar Position and Insolation Values for 40° North Latitude

Date	Solar time AM	Solar time PM	Solar position Alt	Solar position Azm	Normal[c]	Horiz.	30	40	50	60	90
							South facing surface angle with horiz.				
Jan 21	8	4	8.1	55.3	142	28	65	74	81	85	84
	9	3	16.8	44.0	239	83	155	171	182	187	171
	10	2	23.8	30.9	274	127	218	237	249	254	223
	11	1	28.4	16.0	289	154	257	277	290	293	253
	12		30.0	0.0	294	164	270	291	303	306	263
	Surface daily totals				2182	948	1660	1810	1906	1944	1726
Feb 21	7	5	4.8	72.7	69	10	19	21	23	24	22
	8	4	15.4	62.2	224	73	114	122	126	127	107
	9	3	25.0	50.2	274	132	195	205	209	208	167
	10	2	32.8	35.9	295	178	256	267	271	267	210
	11	1	38.1	18.9	305	206	293	306	310	304	236
	12		40.0	0.0	308	216	306	319	323	317	245
	Surface daily totals				2640	1414	2060	2162	2202	2176	1730
Mar 21	7	5	11.4	80.2	171	46	55	55	54	51	35
	8	4	22.5	69.6	250	114	140	141	138	131	89
	9	3	32.8	57.3	282	173	215	217	213	202	138
	10	2	41.6	41.9	297	218	273	276	271	258	176
	11	1	47.7	22.6	305	247	310	313	307	293	200
	12		50.0	0.0	307	257	322	326	320	305	208
	Surface daily totals				2916	1852	2308	2330	2284	2174	1484
Apr 21	6	6	7.4	98.9	89	20	11	8	7	7	4
	7	5	18.9	89.5	206	87	77	70	61	50	12
	8	4	30.3	79.3	252	152	153	145	133	117	53
	9	3	41.3	67.2	274	207	221	213	199	179	93
	10	2	51.2	51.4	286	250	275	267	252	229	126
	11	1	58.7	29.2	292	277	308	301	285	260	147
	12		61.6	0.0	293	287	320	313	296	271	154
	Surface daily totals				3092	2274	2412	2320	2168	1956	1022
May 21	5	7	1.9	114.7	1	0	0	0	0	0	0
	6	6	12.7	105.6	144	49	25	15	14	13	9
	7	5	24.0	96.6	216	214	89	76	60	44	13
	8	4	35.4	87.2	250	175	158	144	125	104	25
	9	3	46.8	76.0	267	227	221	206	186	160	60
	10	2	57.5	60.9	277	267	270	255	233	205	89
	11	1	66.2	37.1	283	293	301	287	264	234	108
	12		70.0	0.0	284	301	312	297	274	243	114
	Surface daily totals				3160	2552	2442	2264	2040	1760	724
Jun 21	5	7	4.2	117.3	22	4	3	3	2	2	1
	6	6	14.8	108.4	155	60	30	18	17	16	10
	7	5	26.0	99.7	216	123	92	77	59	41	14
	8	4	37.4	90.7	246	182	159	142	121	97	16
	9	3	48.8	80.2	263	233	219	202	179	151	47
	10	2	59.8	65.8	272	272	266	248	224	194	74
	11	1	69.2	41.9	277	296	296	278	253	221	92
	12		73.5	0.0	279	304	306	289	263	230	98
	Surface daily totals				3180	2648	2434	2224	1974	1670	610

BTUH/sq. ft. total insolation on surface[b]

Solar Position and Insolation Values for 40° North Latitude

Date	Solar time AM	Solar time PM	Solar position Alt	Solar position Azm	Normal[c]	Horiz.	30	40	50	60	90
							South facing surface angle with horiz.				
Jul 21	5	7	2.3	115.2	2	0	0	0	0	0	0
	6	6	13.1	106.1	138	50	26	17	15	14	9
	7	5	24.3	97.2	208	114	89	75	60	44	14
	8	4	35.8	87.8	241	174	157	142	124	102	24
	9	3	47.2	76.7	259	225	218	203	182	157	58
	10	2	57.9	61.7	269	265	266	251	229	200	86
	11	1	66.7	37.9	275	290	296	281	258	228	104
	12		70.6	0.0	276	298	307	292	269	238	111
	Surface daily totals				3062	2534	2409	2230	2006	1728	702
Aug 21	6	6	7.9	99.5	81	21	12	9	8	7	5
	7	5	19.3	90.9	191	87	76	69	60	49	12
	8	4	30.7	79.9	237	150	150	141	129	113	50
	9	3	41.8	67.9	260	205	216	207	193	173	89
	10	2	51.7	52.1	272	246	267	259	244	221	120
	11	1	59.3	29.7	278	273	300	292	276	252	140
	12		62.3	0.0	280	282	311	303	287	262	147
	Surface daily totals				2916	2244	2354	2258	2104	1894	978
Sep 21	7	5	11.4	80.2	149	43	51	51	49	47	32
	8	4	22.5	69.6	230	109	133	134	131	124	84
	9	3	32.8	57.3	263	167	206	208	203	193	132
	10	2	41.6	41.9	280	211	262	265	260	247	168
	11	1	47.7	22.6	287	239	298	301	295	281	192
	12		50.0	0.0	290	249	310	313	307	292	200
	Surface daily totals				2708	1788	2210	2228	2182	2074	1416
Oct 21	7	5	4.5	72.3	48	7	14	15	17	17	16
	8	4	15.0	61.9	204	68	106	113	117	118	100
	9	3	24.5	49.8	257	126	185	195	200	198	160
	10	2	32.4	35.6	280	170	245	257	261	257	203
	11	1	37.6	18.7	291	199	283	295	299	294	229
	12		39.5	0.0	294	208	295	308	312	306	238
	Surface daily totals				2454	1348	1962	2060	2098	2074	1654
Nov 21	8	4	8.2	55.4	136	28	63	72	78	82	81
	9	3	17.0	44.1	232	82	152	167	178	183	167
	10	2	24.0	31.0	268	126	215	233	245	249	219
	11	1	28.6	16.1	283	153	254	273	285	288	248
	12		30.2	0.0	288	163	267	287	298	301	258
	Surface daily totals				2128	942	1636	1778	1870	1908	1686
Dec 21	8	4	5.5	53.0	89	14	39	45	50	54	56
	9	3	14.0	41.9	217	65	135	152	164	171	163
	10	2	20.,	29.4	261	107	200	221	235	242	221
	11	1	25.0	15.2	280	134	239	262	276	283	252
	12		26.6	0.0	285	143	253	275	290	296	263
	Surface daily totals				1978	782	1480	1634	1740	1796	1646

BTUH/sq. ft. total insolation on surface[b]

Solar Position and Insolation Values for 48° North Latitude

Date	Solar time AM	PM	Solar position Alt	Azm	BTUH/sq. ft. total insolation on surfaces[b] Normal[c]	Horiz.	South facing surface angle with horiz. 38	48	58	68	90
Jan 21	8	4	3.5	54.6	37	4	17	19	21	22	22
	9	3	11.0	42.6	185	46	120	132	140	145	139
	10	2	16.9	29.4	239	83	190	206	216	220	206
	11	1	20.7	15.1	261	107	231	249	260	263	243
	12		22.0	0.0	267	115	245	264	275	278	255
	Surface daily totals				1710	596	1360	1478	1550	1578	1478
Feb 21	7	5	2.4	72.2	12	1	3	4	4	4	4
	8	4	11.6	60.5	188	49	95	102	105	106	96
	9	3	19.7	47.7	251	100	178	187	191	190	167
	10	2	26.2	33.3	278	139	240	251	255	251	217
	11	1	30.5	17.2	290	165	278	290	294	288	247
	12		32.0	0.0	293	173	291	304	307	301	258
	Surface daily totals				2330	1080	1880	1972	2024	1978	1720
Mar 21	7	5	10.0	78.7	153	37	49	49	47	45	35
	8	4	19.5	66.8	236	96	131	132	129	122	96
	9	3	28.2	53.4	270	147	205	207	203	193	152
	10	2	35.4	37.8	287	187	263	266	261	248	195
	11	1	40.3	19.8	295	212	300	303	297	283	223
	12		42.0	0.0	298	220	312	315	309	294	232
	Surface daily totals				2780	1578	2208	2228	2182	2074	1632
Apr 21	6	6	8.6	97.8	108	27	13	9	8	7	5
	7	5	18.6	86.7	205	85	76	69	59	48	21
	8	4	28.5	74.9	247	142	149	141	129	113	69
	9	3	37.8	61.2	268	191	216	208	194	174	115
	10	2	45.8	44.6	280	228	268	260	245	223	152
	11	1	51.5	24.0	286	252	301	294	278	254	177
	12		53.6	0.0	288	260	313	305	289	264	185
	Surface daily totals				3076	2106	2358	2266	2114	1902	1262
May 21	5	7	5.2	114.3	41	9	4	4	4	3	2
	6	6	14.7	103.7	162	61	27	16	15	13	10
	7	5	24.6	93.0	219	118	89	75	60	43	13
	8	4	34.7	81.6	248	171	156	142	123	101	45
	9	3	44.3	68.3	264	217	217	202	182	156	86
	10	2	53.0	51.3	274	252	265	251	229	200	120
	11	1	59.5	28.6	279	274	296	281	258	228	141
	12		62.0	0.0	280	281	306	292	269	238	149
	Surface daily totals				3254	2482	2418	2234	2010	1728	982
Jun 21	5	7	7.9	116.5	77	21	9	9	8	7	5
	6	6	17.2	106.2	172	74	33	19	18	16	12
	7	5	27.0	95.8	220	129	93	77	59	39	15
	8	4	37.1	84.6	246	181	157	140	119	95	35
	9	3	46.9	71.6	261	225	216	198	175	147	74
	10	2	55.8	54.8	269	259	262	244	220	189	105
	11	1	62.7	31.2	274	280	291	273	248	216	126
	12		65.5	0.0	275	287	301	283	258	225	133
	Surface daily totals				3312	2626	2420	2204	1950	1644	874

Solar Position and Insolation Values for 48° North Latitude

Date	Solar time AM	Solar time PM	Solar position Alt	Solar position Azm	Normal[c]	Horiz.	38	48	58	68	90
							\multicolumn{5}{South facing surface angle with horiz.}				
Jul 21	5	7	5.7	114.7	43	10	5	5	4	4	3
	6	6	15.2	104.1	156	62	28	18	16	15	11
	7	5	25.1	93.5	211	118	89	75	59	42	14
	8	4	35.1	82.1	240	171	154	140	121	99	43
	9	3	44.8	68.8	256	215	214	199	178	153	83
	10	2	53.5	51.9	266	250	261	246	224	195	116
	11	1	60.1	29.0	271	272	291	276	253	223	137
	12		62.6	0.0	272	279	301	286	263	232	144
	Surface daily totals				3158	2474	2386	2200	1974	1694	956
Aug 21	6	6	9.1	98.3	99	28	14	10	9	8	6
	7	5	19.1	87.2	190	85	75	67	58	47	20
	8	4	29.0	75.4	232	141	145	137	125	109	65
	9	3	38.4	61.8	254	189	210	201	187	168	110
	10	2	46.4	45.1	266	225	260	252	237	214	146
	11	1	52.2	24.3	272	248	293	285	268	244	169
	12		54.3	0.0	274	256	304	296	279	255	177
	Surface daily totals				2898	2086	2300	2200	2046	1836	1208
Sep 21	7	5	10.0	78.7	131	35	44	44	43	40	31
	8	4	19.5	66.8	215	92	124	124	121	115	90
	9	3	28.2	53.4	251	142	196	197	193	183	143
	10	2	35.4	37.8	269	181	251	254	248	236	185
	11	1	40.3	19.8	278	205	287	289	284	269	212
	12		42.0	0.0	280	213	299	302	296	281	221
	Surface daily totals				2568	1522	2102	2118	2070	1966	1546
Oct 21	7	5	2.0	71.9	4	0	1	1	1	1	1
	8	4	11.2	60.2	165	44	86	91	95	95	87
	9	3	19.3	47.4	233	94	167	176	180	178	157
	10	2	25.7	33.1	262	133	228	239	242	239	207
	11	1	30.0	17.1	274	157	266	277	281	276	237
	12		31.5	0.0	278	166	279	291	294	288	247
	Surface daily totals				2154	1022	1774	1860	1890	1866	1626
Nov 21	8	4	3.6	54.7	36	5	17	19	21	22	22
	9	3	11.2	42.7	179	46	117	129	137	141	135
	10	2	17.1	29.5	233	83	186	202	212	215	201
	11	1	20.9	15.1	255	107	227	245	255	258	238
	12		22.2	0.0	261	115	241	259	270	272	250
	Surface daily totals				1668	596	1336	1448	1518	1544	1442
Dec 21	9	3	8.0	40.9	140	27	87	98	105	110	109
	10	2	13.6	28.2	214	63	164	180	192	197	190
	11	1	17.3	14.4	242	86	207	226	239	244	231
	12		18.6	0.0	250	94	222	241	254	260	244
	Surface daily totals				1444	446	1136	1250	1326	1364	1304

Solar Position and Insolation Values for 56° North Latitude

Date	Solar time AM	Solar time PM	Solar position Alt	Solar position Azm	Normal[c]	Horiz.	South facing surface angle with horiz. 46	56	66	76	90
Jan 21	9	3	5.0	41.8	78	11	50	55	59	60	60
	10	2	9.9	28.5	170	39	135	146	154	156	153
	11	1	12.9	14.5	207	58	183	197	206	208	201
	12		14.0	0.0	217	65	198	214	222	225	217
	Surface daily totals				1126	282	934	1010	1058	1074	1044
Feb 21	8	4	7.6	59.4	129	25	65	69	72	72	69
	9	3	14.2	45.9	214	65	151	159	162	161	151
	10	2	19.4	31.5	250	98	215	225	228	224	208
	11	1	22.8	16.1	266	119	254	265	268	263	243
	12		24.0	0.0	270	126	268	279	282	276	255
	Surface daily totals				1986	740	1640	1716	1742	1716	1598
Mar 21	7	5	8.3	77.5	128	28	40	40	39	37	32
	8	4	16.2	64.4	215	75	119	120	117	111	97
	9	3	23.3	50.3	253	118	192	193	189	180	154
	10	2	29.0	34.9	272	151	249	251	246	234	205
	11	1	32.7	17.9	282	172	285	288	282	268	236
	12		34.0	0.0	284	179	297	300	294	280	246
	Surface daily totals				2586	1268	2066	2084	2040	1938	1700
Apr 21	5	7	1.4	108.8	0	0	0	0	0	0	0
	6	6	9.6	96.5	122	32	14	9	8	7	6
	7	5	18.0	84.1	201	81	74	66	57	46	29
	8	4	26.1	70.9	239	129	143	135	123	108	82
	9	3	33.6	56.3	260	169	208	200	186	167	133
	10	2	39.9	39.7	272	201	259	251	236	214	174
	11	1	44.1	20.7	278	220	292	284	268	245	200
	12		45.6	0.0	280	227	303	295	279	255	209
	Surface daily totals				3024	1892	2282	2186	2038	1830	1458
May 21	4	8	1.2	125.5	0	0	0	0	0	0	0
	5	7	8.5	113.4	93	25	10	9	8	7	6
	6	6	16.5	101.5	175	71	28	17	15	13	11
	7	5	24.8	89.3	219	119	88	74	58	41	16
	8	4	33.1	76.3	244	163	153	138	119	98	63
	9	3	40.9	61.6	259	201	212	197	176	151	109
	10	2	47.6	44.2	268	231	259	244	222	194	146
	11	1	52.3	23.4	273	249	288	274	251	222	170
	12		54.0	0.0	275	255	299	284	261	231	178
	Surface daily totals				3340	2374	2374	2188	1962	1682	1218
Jun 21	4	8	4.2	127.2	21	4	2	2	2	2	1
	5	7	11.4	115.3	122	40	14	13	11	10	8
	6	6	19.3	103.6	185	86	34	19	17	15	12
	7	5	27.6	91.7	222	132	92	76	57	38	15
	8	4	35.9	78.8	243	175	154	137	116	92	55
	9	3	43.8	64.1	257	212	211	193	170	143	98
	10	2	50.7	46.4	265	240	255	238	214	184	133
	11	1	55.6	24.9	269	258	284	267	242	210	156
	12		57.5	0.0	271	264	294	276	251	219	164
	Surface daily totals				3438	2526	2388	2166	1910	1606	1120

Solar Position and Insolation Values for 56° North Latitude

Date	Solar time AM	PM	Solar position Alt	Azm	BTUH/sq. ft. total insolation on surfaces[b] Normal[c]	Horiz.	South facing surface angle with horiz. 46	56	66	76	90
Jul 21	4	8	1.7	125.8	0	0	0	0	0	0	0
	5	7	9.0	113.7	91	27	11	10	9	8	6
	6	6	17.0	101.9	169	72	30	18	16	14	12
	7	5	25.3	89.7	212	119	88	74	58	41	15
	8	4	33.6	76.7	237	163	151	136	117	96	61
	9	3	41.4	62.0	252	201	208	193	173	147	106
	10	2	48.2	44.6	261	230	254	239	217	189	142
	11	1	52.9	23.7	265	248	283	268	245	216	165
	12		54.6	0.0	267	254	293	278	255	225	173
	Surface daily totals				3240	2372	2342	2152	1926	1646	1186
Aug 21	5	7	2.0	109.2	1	0	0	0	0	0	0
	6	6	10.2	97.0	112	34	16	11	10	9	7
	7	5	18.5	84.5	187	82	73	65	56	45	28
	8	4	26.7	71.3	225	128	140	131	119	104	78
	9	3	34.3	56.7	246	168	202	193	179	160	126
	10	2	40.5	40.0	258	199	251	242	227	206	166
	11	1	44.8	20.9	264	218	282	274	258	235	191
	12		46.3	0.0	266	225	293	285	269	245	200
	Surface daily totals				2850	1884	2218	2118	1966	1760	1392
Sep 21	7	5	8.3	77.5	107	25	36	36	34	32	28
	8	4	16.2	64.4	194	72	111	111	108	102	89
	9	3	23.3	50.3	233	114	181	182	178	168	147
	10	2	29.0	34.9	253	146	236	237	232	221	193
	11	1	32.7	17.9	263	166	271	273	267	254	223
	12		34.0	0.0	266	173	283	285	279	265	233
	Surface daily totals				2368	1220	1950	1962	1918	1820	1594
Oct 21	8	4	7.1	59.1	104	20	53	57	59	59	57
	9	3	13.8	45.7	193	60	138	145	148	147	138
	10	2	19.0	31.3	231	92	201	210	213	210	195
	11	1	22.3	16.0	248	112	240	250	253	248	230
	12		23.5	0.0	253	119	253	263	266	261	241
	Surface daily totals				1804	688	1516	1586	1612	1588	1480
Nov 21	9	3	5.2	41.9	76	12	49	54	57	59	58
	10	2	10.0	28.5	165	39	132	143	149	152	148
	11	1	13.1	14.5	201	58	179	193	201	203	196
	12		14.2	0.0	211	65	194	209	217	219	211
	Surface daily totals				1094	284	914	986	1032	1046	1016
Dec 21	9	3	1.9	40.5	5	0	3	4	4	4	4
	10	2	6.6	27.5	113	19	86	95	101	104	103
	11	1	9.5	13.9	166	37	141	154	163	167	164
	12		10.6	0.0	180	43	159	173	182	186	182
	Surface daily totals				748	156	620	678	716	734	722

Solar Position and Insolation Values for 64° North Latitude

Date	Solar time AM	Solar time PM	Solar position Alt	Solar position Azm	Normal[c]	Horiz.	54	64	74	84	90
							South facing surface angle with horiz.				
Jan 21	10	2	2.8	28.1	22	2	17	19	20	20	20
	11	1	5.2	14.1	81	12	72	77	80	81	81
	12		6.0	0.0	100	16	91	98	102	103	103
	Surface daily totals				306	45	268	290	302	306	304
Feb 21	8	4	3.4	58.7	35	4	17	19	19	19	19
	9	3	8.6	44.8	147	31	103	108	111	110	107
	10	2	12.6	30.3	199	55	170	178	181	178	173
	11	1	15.1	15.3	222	71	212	220	223	219	213
	12		16.0	0.0	228	77	225	235	237	232	226
	Surface daily totals				1432	400	1230	1286	1302	1282	1252
Mar 21	7	5	6.5	76.5	95	18	30	29	29	27	25
	8	4	20.7	62.6	185	54	101	102	99	94	89
	9	3	18.1	48.1	227	87	171	172	169	160	153
	10	2	22.3	32.7	249	112	227	229	224	213	203
	11	1	25.1	16.6	260	129	262	265	259	246	235
	12		26.0	0.0	263	134	274	277	271	258	246
	Surface daily totals				2296	932	1856	1870	1830	1736	1656
Apr 21	5	7	4.0	108.5	27	5	2	2	2	1	1
	6	6	10.4	95.1	133	37	15	9	8	7	6
	7	5	17.0	81.6	194	76	70	63	54	43	37
	8	4	23.3	67.5	228	112	136	128	116	102	91
	9	3	29.0	52.3	248	144	197	189	176	158	145
	10	2	33.5	36.0	260	169	246	239	224	203	188
	11	1	36.5	18.4	266	184	278	270	255	233	216
	12		97.6	0.0	268	190	289	281	266	243	225
	Surface daily totals				2982	1644	2176	2082	1936	1736	1594
May 21	4	8	5.8	125.1	51	11	5	4	4	3	3
	5	7	11.6	112.1	132	42	13	11	10	9	8
	6	6	17.9	99.1	185	79	29	16	14	12	11
	7	5	24.5	85.7	218	117	86	72	56	39	28
	8	4	30.9	71.5	239	152	148	133	115	94	80
	9	3	36.8	56.1	252	182	204	190	170	145	128
	10	2	41.6	38.9	261	205	249	235	213	186	167
	11	1	44.9	20.1	265	219	278	264	242	213	193
	12		46.0	0.0	267	224	228	274	251	222	201
	Surface daily totals				3470	2236	2312	2124	1898	1624	1436
Jun 21	3	9	4.2	139.4	21	4	2	2	2	2	1
	4	8	9.0	126.4	93	27	10	9	8	7	6
	5	7	14.7	113.6	154	60	16	15	13	11	10
	6	6	21.0	100.8	194	96	34	19	17	14	13
	7	5	27.5	87.5	221	132	91	74	55	36	23
	8	4	34.0	73.3	239	166	150	133	112	88	73
	9	3	39.9	57.8	251	195	204	187	164	137	119
	10	2	44.9	40.4	258	217	247	230	206	177	157
	11	1	48.3	20.9	262	231	275	258	233	202	181
	12		49.5	0.0	263	235	284	267	242	211	189
	Surface daily totals				3650	2488	2342	2118	1862	1558	1356

Solar Position and Insolation Values for 64° North Latitude

Date	Solar time AM	Solar time PM	Solar position Alt	Solar position Azm	Normal[c]	Horiz.	South facing surface angle with horiz. 54	64	74	84	90
Jul 21	4	8	6.4	125.3	53	13	6	5	5	4	4
	5	7	12.1	112.4	128	44	14	13	11	10	9
	6	6	18.4	99.4	179	81	30	17	16	13	12
	7	5	25.0	86.0	211	118	86	72	56	38	28
	8	4	31.4	71.8	231	152	146	131	113	91	77
	9	3	37.3	56.3	245	182	201	186	166	141	124
	10	2	42.2	39.2	253	204	245	230	208	181	162
	11	1	45.4	20.2	257	218	273	258	236	207	187
	12		46.6	0.0	259	223	282	267	245	216	195
	Surface daily totals				3372	2248	2280	2090	1864	1588	1400
Aug 21	5	7	4.6	108.8	29	6	3	3	2	2	2
	6	6	11.0	95.5	123	39	16	11	10	8	7
	7	5	17.6	81.9	181	77	69	61	52	42	35
	8	4	23.9	67.8	214	113	132	123	112	97	87
	9	3	29.6	52.6	234	144	190	182	169	150	138
	10	2	34.2	36.2	246	168	237	229	215	194	179
	11	1	37.2	18.5	252	183	268	260	244	222	205
	12		38.3	0.0	254	188	278	270	255	232	215
	Surface daily totals				2808	1646	2108	1008	1860	1662	1522
Sep 21	7	5	6.5	76.5	77	16	25	25	24	23	21
	8	4	12.7	72.6	163	51	92	92	90	85	81
	9	3	18.1	48.1	206	83	159	159	156	147	141
	10	2	22.3	32.7	229	108	212	213	209	198	189
	11	1	25.1	16.6	240	124	246	248	243	230	220
	12		26.0	0.0	244	129	258	260	254	241	230
	Surface daily totals				2074	892	1726	1736	1696	1608	1532
Oct 21	8	4	3.0	58.5	17	2	9	9	10	10	10
	9	3	8.1	44.6	122	26	86	91	93	92	90
	10	2	12.1	30.2	176	50	152	159	161	159	155
	11	1	14.6	15.2	201	65	193	201	203	200	195
	12		15.5	0.0	208	71	207	215	217	213	208
	Surface daily totals				1238	358	1088	1136	1152	1134	1106
Nov 21	10	2	3.0	28.1	23	3	18	20	21	21	21
	11	1	5.4	14.2	79	12	70	76	79	80	79
	12		6.2	0.0	97	17	89	96	100	101	100
	Surface daily totals				302	46	266	286	298	302	300
Dec 21	11	1	1.8	13.7	4	0	3	4	4	4	4
	12		2.6	0.0	16	2	14	15	16	17	17
	Surface daily totals				24	2	20	22	24	24	24

CHAPTER 8

Passive Solar Energy

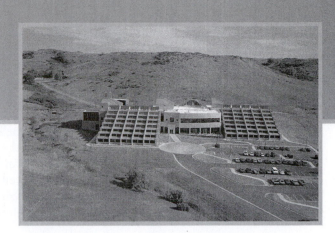

8.1 FUNDAMENTAL CONCEPTS OF PASSIVE SOLAR ENERGY

Passive solar energy is generally defined to include systems in which the flow of solar energy is by passive (or natural) processes, such as natural (free) convection. Passive solar concepts have been known and practiced since ancient times. The following statements are attributed to Socrates (~400 B.C.) and Aristotle (~350 B.C.), respectively:

> In houses with south aspect, the sun's rays penetrate into the porticos in winter, but in summer the path of the sun is right over our heads and above the roofs so that there is shade.
> For well-being and health .... the homestead should be airy in summer and sunny in winter. A homestead promising these qualities would be longer than it is deep and the main front would face south.

In the modern vernacular of passive solar practices, these statements mean that a home should be oriented east-to-west with significant southern exposure and with an overhang such that in the winter the sunlight is incident on the southern exposure but in the summer the sunlight is blocked.

However, even a first-order examination of passive solar concepts requires a more formal examination of fundamental principles. The distinction between an active and a passive system is that a passive system uses no mechanical intervention to distribute energy, but relies on radiation and free convection. A passive solar system generally possesses five components:

1. Collector (or aperture) arrangement
2. Absorber
3. Thermal mass
4. Distribution protocol
5. Control strategy

Additionally, most passive solar systems have a backup energy source for use when passive solar features are insufficient for comfort or utility.

Passive collectors include windows or more exotic features such as a water roof. An absorber is a hard surface with a high absorptivity that is in the direct path of the solar irradiation entering through the collector. The passive thermal storage mass can be provided by external or internal walls, floors, or water walls. These thermal masses receive energy for solar irradiation in the winter, but are blocked in the summer. The distribution protocol is the method by which the thermal energy is stored, retrieved, and circulated to different areas of the structure. Controls are provided to enhance distribution from the collector–absorber–thermal mass arrangement. In many residences, passive controls include human participation in such tasks as moving panels or opening and closing dampers, as well as passive features such as overhangs.

If a structure has south-facing glazing but no thermal storage mass, the structure is called sun-tempered. The use of extensive south-facing glazing, up to 7 percent of

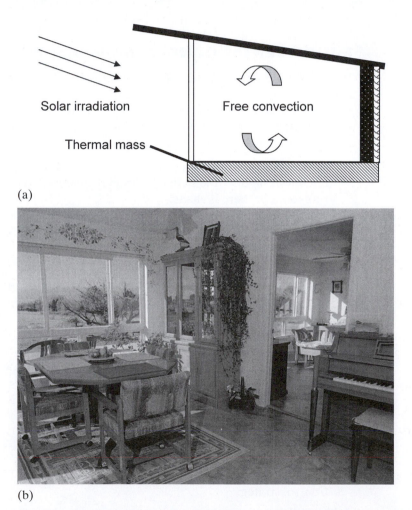

(a)

(b)

Figure 8.1 Direct-gain passive solar system. (a) Schematic. (b) Implementation example (NREL 08427).

the floor area, can markedly reduce the heating energy requirement, especially in a cold climate. If the south-facing glazing exceeds roughly 7 percent of the floor area, then a thermal storage mass is needed to effectively utilize the solar irradiation incident on the interior of the structure.

Passive solar heating configurations are categorized as direct gain, indirect gain, or isolated gain. As the name implies, in direct-gain solar systems, the solar irradiation is directly incident on the interior of the structure. Figure 8.1(a) is a schematic of a direct-gain system, and Figure 8.1(b) illustrates a direct-gain implementation in a residence. The thermal mass thickness should not exceed 6 inches, and the thermal mass should not be carpeted. Two well-accepted rules of thumb are that 150 lbm (pound-mass) of masonry is needed for every square foot of south-facing glazing, and that 9 ft^2 of thermal mass storage area is needed for every square foot of south-facing glazing.

Indirect-gain systems are characterized by having the thermal mass located between the sun and the interior space. Figures 8.2(a) and 8.2(b) present a schematic

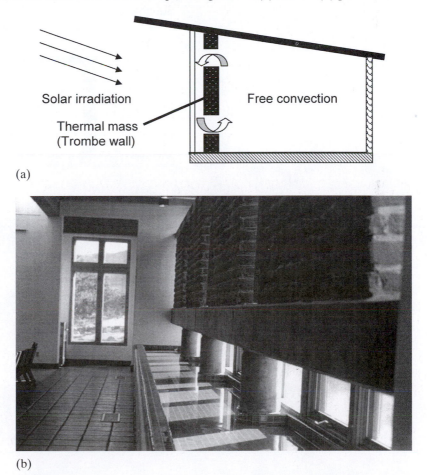

(a)

(b)

Figure 8.2 Indirect-gain solar system with a Trombe wall. (a) Schematic. (b) Implementation example (NREL 02929).

and an example of an indirect-gain passive solar system with a Trombe wall. A Trombe wall is a thick wall, usually constructed of masonry, designed to absorb solar irradiation during the day, store it, and radiate heat during the night or on overcast days. A rule of thumb is that 0.2 ft^2 of thermal mass wall is required for every square foot of floor area. The recommended Trombe wall thickness is 12–18 inches of concrete or 8–12 inches of adobe.

Figures 8.3(a) and 8.3(b) illustrate a schematic and an example of an isolated-gain solar system, a sunspace. An isolated-gain system is separate from the main building, but thermal mass is provided for energy storage. For a masonry thermal wall, the recommended thicknesses are the same as for a Trombe wall, but 0.3 ft^2 of south-facing glazing is needed for every square foot of floor area. Ventilation and shading are generally required during the summer months.

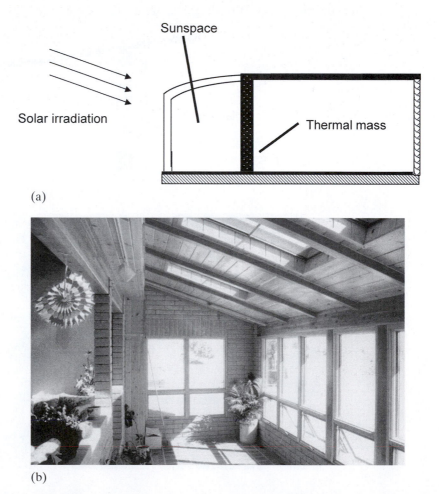

(a)

(b)

Figure 8.3 Isolated solar system (sunspace). (a) Schematic. (b) Implementation example (NREL 08427).

Figure 8.4 Daylighting example (NREL 09536).

The document "Passive Solar Guidelines," on the website www.green builder.com, contains recommendations and rules of thumb for the passive solar features presented in Figures 8.1–8.3. Additionally, "Passive Solar Design for the Home," from the U.S. Department of Energy, lists recommendations for passive features in residences.

Another important feature often implemented in passive solar is daylighting—the use of radiant energy from the sun for interior illumination so that the need for artificial illumination during daylight hours is reduced. Figure 8.4 illustrates the use of daylighting for the stacks of a library.

Passive solar features can be implemented in virtually any structure regardless of size. Figures 8.5(a) and 8.5(b) reproduce photographs of the NREL Solar Energy Research Facility (SERF) and a residence, respectively, illustrating the implementation of many passive (and some active) solar features.

SERF is a state-of-the-art research facility that was completed in 1993. The facility faces 15° east of due south and is built into the site slope to take advantage of earth sheltering. SERF incorporates a number of energy-saving features, including daylighting, window shades, evaporative cooling, and heat recovery systems. Annual energy usage and costs are 30–40 percent lower than for a similar building designed to meet federal standards. SERF costs less than half of what a typical research facility costs to operate. Photovoltaic arrays, which are not passive solar elements, have been installed on the roof of the east and west wings.

(a)

(b)

Figure 8.5 Passive solar building examples. (a) SERF (NREL 01257). (b) Residence (NREL 11050).

Figures 8.1 to 8.3 illustrate the use of south-facing glazing to capture solar irradiation for heating. An equally important passive solar feature, especially in warm climates, is overhang on the south-facing wall to block out solar irradiation in the summer. The residence in Figure 8.5(b) shows overhang that blocks the sun in summer (high sun altitude angle) but admits solar irradiation in winter (low sun altitude angle).

The preceding schematics and photos as well as the accompanying text establish the fundamentals of passive solar energy, but do little to quantify expected performance and energy savings. The next section delineates a first-order process and a second-order process to quantify the effects of various passive solar features.

8.2   QUANTIFYING PASSIVE SOLAR FEATURES

The first- and second-order procedures for estimating or quantifying the effects of various passive solar features are based on the correlated results of a large number of computer simulations. The basis of the passive solar estimates is analogous to the *f*-chart procedure explored in the previous chapter. Perhaps the single most referenced document for passive solar energy is the *Passive Solar Design Handbook*, typically referred to as PSDH (1980, 1984). Additional information is available in *The Passive Solar Design and Construction Handbook* (Crosbie 1997). A common set of definitions used to quantify the performance of passive solar features is examined next.

The most common single-variable method used to express climatological effects on building energy usage is the "degree-day." The degree-day is built around the concepts of a no-load (balance point) temperature and an average daily temperature. The no-load temperature is the ambient (outdoor) temperature for which no heating or cooling energy would be required for human comfort in a building. Traditionally, the no-load temperature has been taken to be 65°F. Thus, for ambient temperatures less than 65°F, heating would be required, and for ambient temperatures greater than 65°F, cooling would be required. The average daily temperature is defined as the average of the minimum and maximum temperatures for a given day. The difference between the no-load temperature and the average temperature is defined as the number of degree-days for that day. Hence, if the average temperature is 45°F during a particular day, that day has 20 heating degree-days. If the average temperature is 85°F, the day has 20 cooling degree-days. The total heating degree-days (HDD) for a given location is the sum of all the daily heating degree-days, and the total cooling degree-days (CDD) is the sum of the daily cooling degree-days for the location. The total values are a useful metric for estimating energy usage to heat and cool a building since the energy required is essentially proportional to the HDD and CDD. Energy usage estimated from the HDD is generally more accurate than energy usage estimated from the CDD since cooling degree-days do not consider humidity effects, which can be a significant cooling load in a humid climate.

HDD and CDD data are available from the National Weather Service and are archived by the National Climatic Data Center. The *Solar Radiation Data Manual for Flat-Plate and Concentrating Collectors* (1994), discussed in the previous chapter, also tabulates HDD and the CDD values. Figure 7.12 indicates that Meridian, MS, has 1358 Celsius heating degree-days (C-HDD). In the NREL data in Figure 7.12, the Celsius degree-days are referenced to a no-load temperature of 18.3°C, which is 65°F. Using this basis, conversion from Celsius (C) degree-days to Fahrenheit (F) degree-days is simple: the conversion factor is 5/9. Care needs to be exercised in using SI units for degree-days since some are tabulated on the basis of 19.0°C, for which the 5/9 conversion factor is slightly off. Additionally, many heating, ventilating, and air conditioning (HVAC) and solar energy engineering textbooks provide tabulations for selected cities as well as maps containing contour lines representing constant HDD and CDD regions. For the purposes of this textbook, a contour map will suffice. Figure 8.6 presents a contour map of F heating degree-days for the continental United States.

The higher the HDD, the more severe the winter. In Figure 8.6, the HDD values in the north are much larger than those in the south. The HDD numbers in the mountains of the west show the effects of elevation and latitude. More detailed contour maps are available from the National Climatic Data Center (NCDC) at www.ncdc.noaa.gov.

In addition to the degree-days concept, many passive solar energy applications utilize a common nomenclature. This includes the following terms:

A_p—vertically projected net south-facing solar glazing, ft^2 or m^2

NLC—net building load coefficient: the net heating load of the non-solar portion of the structure per heating degree-day, kJ/C-HDD or Btu/F-HDD

Q_{net}—net heating load: Q_{net} = NLC · HDD, kWh or Btu

Q_{solar}—heating load provided by solar energy, kWh or Btu

Q_{aux}—heating load not provided by solar energy, kWh or Btu

LCR—load collector ratio: LCR = NLC/A_p, kJ/m^2 C-HDD or Btu/ft^2 F-HDD

SSF—solar savings fraction: the fraction of the total heating energy provided by solar, Q_{solar}/Q_{net}

The net heating energy required is the sum of the solar contribution and the auxiliary, or

$$Q_{net} = Q_{solar} + Q_{aux} \tag{8-1}$$

Dividing by Q_{net} and solving for Q_{solar}/Q_{net}, the solar savings fraction (SSF) is

$$\frac{Q_{solar}}{Q_{net}} = SSF = 1 - \frac{Q_{aux}}{Q_{net}} \tag{8-2}$$

from which the auxiliary heating energy is found as

$$Q_{aux} = (1 - SSF) \cdot NLC \cdot HDD \tag{8-3}$$

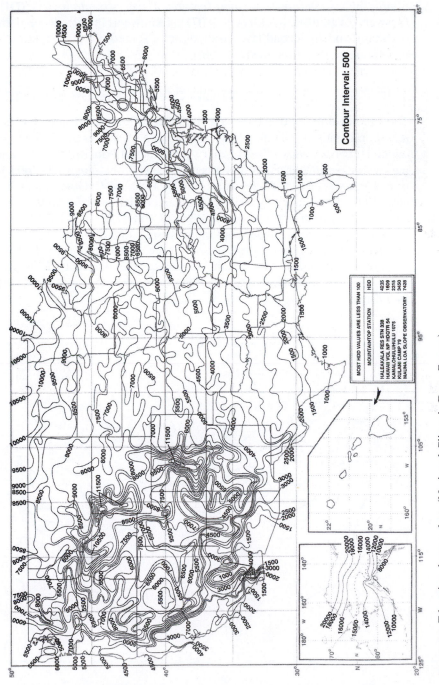

Contour Interval: 500

MOST HDD VALUES ARE LESS THAN 100	
MOUNTAINTOP STATION	HDD
HALEAKALA RES STN 338	4235
HAWAII VOL NP HDQTR 54	1609
KANALOHULUHULU 1075	2315
KULANI CAMP 79	3450
MAUNA LOA SLOPE OBSERVATORY	7628

Figure 8.6 F heating degree-days (National Climatic Data Center).

199

The preceding definitions are used in the two most common procedures for quantifying the effects of passive solar features. These two methods (PSDH 1980, 1984; Goswami et al. 2000; Kreith et al. 1997) are known as the first-level method (or rules of thumb) and the second-level method (or LCR method). Each method's procedures will be examined in the next sections.

8.3 ## THE FIRST-LEVEL METHOD OR RULES OF THUMB

The first-level method is based on the observation that residences that effectively incorporate passive solar features have similar NLC values per floor area and climate-dependent SSF values. The NLC per square feet of floor area in conventional buildings with energy-conserving features is typically in the range 120–160 kJ/C-HDD m^2 or 6–8 Btu/F-HDD ft^2. These values (PSDH 1980, 1984); Goswami et al. 2000; Kreith et al. 1997) are reduced by 20 percent when passive solar features are added, resulting in NLC/floor area values of 100–130 kJ/C-HDD m^2 or 4.8–6.4 Btu/F-HDD ft^2. If climate-dependent, cost-based optimum passive solar features are used, the expected SSF for various regions of the United States can be found in Figure 8.7.

For well-designed passive solar features, the expected LCR values are climate dependent and are defined (PSDH 1980, 1984; Goswami et al. 2000; Kreith et al. 1997) as follows:

	Warm Climate	*Cold Climate*
LCR, kJ/C-HDD m^2 (Btu/F-HDD ft^2)	610 (30)	410 (20)

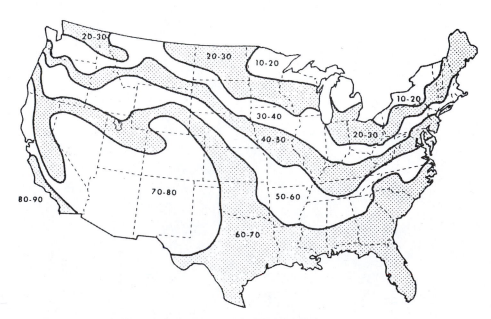

Figure 8.7 Expected SSF values (percentages) (PDSH 1980).

When the thermal mass directly absorbs the solar irradiation, sufficient thermal mass to store 610 kJ/°C m^2 or 30 Btu/°F ft^2 per unit area of glazing must be available (PSDH 1980, 1984; Goswami et al. 2000; Kreith et al. 1997). If the storage material is heated by room air, then four times as much mass is needed. A reasonable value for the specific heat of typical masonry material is 0.2 Btu/lbm F (837 J/kg C) with a density of 150 lbm/ft^3 (2400 kg/m^3).

Example 8.1

For 2500-ft^2 house located in Meridian, MS, the estimated NLC/ft^2 of floor area is 7.5 Btu/F-HDD ft^2. Estimate the characteristics of an optimum passive solar design and the auxiliary heating energy required. Also estimate the thermal storage mass required.

Solution   Figure 7.12 shows that Meridian, MS, has 1358 C-HDD, which converts (multiplying by 9/5) to 2444 F-HDD. A similar number could be estimated from Figure 8.6. The NLC is thus

$$NLC = 7.5 \frac{Btu}{F\text{-}HDD\ ft^2} \cdot 2500\ ft^2 = 18{,}750 \frac{Btu}{F\text{-}HDD}$$

The LCR for a warm climate is 30 Btu/F-HDD ft^2. The projected amount of south-facing glazing is computed as

$$A_p = \frac{NLC}{LCR} = 18{,}750 \frac{Btu}{F\text{-}HDD} \frac{F\text{-}HDD \cdot ft^2}{30\ Btu} = 625\ ft^2$$

The SSF is estimated from Figure 8.7 as 0.6, so Q_{aux} becomes

$$Q_{aux} = (1 - 0.6) \cdot 18{,}750 \frac{Btu}{F\text{-}HDD} \cdot 2444\ F\text{-}HDD = 18.3 \times 10^6\ Btu$$

The thermal mass storage requirement can be estimated by equating the storage requirement per Btu/°F to the product of the mass times the specific heat, or

$$30 \frac{Btu}{ft^2\ °F} \cdot 625\ ft^2 = mass \cdot 0.2 \frac{Btu}{lbm\ °F}$$

from which the mass storage requirement is found to be 94,750 lbm, or 625 ft^3 with the typical masonry density.

The results of the preceding example problem give some indication of the effectiveness of the optimum use of passive solar features, but the solution does not include any details of the solar features. The second-level method incorporates some of the details of the solar features and is examined next.

8.4 | THE SECOND-LEVEL METHOD (LCR METHOD)

Instead of rules of thumb, the LCR method is based on extensive computer-based analyses of various passive solar features in a variety of cities (or climates). The LCR method is philosophically similar to the f-chart approach for active solar features discussed in Chapter 7. Ninety-four different passive solar systems were analyzed to develop the LCR method (PSDH 1980, 1984; Goswami et al. 2000; Kreith et al. 1997). The LCR method is specific to a given structure in that the NLC and projected solar area, A_p, are required. From these two quantities, the LCR can be determined by the definition given earlier:

$$LCR = \frac{NLC}{A_p} \qquad (8\text{-}4)$$

Once the LCR is known, the SSF can be found for a given city by matching the LCR with specific passive solar features.

The five different passive solar configurations covered by the LCR method are as follows:

1. Direct gain (DG)
2. Vented Trombe wall (TW)
3. Unvented Trombe wall (TW)
4. Waterwall (WW)
5. Sunspace (SS)

Within each configuration, different values of various feature parameters are used. These parameters include thermal storage capacity, wall thickness, number of glazings, nighttime insulation, ρck value, and sunspace details. The parameter variations for each of the five types lead to the 94 distinct passive solar energy systems. For a given city, the solar savings fraction (SSF) is tabulated as a function of the LCR. Appendix 8A, tabulated in Goswami et al. (2000) from data by PSDH (1980, 1984), is convenient and useful with the LCR method.

Appendix 8A starts with the common system assumptions—masonry density, for example—incorporated into each analysis. Variations of each of the five configuration types are tabulated and assigned a designation—such as TWA1 for the first vented Trombe wall system. Tables for Albuquerque, Boston, Madison, Medford, Nashville, and Santa Maria are provided in which LCR values for each system variation are provided, and from which the SSF can be estimated. An example is in order.

Example 8.2

A 2000-ft^2 home with passive solar features is located in Nashville. A heating load analysis determines that the NLC is 12,000 Btu/F-HDD. The home has 150 ft^2 of direct-gain solar glazing. Characteristics of the direct-gain features include double glazing, nighttime insulation, and 30 Btu/ft^2 °F thermal storage. Determine the SSF and the auxiliary heating required.

Solution   The LCR is first calculated:

$$LCR = \frac{NLC}{A_p} = \frac{12{,}000\dfrac{Btu}{F\text{-}HDD}}{150 \ ft^2} = 80 \ \frac{Btu}{F\text{-}HDD \ ft^2}$$

The system designation is DGA3, for a direct-gain system with double glazing, night-time insulation, and 30 Btu/ft^2 °F thermal storage. Entering the LCR table (Appendix 8A) for Nashville with an LCR = 80 Btu/F-HDD ft^2 and the system designator DGA3, an SSF of 0.2 is estimated. The table also gives the F-HDD as 3696. The auxiliary heating, Q_{aux}, is calculated as

$$Q_{aux} = (1 - 0.2) \cdot 12{,}000\frac{Btu}{F\text{-}HDD} \cdot 3696 \ F\text{-}HDD = 35.5 \times 10^6 \ Btu$$

The thermal mass can also be estimated since system DGA3 has a mass area to glazing area ratio of 6 and a thickness of 2 inches. Thus,

$$mass = volume \cdot \rho = 6 \cdot 150 \ ft^2 \cdot 2 \ in \cdot \frac{ft}{12 \ in} \cdot 150\frac{lbm}{ft^3} = 22{,}500 \ lbm$$

The utility of the LCR method is that once a baseline structure is described, the method provides an easy procedure for ascertaining the effects of changing passive solar features within the 94 systems in the original computational base.

8.5   DAYLIGHTING

Figure 8.4 demonstrates the effectiveness of daylighting—the use of natural lighting, not artificial lighting (incandescent or fluorescent, for example), to illuminate the interior of a structure. An examination of the salient features of daylighting is appropriate since significant energy savings are possible with the use of properly designed and effectively integrated daylighting implementations.

Spaces that are occupied during daylight hours have a high potential for benefiting from daylighting. For a new building, maximizing the use of daylighting will likely allow a higher skin-to-volume ratio than for a conventional building. A standard window produces useful illumination to a depth of 1.5 times the height of the window. Using additional daylighting features, illumination penetration to 2.0 times the window height is possible. The higher the placement of the window, the deeper the penetration. The illumination of daylighting is composed of three components: (1) an exterior reflected component, (2) a direct sun/sky component, and (3) an internal reflected component. The ceiling is the most important internal reflected component; ceiling reflectivities above 80 percent are recommended. Wall reflectivities between 50 and 70 percent and floor reflectivities between 20 and 40 percent are recommended. The proportions of a room are more important than the dimensions for daylighting considerations.

One method of quantifying the relationship between the visible light and the size of a window (glazing) is the effective-aperture method. The effective aperture is the product of the window-to-wall ratio and the window transmissivity. The window-to-wall ratio is the window area divided by the area of the wall where the window is located. Effective apertures in the range 0.2 to 0.3 are considered good. Consider an example.

| **Example 8.3** | A south-facing wall has a total area of 300 ft^2 and a window area of 100 ft^2. If the transmissivity of the glass is 0.7, determine the effective aperture. Comment on the appropriateness of this configuration for daylighting. |

Solution   The effective aperture is calculated as

$$\text{effective aperture} = \frac{A_{\text{window}}}{A_{\text{wall}}} \tau_{\text{window}} = \frac{100 \text{ ft}^2}{300 \text{ ft}^2} \cdot 0.7 = 0.233$$

Since the effective aperture, 0.233, is between 0.2 and 0.3, the configuration is appropriate for daylighting.

One way to increase the illumination level without increasing the brightness is to utilize light shelves. A light shelf is a horizontal light-reflecting overhang placed above eye level with a transom window above the light shelf. Light is reflected from the light shelf into the interior of the building, and the ceiling is used to distribute the light throughout the room.

The effectiveness of daylighting can be increased by the use of toplighting strategies. Toplighting features include skylights, clerestory windows, roof monitors, and sawtooth roofs. However, some of these features have problems and therefore are sparingly used. Skylights seem a natural for daylighting, but skylights receive the maximum solar gain at solar noon, which can lead to increased cooling loads as well as severe variations in illumination between the morning and afternoon. Clerestory windows are vertical glazings located high on an exterior wall. North-south-oriented clerestory windows are very effective daylighting features; east-west-oriented clerestory windows can lead to high heat gains and major variations in interior illumination. A roof monitor is a flat roof section raised above the adjacent roofline with vertical glazing on all sides. Monitors can result in excessive glazing areas, high heat losses and gains, and interior glare and shading problems during the day. Some older industrial buildings have sawtooth roofs. The sloped surface is opaque, and the vertical surface is glazed. South-facing sawtooth roofs may result in large heat gains during the summer. Thus, taking into account all of these factors, daylighting features must be carefully considered if satisfactory illumination, without significant heat gains and losses, is to result.

An important part of any daylighting implementation is control of the electrical lighting within a daylighted area. Light-level sensors are effective in cutting off or dimming electrical lights when illumination for daylighting features reaches the desired level.

A number of computer programs are available to model and predict daylighting illumination levels in a specific building at a specific site. The U.S. DOE points out that Lumen-Micro, Radiance, and Lightscape include daylighting calculations.

8.6 PASSIVE SOLAR SOFTWARE

There is a plethora of passive solar software elements and programs available. Many of the common energy analysis programs incorporate passive solar calculations that are more accurate than the methods previewed in this chapter. Indeed, many organizations have web-based passive solar energy software that can be executed from the website. But in many of the web-based passive solar programs, the assumptions and numerical procedures are not clearly defined or explained, so the user should exercise caution.

A frequently referenced, widely used, and relatively inexpensive commercial passive solar computer program is Energy-10. Energy-10 is marketed and maintained by the Sustainable Buildings Industry Council, or SBIC. The SBIC web address is www.sbicouncil.org. Details about Energy-10 are available on the website. Essentially, Energy-10 analyzes and quantifies the energy and cost savings for a number of sustainable design strategies, including a nearly complete suite of passive solar features. Energy-10 explicitly simulates the following systems:

1. Passive solar heating
2. Passive solar cooling
3. Photovoltaic
4. Solar domestic hot water
5. Daylighting
6. Natural ventilation
7. High-performance windows

The program is set up to answer "what if" questions concerning various solar features. In Energy-10, a base structure with conventional construction can be specified, and then additional features, including many passive solar options, can be "added" and the energy and cost savings evaluated for the added features. The program can be used for a variety of small commercial building types and is not restricted to residences. An example of the potential savings for a small commercial structure with extensive passive solar features is provided on the SBIC website.

Figure 8.8, taken from the Energy-10 information on the SBIC website, presents a graphical comparison of a conventional building and the same building with

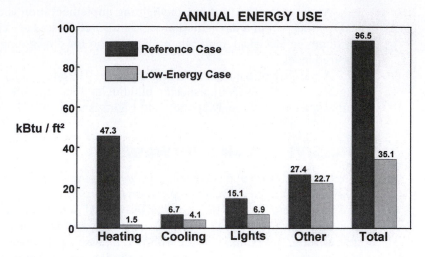

Figure 8.8 Energy-10 output showing end-point energy usage differences (image provided courtesy of the Sustainable Buildings Industry Council).

extensive energy conservation features, including passive solar components. As is evident from the figure, the energy consumption of the "low-energy" case is about one-third that of the base case. Of particular interest is the decrease in heating energy required—from 47.3 kBut/ft^2 to only 1.5 kBtu/ft^2. Energy savings are also indicated for the cooling and lighting functions.

Computer program similar in capability to Energy-10 are necessary for engineering endeavors involving passive solar systems. Much of the passive solar information available on the web is qualitative, not quantitative. In reality, building owners/operators and home owners need quantified energy and cost savings information to justify the use of passive solar implementations.

8.7 CLOSURE

This chapter has reviewed the fundamentals of passive solar energy. Many passive solar features can be added with little additional cost for a new construction. Retrofits of passive solar features to an existing building may be more expensive, but are possible with careful consideration. Additional information is available in the references and on the web. Indeed, a Google search for "passive solar energy" will produce many hits. As with all web information, however, care must be exercised in assessing the validity of the information. In the case of passive solar energy, the U.S. Department of Energy websites provide a wide range of validated information.

REVIEW QUESTIONS

1. A 60-ft long by 25-ft wide house has mostly glass on one side. Sketch the house orientation necessary to take advantage of passive solar principles. Be sure to indicate all directions (N, S, E, W) and the side of the house that is mostly glass.

2. Why is thermal storage needed in a passive solar design?

3. What is the SSF?

4. What SSF would be expected for a house with well-designed passive solar features in Jackson, MS?

5. Describe the features that a passive solar design might include.

6. What is the PSDH?

7. Define and explain NLC and LCR.

8. What is the first-level method, or rules of thumb, for estimating passive solar potential?

9. What is the second-level method (LCR method) for estimating the effects of passive solar features? How do the first-level and second-level methods differ in terms of accuracy?

10. What is daylighting?

11. What is Energy-10?

EXERCISES

1. For a properly designed 2500-ft^2 passive solar house in Meridian, MS, how much auxiliary energy is needed? How much solar energy is captured? Use the rules of thumb (first-level estimate).

2. A 2500-ft^2 house has an NLC = 10,000 Btu/F-HDD and 250 ft^2 of south-facing double glazing. The passive solar energy system consists of a 12-inch thick vented Trombe wall with a selective wall surface and nighttime insulation. For the same house located in Albuquerque, Boston, Madison, Medford, Nashville, and Santa Maria, estimate how much auxiliary heat is needed and how much solar energy is captured in each case. Discuss the effectiveness of these passive solar features for the different climates and locations. Use the second-level (LCR) method.

3. Repeat Exercise 2 for the case where the house has a 6-inch vented Trombe wall with no nighttime insulation. Contrast the effectiveness of the different passive solar features in the different locations.

4. For a properly designed 3000-ft^2 passive solar house in a specified location, how much auxiliary energy is needed? How much solar energy is captured? Use the rules of thumb (first-level estimate).

5. A 3000-ft^2 house in a specified location is being designed with NLC = 15,000 Btu/F-HDD and 350 ft^2 of south-facing double glazing. The passive solar energy system consists of a 12-inch thick vented Trombe wall with a selective wall surface and nighttime insulation. How much auxiliary heat is needed and how much solar energy is captured? Use the second-level (LCR) method.

6. A 2400-ft^2 house in Nashville, TN, is being designed with NLC = 8000 Btu/F-HDD and 250 ft^2 of south-facing double glazing.

   (a) Using the first-level method, estimate the solar savings fraction for a well-designed passive solar home.

   (b) If the passive solar energy system consists of a 12-inch vented Trombe wall with a normal wall surface and nighttime insulation, estimate the solar savings fraction using the second-level method.

   (c) Comment on the relative accuracy of the two methods.

7. A residence in Nashville is constructed with a direct-gain passive solar energy system with 60 Btu/ft^2 °F thermal capacity, double glazing, and no nighttime insulation. What is the expected solar savings fraction (SSF) if the LCR is 17 Btu/F-HDD ft^2?

REFERENCES

Crosbie, M. J. 1997. *The Passive Solar Design and Construction Handbook.* New York: Wiley.

Duffie, J. A., and Beckman, W. A. 2006. *Solar Engineering of Thermal Processes,* 3rd ed. New York: Wiley.

Goswami, Y., Kreith, F., and Kreider, J. F. 2000. *Principles of Solar Engineering,* 2nd ed. New York: Taylor and Francis.

Kreith, F., and West, R. E. 1997. *CRC Handbook of Energy Efficiency.* Boca Raton, FL: CRC Press.

PSDH. 1980. *Passive Solar Design Handbook,* 2 vols. DOE/CS-0127/1 and DOE/CS-0127/2. Washington, DC: U.S. DOE.

PSDH. 1984. *Passive Solar Design Handbook.* New York: Van Nostrand Reinhold.

Sustainable Sources. 2007. *Passive Solar Guidelines.* Available at www.greenbuilder.com.

U.S. Dept. of Energy. 2001. *Passive Solar Design for the Home.* DOE/GO-102001-1105. Available at www.doe.gov.

U.S. Dept. of Energy, National Renewable Energy Laboratory. 1994. *Solar Radiation Data Manual for Flat-Plate and Concentrating Collectors.* Golden, CO: NREL.

APPENDIX A

(Reproduced from Goswami et al., 2000)

Designation and Characteristics for 94 Referenced Passive Systems

(a) Overall System Characteristics

Masonry properties

thermal conductivity (k)	
sunspace floor	0.5 Btu/hr/ft/°F
all other masonry	1.0 Btu/hr/ft/°F
density (Q)	150 lb/ft³
specific heat (c)	0.2 Btu/lb/°F
infrared emittance of normal surface	0.9
infrared emittance of selective surface	0.1

Solar absorptances

waterwall	1.0
masonry, Trombe wall	1.0
direct gain and sunspace	0.8
sunspace: water containers	0.9
lightweight common wall	0.7
other lightweight surfaces	0.3

Glazing properties

transmission characteristics	diffuse
orientation	due south
index of refraction	1.526
extinction coefficient	0.5 inch⁻¹
thickness of each pane	one-eighth inch
gap between panes	one-half inch
ared emittance	0.9

Control range

room temperature	65 to 75°F
sunspace temperature	45 to 95°F
internal heat generation	0

Designation and Characteristics for 94 Referenced Passive Systems

(a) Overall System Characteristics

Thermocirculation vents (when used)

vent area/projected area (sum of both upper and lower vents)	0.06
height between vents	8 ft
reverse flow	none

Nighttime insulation (when used)

thermal resistance	R9
in place, solar time	5:30 P.M. to 7:30 A.M.

Solar radiation assumptions

shading	none
ground diffuse reflectance	0.3

(b) Direct-Gain (DG) System Types

Designation	Thermal Storage Capacity* (in Btu/ft²/°F)	Mass Thickness* (inches)	Mass-Area-to-Glazing-Area Ratio	No. of Glazings	Nighttime Insulation
A1	30	2	6	2	no
A2	30	2	6	3	no
A3	30	2	6	2	yes
B1	45	6	3	2	no
B2	45	6	3	3	no
B3	45	6	3	2	yes
C1	60	4	6	2	no
C2	60	4	6	3	no
C3	60	4	6	2	yes

(c) Vented Trombe-Wall (TW) System Types

Designation	Thermal Storage Capacity* (Btu/ft²/°F)	Wall Thickness* (inches)	ρck Btu²/hr/ft⁴/°F²)	No. of Glazings	Wall Surface	Nighttime Insulation
A1	15	6	30	2	normal	no
A2	22.5	9	30	2	normal	no
A3	30	12	30	2	normal	no
A4	45	18	30	2	normal	no
B1	15	6	15	2	normal	no
B2	22.5	9	15	2	normal	no
B3	30	12	15	2	normal	no
B4	45	18	15	2	normal	no
C1	15	6	7.5	2	normal	no
C2	22.5	9	7.5	2	normal	no
C3	30	12	7.5	2	normal	no
C4	45	18	7.5	2	normal	no

Designation and Characteristics for 94 Referenced Passive Systems

(c) Vented Trombe-Wall (TW) System Types

D1	30	12	30	1	normal	no
D2	30	12	30	3	normal	no
D3	30	12	30	1	normal	yes
D4	30	12	30	2	normal	yes
D5	30	12	30	3	normal	yes
E1	30	12	30	1	selective	no
E2	30	12	30	2	selective	no
E3	30	12	30	1	selective	yes
E4	30	12	30	2	selective	yes

(d) Unvented Trombe-Wall (TW) System Types

Designation	Thermal Storage Capacity* ($Btu/ft^2/°F$)	Wall Thickness* (inches)	ρck ($Btu^2/hr/ft^4/°F^2$)	No. of Glazings	Wall Surface	Nighttime Insulation
F1	15	6	30	2	normal	no
F2	22.5	9	30	2	normal	no
F3	30	12	30	2	normal	no
F4	45	18	30	2	normal	no
G1	15	6	15	2	normal	no
G2	22.5	9	15	2	normal	no
G3	30	12	15	2	normal	no
G4	45	18	15	2	normal	no
H1	15	6	7.5	2	normal	no
H2	22.5	9	7.5	2	normal	no
H3	30	12	7.5	2	normal	no
H4	45	18	7.5	2	normal	no
I1	30	12	30	1	normal	no
I2	30	12	30	3	normal	no
I3	30	12	30	1	normal	yes
I4	30	12	30	2	normal	yes
I5	30	12	30	3	normal	yes
J1	30	12	30	1	selective	no
J2	30	12	30	2	selective	no
J3	30	12	30	1	selective	yes
J4	30	12	30	2	selective	yes

(e) Waterwall (WW) System Types

Designation	Thermal Storage Capacity* (in $Btu/ft^2/°F$)	Wall Thickness (inches)	No. of Glazings	Wall Surface	Nighttime Insulation
A1	15.6	3	2	normal	no
A2	31.2	6	2	normal	no
A3	46.8	9	2	normal	no
A4	62.4	12	2	normal	no

Designation and Characteristics for 94 Referenced Passive Systems

(e) Waterwall (WW) System Types

A5	93.6	18	2	normal	no
A6	124.8	24	2	normal	no
B1	46.8	9	1	normal	no
B2	46.8	9	3	normal	no
B3	46.8	9	1	normal	yes
B4	46.8	9	2	normal	yes
B5	46.8	9	3	normal	yes
C1	46.8	9	1	selective	no
C2	46.8	9	2	selective	no
C3	46.8	9	1	selective	yes
C4	46.8	9	2	selective	yes

(f) Sunspace (SS) System Types

Designation	Type	Tilt (degrees)	Common Wall	End Walls	Nighttime Insulation
A1	attached	50	masonry	opaque	no
A2	attached	50	masonry	opaque	yes
A3	attached	50	masonry	glazed	no
A4	attached	50	masonry	glazed	yes
A5	attached	50	insulated	opaque	no
A6	attached	50	insulated	opaque	yes
A7	attached	50	insulated	glazed	no
A8	attached	50	insulated	glazed	yes
B1	attached	90/30	masonry	opaque	no
B2	attached	90/30	masonry	opaque	yes
B3	attached	90/30	masonry	glazed	no
B4	attached	90/30	masonry	glazed	yes
B5	attached	90/30	insulated	opaque	no
B6	attached	90/30	insulated	opaque	yes
B7	attached	90/30	insulated	glazed	no
B8	attached	90/30	insulated	glazed	yes
C1	semienclosed	90	masonry	common	no
C2	semienclosed	90	masonry	common	yes
C3	semienclosed	90	insulated	common	no
C4	semienclosed	90	insulated	common	yes
D1	semienclosed	50	masonry	common	no
D2	semienclosed	50	masonry	common	yes
D3	semienclosed	50	insulated	common	no
D4	semienclosed	50	insulated	common	yes
E1	semienclosed	90/30	masonry	common	no
E2	semienclosed	90/30	masonry	common	yes
E3	semienclosed	90/30	insulated	common	no
E4	semienclosed	90/30	insulated	common	yes

LCR Table for Santa Maria, California

SSF	.10	.20	.30	.40	.50	.60	.70	.80	.90
Santa Maria, California									3053 DD
WW A1	1776	240	119	73	50	35	25	18	12
WW A2	617	259	154	103	74	54	39	28	19
WW A3	523	261	164	114	82	61	45	33	22
WW A4	482	260	169	119	87	65	48	35	24
WW A5	461	263	175	125	92	69	52	38	26
WW A6	447	263	177	128	95	72	54	40	27
WW B1	556	220	128	85	60	43	32	23	15
WW B2	462	256	168	119	88	66	49	36	25
WW B3	542	315	211	151	112	85	64	47	32
WW B4	455	283	197	144	109	83	63	47	32
WW B5	414	263	184	136	103	79	60	45	31
WW C1	569	330	221	159	118	89	67	49	33
WW C2	478	288	197	143	107	81	61	45	31
WW C3	483	318	228	170	130	100	77	57	40
WW C4	426	280	200	149	114	88	68	51	35
TW A1	1515	227	113	70	48	34	24	17	11
TW A2	625	234	134	89	63	46	33	24	16
TW A3	508	231	140	95	68	50	37	27	18
TW A4	431	217	137	95	69	51	38	28	19
TW B1	859	212	112	71	49	35	25	18	12
TW B2	502	209	124	83	59	43	32	23	15
TW B3	438	201	123	84	60	44	33	24	16
TW B4	400	184	112	76	55	40	30	22	14
TW C1	568	188	105	69	48	35	25	18	12
TW C2	435	178	105	70	50	36	27	19	13
TW C3	413	165	97	64	46	33	25	18	12
TW C4	426	146	82	54	38	27	20	14	10
TW D1	403	170	101	67	48	35	25	18	12
TW D2	488	242	152	105	76	57	42	31	21
TW D3	509	271	175	123	90	67	50	36	25
TW D4	464	266	177	127	94	71	53	39	27
TW D5	425	250	169	122	91	69	52	38	26
TW E1	581	309	199	140	102	76	57	42	28
TW E2	512	283	186	132	97	73	55	40	27
TW E3	537	328	225	164	123	94	71	53	36
TW E4	466	287	199	145	109	83	63	47	32
TW F1	713	198	107	68	47	34	25	18	12
TW F2	455	199	120	81	58	42	31	22	15
TW F3	378	190	120	83	60	45	33	24	16
TW F4	311	169	110	77	57	42	32	23	16
TW G1	450	170	98	65	46	33	24	17	12
TW G2	331	163	102	70	51	38	28	20	14
TW G3	278	147	94	66	48	36	27	20	13
TW G4	222	120	78	55	40	30	22	16	11
TW H1	295	137	84	57	41	30	22	16	11
TW H2	226	118	75	52	38	28	21	15	10
TW H3	187	99	64	44	33	24	18	13	9

LCR Table for Santa Maria, California

SSF	.10	.20	.30	.40	.50	.60	.70	.80	.90
TW H4	143	75	48	33	24	18	14	10	7
TW I1	318	144	88	59	42	31	23	16	11
TW I2	377	203	132	93	68	51	38	28	19
TW I3	404	226	149	106	78	58	44	32	22
TW I4	387	230	156	113	84	64	48	36	24
TW I5	370	226	155	113	85	65	49	36	25
TW J1	483	271	179	127	94	71	53	39	26
TW J2	422	246	165	119	88	67	50	37	25
TW J3	446	283	199	146	111	85	65	48	33
TW J4	400	254	178	132	100	77	58	43	30
DG A1	392	188	117	79	55	38	26	16	7
DG A2	389	190	121	85	61	45	32	22	14
DG A3	443	220	142	102	77	58	44	31	19
DG B1	384	191	122	86	64	48	35	24	13
DG B2	394	196	127	91	69	53	40	29	19
DG B3	445	222	145	105	80	62	49	37	25
DG C1	451	225	146	104	78	61	47	34	21
DG C2	453	226	148	106	80	63	49	37	25
DG C3	509	254	167	121	92	73	58	45	31
SS A1	1171	396	220	142	98	69	49	34	22
SS A2	1028	468	283	190	135	98	71	50	33
SS A3	1174	380	209	133	91	64	45	31	20
SS A4	1077	481	289	193	136	98	71	50	32
SS A5	1896	400	204	127	86	60	42	29	18
SS A6	1030	468	283	190	135	97	71	50	32
SS A7	2199	359	178	109	72	50	35	24	15
SS A8	1089	478	285	190	133	96	69	48	31
SS B1	802	298	170	111	77	55	40	28	18
SS B2	785	366	224	152	108	79	57	41	27
SS B3	770	287	163	106	74	52	37	26	17
SS B4	790	368	224	152	108	78	57	40	26
SS B5	1022	271	144	91	62	44	31	22	14
SS B6	750	356	219	149	106	77	56	40	26
SS B7	937	242	127	80	54	38	27	19	12
SS B8	750	352	215	146	103	75	55	39	25
SS C1	481	232	144	99	71	52	39	28	19
SS C2	482	262	170	120	88	66	49	36	24
SS C3	487	185	107	71	50	36	27	19	13
SS C4	473	235	147	102	74	55	41	30	20
SS D1	1107	477	282	188	132	95	68	48	31
SS D2	928	511	332	232	169	125	92	66	43
SS D3	1353	449	248	160	110	78	56	39	25
SS D4	946	500	319	222	160	117	86	61	40
SS E1	838	378	227	153	108	78	56	40	26
SS E2	766	419	272	190	138	102	75	54	36
SS E3	973	322	178	115	79	56	40	28	18
SS E4	780	393	247	170	122	89	65	47	31

LCR Table for Albuquerque, New Mexico

SSF	.10	.20	.30	.40	.50	.60	.70	.80	.90
Albuquerque, New Mexico								4292 DD	
WW A1	1052	130	62	38	25	18	13	9	6
WW A2	354	144	84	56	39	29	21	15	10
WW A3	300	146	90	62	45	33	24	18	12
WW A4	276	146	93	65	47	35	26	19	13
WW A5	264	148	97	69	50	38	28	21	14
WW A6	256	148	99	70	52	39	30	22	15
WW B1	293	111	63	41	28	20	15	11	7
WW B2	270	147	96	67	49	37	28	20	14
WW B3	314	179	119	84	62	47	35	26	18
WW B4	275	169	116	85	64	49	37	28	19
WW B5	252	159	110	81	61	47	36	27	19
WW C1	333	190	126	89	66	50	38	28	19
WW C2	287	171	115	83	62	47	36	27	18
WW C3	293	191	136	101	77	59	46	34	24
WW C4	264	172	122	91	69	54	41	31	22
TW A1	900	124	60	37	25	17	12	9	6
TW A2	361	130	73	48	33	24	18	13	8
TW A3	293	129	77	52	37	27	20	15	10
TW A4	249	123	76	52	38	28	21	15	10
TW B1	502	117	60	38	26	18	13	9	6
TW B2	291	118	68	45	32	23	17	12	8
TW B3	254	114	68	46	33	24	18	13	9
TW B4	233	104	63	42	30	22	16	12	8
TW C1	332	106	58	37	26	19	14	10	6
TW C2	255	101	58	39	27	20	15	11	7
TW C3	243	94	54	36	25	18	13	10	7
TW C4	254	84	46	30	21	15	11	8	5
TW D1	213	86	50	33	23	17	12	9	6
TW D2	287	139	86	59	43	32	24	17	12
TW D3	294	153	97	68	49	37	27	20	14
TW D4	281	158	104	74	55	41	31	23	16
TW D5	260	151	101	73	54	41	31	23	16
TW E1	339	177	113	78	57	43	32	23	16
TW E2	308	168	109	77	56	42	32	23	16
TW E3	323	195	133	96	72	55	42	31	21
TW E4	287	175	120	88	66	50	38	28	20
TW F1	409	108	57	36	24	17	13	9	6
TW F2	260	110	65	43	31	22	17	12	8
TW F3	216	106	66	45	33	24	10	13	9
TW F4	178	95	61	42	31	23	17	13	9
TW G1	256	93	53	34	24	17	13	9	6
TW G2	189	91	56	38	27	20	15	11	7
TW G3	159	82	52	36	26	20	15	11	7
TW G4	128	68	43	30	22	16	12	9	6
TW H1	168	76	45	31	22	16	12	9	6
TW H2	130	66	41	29	21	15	11	8	6

LCR Table for Albuquerque, New Mexico

SSF	.10	.20	.30	.40	.50	.60	.70	.80	.90
TW H3	108	56	35	25	8	13	10	7	5
TW H4	83	42	27	19	13	10	7	5	4
TW I1	166	73	43	29	20	15	11	8	5
TW I2	221	117	75	52	30	28	21	16	11
TW I3	234	128	83	59	43	32	24	10	12
TW I4	234	137	92	66	49	37	28	21	14
TW I5	226	136	93	67	50	38	29	22	15
TW J1	282	156	102	72	53	40	30	22	15
TW J2	254	146	97	69	51	39	29	22	15
TW J3	269	169	118	86	65	50	38	29	20
TW J4	247	155	106	80	60	46	35	26	18
DG A1	211	97	57	36	22	13	5	—	—
DG A2	227	107	67	46	32	23	16	10	5
DG A3	274	131	83	59	44	34	25	18	10
DG B1	210	97	60	42	30	21	13	6	—
DG B2	232	110	69	49	37	28	21	14	8
DG B3	277	134	85	61	47	37	28	21	14
DG C1	253	120	74	53	39	30	22	14	—
DG C2	271	130	82	59	45	35	26	19	12
DG C3	318	155	96	71	54	43	34	26	18
SS A1	591	187	101	64	44	31	22	16	10
SS A2	531	232	137	92	65	47	34	25	16
SS A3	566	170	90	56	38	27	19	13	8
SS A4	537	230	135	89	63	45	33	23	15
SS A5	980	187	92	56	37	26	18	13	8
SS A6	529	231	136	91	64	47	34	24	16
SS A7	1103	158	74	44	29	20	14	10	6
SS A8	540	226	131	87	61	44	32	23	15
SS B1	403	141	78	50	35	25	18	13	8
SS B2	412	186	111	75	53	39	28	20	14
SS B3	372	130	71	46	31	22	16	11	7
SS B4	403	181	106	72	51	37	27	20	13
SS B5	518	127	65	40	27	19	13	9	6
SS B6	390	179	106	73	52	38	28	20	13
SS B7	457	108	54	33	22	16	11	8	5
SS B8	379	171	102	69	49	35	26	19	12
SS C1	270	126	77	52	37	27	20	15	10
SS C2	282	150	97	68	49	37	28	20	14
SS C3	276	101	57	37	26	19	14	10	7
SS C4	277	135	83	57	41	31	23	17	11
SS D1	548	225	130	85	59	43	31	22	14
SS D2	474	253	162	113	82	61	45	33	22
SS D3	683	212	113	72	49	35	25	17	11
SS D4	484	248	156	107	77	57	42	30	20
SS E1	410	176	103	68	48	35	25	18	12
SS E2	390	208	133	92	67	50	37	27	18
SS E3	487	151	80	51	35	25	18	12	8
SS E4	400	195	120	82	59	43	32	23	15

LCR Table for Nashville, Tennessee

SSF	.10	.20	.30	.40	.50	.60	.70	.80	.90
Nashville, Tennessee									3696 DD
WW A1	588	60	24	13	8	5	3	2	1
WW A2	192	70	38	23	15	11	7	5	3
WW A3	161	72	42	27	18	13	9	6	4
WW A4	148	72	43	29	20	14	10	7	5
WW A5	141	74	46	31	22	16	11	8	5
WW A6	137	74	47	32	22	16	12	8	5
WW B1	135	41	19	10	6	3	2	—	—
WW B2	152	78	48	33	23	17	12	9	6
WW B3	179	97	61	42	30	22	16	12	8
WW B4	164	97	65	46	34	25	19	14	9
WW B5	153	93	63	45	33	25	19	14	9
WW C1	193	105	67	46	33	24	18	13	8
WW C2	169	97	63	44	32	24	18	13	8
WW C3	181	115	79	58	43	33	25	18	12
WW C4	164	104	72	53	39	30	23	17	11
TW A1	509	59	25	13	8	5	3	2	1
TW A2	199	64	33	20	13	9	6	4	3
TW A3	160	65	36	23	15	11	8	5	3
TW A4	136	62	36	23	16	11	8	6	4
TW B1	282	57	26	15	9	6	4	3	2
TW B2	161	59	32	20	13	9	6	4	3
TW B3	141	58	32	21	14	10	7	5	3
TW B4	131	54	30	19	13	9	7	5	3
TW C1	188	53	27	16	10	7	5	3	2
TW C2	144	52	28	18	12	8	6	4	2
TW C3	139	49	27	17	11	8	5	4	2
TW C4	149	45	23	14	9	7	5	3	2
TW D1	99	33	16	9	5	3	2	1	—
TW D2	164	75	44	29	20	14	10	7	5
TW D3	167	82	49	33	23	17	12	8	5
TW D4	168	91	58	40	29	21	15	11	7
TW D5	160	89	58	40	29	22	16	12	8
TW E1	198	98	59	40	28	20	15	10	7
TW E2	182	95	59	40	29	21	15	11	7
TW E3	197	115	76	54	39	29	22	16	11
TW E4	178	105	70	50	37	27	20	15	10
TW F1	221	50	23	13	8	5	4	2	1
TW F2	139	53	29	18	12	8	6	4	2
TW F3	116	52	30	19	13	9	7	5	3
TW F4	96	47	28	19	13	9	7	5	3
TW G1	137	44	22	13	9	6	4	3	2
TW G2	101	44	25	16	11	8	5	4	2
TW G3	86	41	24	16	11	8	6	4	2
TW G4	69	34	21	14	10	7	5	3	2
TW H1	89	36	20	13	8	6	4	3	2
TW H2	69	33	19	12	9	6	4	3	2
TW H3	59	28	17	11	8	5	4	3	2

LCR Table for Nashville, Tennessee

SSF	.10	.20	.30	.40	.50	.60	.70	.80	.90
TW H4	46	22	13	9	6	4	3	2	1
TW I1	74	26	13	7	4	2	1	—	—
TW I2	125	62	38	25	18	13	9	7	4
TW I3	133	69	43	29	20	15	11	8	5
TW I4	139	78	51	35	26	19	14	10	7
TW I5	137	80	53	37	27	20	15	11	7
TW J1	164	86	54	36	26	19	14	10	6
TW J2	150	82	53	36	26	19	14	10	7
TW J3	165	101	68	49	36	27	20	15	10
TW J4	153	93	63	46	34	25	19	14	10
DG A1	98	34	—	—	—	—	—	—	—
DG A2	130	55	31	19	11	6	—	—	—
DG A3	173	78	47	32	23	16	11	7	2
DG B1	100	36	17	—	—	—	—	—	—
DG B2	134	58	33	22	15	10	6	—	—
DG B3	177	81	49	33	24	18	14	10	6
DG C1	131	52	28	17	9	—	—	—	—
DG C2	161	71	42	28	20	14	10	6	—
DG C3	205	94	57	39	29	22	17	12	8
SS A1	351	100	50	29	19	13	9	6	4
SS A2	328	135	76	49	33	24	17	12	8
SS A3	330	87	41	24	15	10	6	4	2
SS A4	331	133	74	47	32	22	16	11	7
SS A5	595	98	43	24	15	10	7	4	2
SS A6	324	132	75	48	32	23	16	11	7
SS A7	668	79	32	17	10	6	4	2	1
SS A8	330	129	71	45	30	21	15	10	6
SS B1	236	74	38	23	15	10	7	5	3
SS B2	258	110	63	41	28	20	14	10	6
SS B3	212	65	32	19	12	8	5	3	2
SS B4	251	105	60	39	27	19	13	9	6
SS B5	307	65	30	17	10	7	4	3	2
SS B6	241	104	60	39	27	19	14	10	6
SS B7	264	52	23	12	7	5	3	2	—
SS B8	233	98	56	36	25	17	12	9	5
SS C1	141	60	33	21	14	10	7	5	3
SS C2	161	81	50	33	23	17	12	9	6
SS C3	149	48	25	15	10	7	4	3	2
SS C4	160	73	43	28	19	14	10	7	5
SS D1	317	119	64	39	26	18	13	8	5
SS D2	287	147	90	61	43	31	23	16	10
SS D3	405	113	55	33	21	14	10	6	4
SS D4	295	144	87	58	40	29	21	15	10
SS E1	229	89	48	29	19	13	9	6	4
SS E2	233	118	72	48	34	24	18	12	8
SS E3	283	77	37	22	14	9	6	4	2
SS E4	242	111	65	43	29	21	15	11	7

LCR Table for Medford, Oregon

SSF	.10	.20	.30	.40	.50	.60	.70	.80	.90
Medford, Oregon								4930 DD	
WW A1	708	64	24	11	—	—	—	—	—
WW A2	212	73	38	22	13	7	3	—	—
WW A3	174	75	41	25	16	9	5	2	—
WW A4	158	74	43	27	17	11	6	3	1
WW A5	149	75	45	29	19	12	7	4	2
WW A6	144	75	46	30	20	13	8	4	2
WW B1	154	43	16	—	—	—	—	—	—
WW B2	162	80	48	31	21	14	9	6	3
WW B3	190	100	62	41	28	19	13	8	5
WW B4	171	99	65	45	32	23	16	11	7
WW B5	160	95	63	45	32	23	17	12	7
WW C1	205	108	67	45	31	21	15	10	6
WW C2	178	99	63	43	30	22	15	10	6
WW C3	189	117	80	57	42	31	23	16	10
WW C4	170	106	72	52	38	28	21	15	9
TW A1	607	63	25	12	5	—	—	—	—
TW A2	222	68	33	19	11	6	2	—	—
TW A3	175	67	36	21	13	8	4	2	—
TW A4	147	64	36	22	14	9	5	3	1
TW B1	327	61	27	14	7	3	—	—	—
TW B2	178	62	32	19	12	7	4	2	—
TW B3	154	60	33	20	12	8	4	2	1
TW B4	143	56	31	19	12	8	5	2	1
TW C1	212	56	27	15	9	5	2	—	—
TW C2	159	55	28	17	11	7	4	2	—
TW C3	154	52	27	16	10	6	4	2	1
TW C4	167	48	24	14	9	5	3	2	—
TW D1	112	34	14	—	—	—	—	—	—
TW D2	177	77	44	28	18	12	8	5	3
TW D3	180	85	50	32	21	14	9	6	3
TW D4	177	93	58	39	27	19	13	9	5
TW D5	168	92	58	40	28	20	14	10	6
TW E1	213	101	60	39	26	18	12	8	4
TW E2	194	98	59	39	27	19	13	9	5
TW E3	208	118	77	53	38	27	20	13	8
TW E4	186	108	71	49	36	26	19	13	8
TW F1	256	53	23	12	5	—	—	—	—
TW F2	153	56	29	17	10	5	2	—	—
TW F3	125	54	30	18	11	7	3	1	—
TW F4	102	48	28	18	11	7	4	2	1
TW G1	153	46	22	12	7	—	—	—	—
TW G2	109	46	25	15	9	5	3	1	—
TW G3	92	42	24	15	9	6	3	2	—
TW G4	74	35	20	13	8	5	3	2	—
TW H1	97	38	20	12	7	4	1	—	—
TW H2	75	34	19	12	7	5	3	1	—

LCR Table for Medford, Oregon

SSF	.10	.20	.30	.40	.50	.60	.70	.80	.90
TW H3	63	29	17	10	7	4	3	1	—
TW H4	49	23	13	8	5	3	2	1	—
TW I1	83	27	10	—	—	—	—	—	—
TW I2	133	64	38	24	16	11	7	4	2
TW I3	142	71	43	28	19	13	9	5	3
TW I4	146	80	51	35	25	17	12	8	5
TW I5	144	82	53	37	26	19	13	9	6
TW J1	175	89	54	36	24	17	11	7	4
TW J2	158	85	53	36	25	18	12	8	5
TW J3	173	103	69	48	35	26	18	13	8
TW J4	160	96	64	45	33	24	17	12	8
DG A1	110	35	—	—	—	—	—	—	—
DG A2	142	58	32	18	9	—	—	—	—
DG A3	187	82	48	32	22	15	9	5	—
DG B1	110	40	15	—	—	—	—	—	—
DG B2	146	61	35	21	13	7	—	—	—
DG B3	193	84	51	34	24	17	12	7	3
DG C1	144	57	29	13	—	—	—	—	—
DG C2	177	75	44	28	19	12	6	—	—
DG C3	224	98	60	41	29	21	14	10	5
SS A1	415	110	51	28	16	9	4	2	—
SS A2	372	146	79	48	31	21	14	8	5
SS A3	397	96	42	21	10	—	—	—	—
SS A4	379	144	76	46	29	19	12	7	4
SS A5	732	111	45	23	12	5	—	—	—
SS A6	368	143	77	47	30	20	13	8	4
SS A7	846	90	33	14	—	—	—	—	—
SS A8	379	140	73	44	27	17	11	6	3
SS B1	274	81	38	21	12	6	3	—	—
SS B2	288	117	65	40	26	18	12	7	4
SS B3	249	71	33	17	8	—	—	—	—
SS B4	282	113	62	38	25	16	11	7	4
SS B5	368	72	30	15	7	—	—	—	—
SS B6	269	111	62	30	25	17	11	7	4
SS B7	323	58	23	10	—	—	—	—	—
SS B8	262	106	57	35	23	15	9	6	3
SS C1	153	62	33	19	11	5	—	—	—
SS C2	172	83	50	32	22	15	10	6	3
SS C3	166	51	24	13	7	3	—	—	—
SS C4	173	76	43	27	18	12	8	5	3
SS D1	367	129	65	37	22	13	7	3	1
SS D2	318	156	92	60	40	27	18	12	7
SS D3	480	124	57	31	18	10	5	2	—
SS D4	328	153	89	57	38	26	17	11	6
SS E1	262	95	48	27	15	7	—	—	—
SS E2	257	124	73	47	31	21	14	9	5
SS E3	334	84	38	20	10	4	—	—	—
SS E4	269	118	67	42	27	18	12	7	4

LCR Table for Boston, Massachusetts

SSF	.10	.20	.30	.40	.50	.60	.70	.80	.90
Boston, Massachusetts									5621 DD
WW A1	368	28	9	—	—	—	—	—	—
WW A2	119	41	20	12	7	5	3	2	—
WW A3	101	43	24	15	10	6	4	3	1
WW A4	93	44	26	16	11	7	5	3	2
WW A5	89	45	27	18	12	8	6	4	2
WW A6	87	46	28	19	13	9	6	4	3
WW B1	59	—	—	—	—	—	—	—	—
WW B2	103	52	31	21	15	10	7	5	3
WW B3	123	66	41	28	20	14	10	7	5
WW B4	118	70	46	33	24	18	13	9	6
WW B5	113	69	46	33	25	18	14	10	7
WW C1	135	72	46	31	22	16	12	8	5
WW C2	121	68	44	31	22	16	12	9	6
WW C3	136	86	60	44	33	25	19	14	9
WW C4	124	78	54	40	30	23	17	12	8
TW A1	324	30	11	4	—	—	—	—	—
TW A2	126	37	18	10	6	4	2	1	—
TW A3	102	39	21	13	8	5	3	2	1
TW A4	88	38	22	14	9	6	4	3	2
TW B1	180	32	13	7	4	2	—	—	—
TW B2	104	36	19	11	7	5	3	2	1
TW B3	92	36	19	12	8	5	3	2	1
TW B4	86	34	19	12	8	5	4	2	1
TW C1	122	32	15	9	5	3	2	1	—
TW C2	95	33	17	10	7	4	3	2	1
TW C3	93	31	16	10	6	4	3	2	1
TW C4	102	29	15	9	6	4	3	2	1
TW D1	45	—	—	—	—	—	—	—	—
TW D2	112	49	28	18	12	9	6	4	3
TW D3	113	54	32	21	15	10	7	5	3
TW D4	121	64	41	28	20	15	11	8	5
TW D5	118	66	42	30	21	16	12	8	6
TW E1	138	67	40	27	18	13	9	7	4
TW E2	130	66	41	28	20	14	10	7	5
TW E3	146	84	56	39	29	21	16	11	8
TW E4	133	78	52	37	27	20	15	11	7
TW F1	134	25	10	4	—	—	—	—	—
TW F2	86	30	16	9	5	3	2	1	—
TW F3	72	31	17	11	7	4	3	2	1
TW F4	61	29	17	11	7	5	3	2	1
TW G1	83	24	11	6	3	2	—	—	—
TW G2	63	26	14	9	5	4	2	1	—
TW G3	54	25	14	9	6	4	3	2	1
TW G4	45	21	12	8	5	4	3	2	1
TW H1	54	21	11	6	4	2	1	—	—
TW H2	44	20	11	7	5	3	2	1	—
TW H3	38	17	10	6	4	3	2	1	—

LCR Table for Boston, Massachusetts

SSF	.10	.20	.30	.40	.50	.60	.70	.80	.90
TW H4	30	14	8	5	3	2	2	1	—
TW I1	30	—	—	—	—	—	—	—	—
TW I2	84	41	24	16	11	8	6	4	2
TW I3	91	46	28	19	13	9	7	5	3
TW I4	100	56	36	25	18	13	10	7	5
TW I5	101	58	38	27	20	15	11	8	5
TW J1	114	59	37	25	17	12	9	6	4
TW J2	107	58	37	25	18	13	10	7	4
TW J3	123	75	51	36	27	20	15	11	7
TW J4	115	70	47	34	25	19	14	10	7
DG A1	43	—	—	—	—	—	—	—	—
DG A2	85	34	18	9	—	—	—	—	—
DG A3	125	56	33	22	16	11	7	4	—
DG B1	44	—	—	—	—	—	—	—	—
DG B2	87	36	20	12	7	—	—	—	—
DG B3	129	58	35	24	17	13	9	6	3
DG C1	71	23	—	—	—	—	—	—	—
DG C2	109	47	27	17	12	8	4	—	—
DG C3	151	68	41	28	21	16	12	8	5
SS A1	230	61	29	16	10	6	4	2	1
SS A2	231	93	52	33	22	15	11	7	5
SS A3	205	48	20	10	4	—	—	—	—
SS A4	229	90	49	31	20	14	9	6	4
SS A5	389	58	23	11	6	3	—	—	—
SS A6	226	91	50	32	21	15	10	7	4
SS A7	420	40	12	—	—	—	—	—	—
SS A8	226	86	46	28	19	12	8	6	3
SS B1	151	44	21	12	7	4	2	1	—
SS B2	183	77	43	28	19	13	9	6	4
SS B3	129	36	16	8	3	—	—	—	—
SS B4	176	73	41	26	17	12	8	6	4
SS B5	193	36	15	7	3	—	—	—	—
SS B6	169	72	41	26	18	12	9	6	4
SS B7	157	25	7	—	—	—	—	—	—
SS B8	160	66	37	23	16	11	7	5	3
SS C1	84	33	17	10	6	4	2	1	—
SS C2	110	54	33	22	15	11	8	5	3
SS C3	91	26	12	7	4	2	—	—	—
SS C4	109	48	28	18	12	9	6	4	3
SS D1	206	73	38	22	14	9	5	3	2
SS D2	203	103	63	42	29	21	15	10	6
SS D3	264	69	32	18	10	6	4	2	1
SS D4	208	100	60	39	27	19	14	9	6
SS E1	140	51	25	14	8	4	2	—	—
SS E2	161	80	48	32	22	15	11	7	5
SS E3	177	44	19	10	5	2	—	—	—
SS E4	166	75	43	28	19	13	9	6	4

LCR Table for Madison, Wisconsin

SSF	.10	.20	.30	.40	.50	.60	.70	.80	.90
Madison, Wisconsin								7730 DD	
WW A1	278	—	—	—	—	—	—	—	—
WW A2	91	27	12	—	—	—	—	—	—
WW A3	77	30	15	8	3	—	—	—	—
WW A4	72	32	17	10	5	—	—	—	—
WW A5	69	33	19	11	7	4	—	—	—
WW A6	67	34	19	12	7	4	2	—	—
WW B1	—	—	—	—	—	—	—	—	—
WW B2	84	41	24	15	10	7	5	3	2
WW B3	102	53	32	21	15	10	7	5	3
WW B4	101	59	39	27	19	14	10	7	5
WW B5	98	59	39	28	20	15	11	8	5
WW C1	113	59	37	25	17	12	8	6	3
WW C2	103	57	37	25	18	13	9	6	4
WW C3	119	75	51	37	28	21	15	11	7
WW C4	109	68	47	34	25	19	14	10	7
TW A1	249	16	—	—	—	—	—	—	—
TW A2	97	26	11	4	—	—	—	—	—
TW A3	79	28	13	7	3	—	—	—	—
TW A4	69	28	15	9	5	3	—	—	—
TW B1	139	20	5	—	—	—	—	—	—
TW B2	81	26	12	6	3	—	—	—	—
TW B3	72	27	13	7	4	2	—	—	—
TW B4	69	26	13	8	5	3	1	—	—
TW C1	96	23	10	4	—	—	—	—	—
TW C2	76	25	12	7	4	2	—	—	—
TW C3	75	24	12	7	4	2	1	—	—
TW C4	84	23	11	6	4	2	1	—	—
TW D1	—	—	—	—	—	—	—	—	—
TW D2	91	39	22	13	9	6	4	2	1
TW D3	93	43	25	16	10	7	5	3	1
TW D4	103	54	34	23	16	12	8	6	4
TW D5	102	56	36	25	18	13	10	7	4
TW E1	115	54	32	21	14	10	7	4	3
TW E2	110	55	34	22	16	11	8	5	3
TW E3	126	72	47	33	24	18	13	9	6
TW E4	116	68	45	32	23	17	13	9	6
TW F1	99	13	—	—	—	—	—	—	—
TW F2	65	20	8	—	—	—	—	—	—
TW F3	55	22	11	5	—	—	—	—	—
TW F4	47	21	11	7	4	2	—	—	—
TW G1	61	14	—	—	—	—	—	—	—
TW G2	47	18	8	4	—	—	—	—	—
TW G3	42	18	9	5	3	—	—	—	—
TW G4	35	16	9	5	3	2	—	—	—
TW H1	41	13	6	—	—	—	—	—	—
TW H2	34	14	7	4	2	—	—	—	—

LCR Table for Madison, Wisconsin

SSF	.10	.20	.30	.40	.50	.60	.70	.80	.90
TW H3	29	13	7	4	2	1	—	—	—
TW H4	24	10	6	3	2	1	—	—	—
TW I1	—	—	—	—	—	—	—	—	—
TW I2	68	32	18	12	8	5	3	2	1
TW I3	75	37	22	14	10	7	4	3	2
TW I4	85	47	30	21	15	11	8	5	3
TW I5	87	50	33	23	16	12	9	6	4
TW J1	95	48	29	19	13	9	6	4	3
TW J2	91	48	30	21	14	10	7	5	3
TW J3	106	65	43	31	23	17	12	9	6
TW J4	100	61	41	29	21	16	12	9	6
DG A1	—	—	—	—	—	—	—	—	—
DG A2	68	25	11	—	—	—	—	—	—
DG A3	109	47	28	18	12	8	5	—	—
DG B1	—	—	—	—	—	—	—	—	—
DG B2	70	27	14	6	—	—	—	—	—
DG B3	114	50	30	20	14	10	7	4	—
DG C1	47	—	—	—	—	—	—	—	—
DG C2	91	37	21	13	7	—	—	—	—
DG C3	133	59	35	24	17	13	9	6	3
SS A1	192	47	20	9	3	—	—	—	—
SS A2	200	78	42	26	17	12	8	5	3
SS A3	166	32	—	—	—	—	—	—	—
SS A4	197	74	39	23	15	10	6	4	2
SS A5	329	42	13	—	—	—	—	—	—
SS A6	195	75	40	25	16	11	7	5	3
SS A7	349	22	—	—	—	—	—	—	—
SS A8	192	69	36	21	13	8	5	3	2
SS B1	122	32	13	5	—	—	—	—	—
SS B2	158	64	36	22	15	10	7	5	3
SS B3	100	22	—	—	—	—	—	—	—
SS B4	150	60	33	29	13	9	6	4	2
SS B5	156	24	—	—	—	—	—	—	—
SS B6	145	59	33	20	13	9	6	4	2
SS B7	122	—	—	—	—	—	—	—	—
SS B8	136	54	29	18	11	7	5	3	2
SS C1	61	20	7	—	—	—	—	—	—
SS C2	90	43	25	16	11	7	5	3	2
SS C3	67	16	—	—	—	—	—	—	—
SS C4	90	38	22	13	9	6	4	2	1
SS D1	169	56	26	13	6	—	—	—	—
SS D2	175	86	51	34	23	16	11	7	5
SS D3	221	52	21	10	—	—	—	—	—
SS D4	179	84	49	32	21	15	10	7	4
SS E1	108	34	12	—	—	—	—	—	—
SS E2	135	65	38	24	16	11	7	5	3
SS E3	141	29	8	—	—	—	—	—	—
SS E4	140	61	34	21	14	9	6	4	2

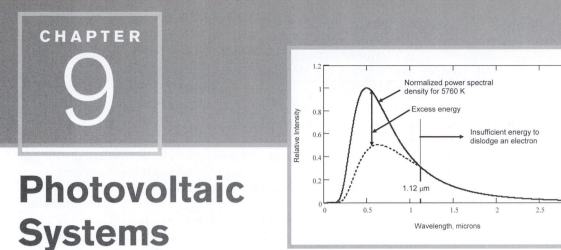

Photovoltaic Systems

INTRODUCTION

"Photovoltaic" refers to the direct generation of electricity by solar irradiation. Like flat-plate solar collectors for thermal solar systems, discussed in Chapter 7, "solar cells" are well-recognized icons of photovoltaic systems. As an example of a photovoltaic system, Figure 9.1 presents a photograph of an 18-kW, flat-plate photovoltaic array with dual axis tracking. The system shown in the figure was installed by the Nevada Power Company and is an example of a very sophisticated application of

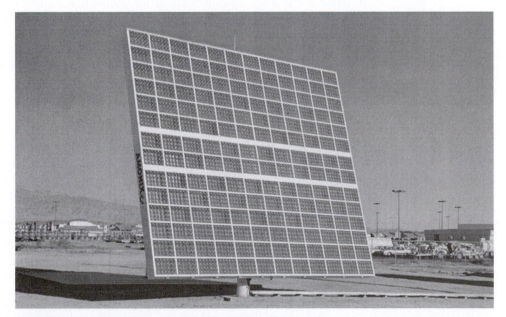

Figure 9.1 Photovoltaic system example (NREL 08851).

photovoltaics (PV). However, no matter how sophisticated the PV system, the solar cell is the fundamental building block. The next section develops the fundamentals of photovoltaic cell operation.

9.2   PHOTOVOLTAIC CELL FUNDAMENTALS

Atomic theory and semiconductor theory are good places to start an examination of the operation of a photovoltaic cell. An atom is composed of a nucleus containing protons and neutrons, with electrons in orbits about the nucleus. In an atom, the number of protons is equal to the number of electrons, and that number is the atomic number of the element. For silicon, a common material extensively used in PV cells, the atomic number is 14. The electrons are positioned in orbitals or bands about the nucleus. All inner bands must be filled before any band farther out can be partially filled. The energy of an electron is determined by its position in one of the various bands. The number of electrons in the outermost band determines the chemical characteristics of the element. The outermost band an electron can occupy and maintain its association with the atom is called the valence band. In silicon, the valence band is partially filled and contains four electrons; it must contain eight electrons to be filled.

If the valence band is filled, the element is chemically inert. If electrons in an unfilled valence band are lightly bound, they may attach themselves to a neighboring atom, which gives that atom a negative charge, resulting in the donor atom's becoming a positively charged ion. The positively and negatively charged ions form an ionic bond. If the electrons in an unfilled valence band are more strongly attached, they configure themselves so that neighboring atoms can share electrons to keep the valence bands filled. Such sharing of electrons results in a covalent bond.

Electrons in the valence band can become so energetic that they jump into a band that is far removed from the nucleus. This remote band is called the conduction band. The difference in energy between an electron in the valence band and one in the conduction band is called the band gap energy. Electrons in the conduction band require only a small amount of energy to move away from the atom, and this is responsible for heat and electrical conduction. Band gap energies are usually expressed in electron-volts, $eV(1\,eV = 1.6 \times 10^{-16}\,J)$. Insulators are materials whose atoms possess full valence bands and have high band gap energies, >3 eV. Conductors are materials whose atoms have relatively empty valence bands. Materials whose atoms have relatively full valence bands are semiconductors with band gap energies <3 eV. Band gap energies for some typical materials used in photovoltaic cells are presented in Table 9.1.

Silicon is an example of a semiconductor (four electrons in the valence band). Pure silicon is called an intrinsic semiconductor, but if a small amount of impurities, usually called the dopant, is combined with the pure silicon, an extrinsic semiconductor results. An n-type semiconductor results if the dopant has more electrons in the valence band than the base material. An n-type semiconductor seems to have an excess of electrons even though the semiconductor is electrically neutral. A p-type semiconductor results if the dopant has fewer electrons in the valence band than the base material. The p-type semiconductor appears to have a deficit of electrons, or an excess of "holes," although it is electrically neutral. The reason for doping a pure

TABLE 9.1 Representative photovoltaic material band gap energies

Material	Band Gap Energy (eV)
Si, silicon	1.11
CdTe, cadmium telluride	1.44
CdS, cadmium sulfide	2.42
CuInSe$_2$, copper indium diselenide	1.01
GaAs, gallium arsenide	1.40
GaP, gallium phosphide	2.24
InP, indium phosphide	1.27

semiconductor such as silicon is that a junction formed with *n*-type and *p*-type semiconductor materials enhances the flow of electrons and holes.

Consider the schematic of a photovoltaic cell in Figure 9.2. The PV cell is shown as a *p-n* junction with an incident photon. An external load is connected. If the incident photon is energetic enough to dislodge a valence electron, the electron will jump to the conduction band and initiate a current flow. For an electron to be forced from the valence band to the conduction band, the electron must possess at least the band gap energy.

The energy of a photon is given by

$$E = h\nu \tag{9-1}$$

with $h = 6.625 \cdot 10^{-34}$ J sec (Planck's constant) and ν the frequency. The frequency, wavelength (λ), and speed of light ($c = 3 \times 10^8$ m/sec) are related as

$$\nu = \frac{c}{\lambda} \tag{9-2}$$

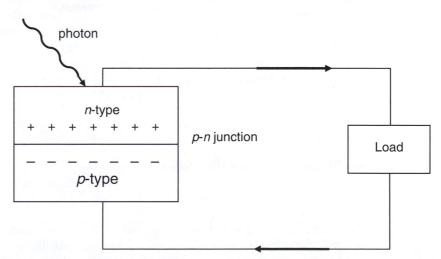

Figure 9.2 Photovoltaic cell schematic.

Example 9.1	Determine the wavelength of light that corresponds to the band gap energy of silicon, 1.11 eV.

Solution Combining Eqs. (9-1) and (9-2) and solving for the wavelength yields

$$\lambda = \frac{hc}{E}$$

$$= 6.625 \cdot 10^{-34} \cdot J\,sec \cdot 3 \cdot 10^8 \cdot \frac{m}{sec} \cdot \frac{1}{1.11\ eV} \cdot \frac{eV}{1.6 \cdot 10^{-19}\ J}$$

$$= 1.1183 \cdot 10^{-6}\ m = 1.12\ \mu m$$

Hence, light with a wavelength of 1.12 μm has sufficient energy to dislodge a valence electron in silicon.

Example 9.1 has important implications for the operation and efficiency of photovoltaic devices. Photons with a wavelength greater than 1.12 μm contain insufficient energy to dislodge a valence electron in silicon; hence, photons with wavelengths greater than 1.12 μm will induce no photovoltaic effect in silicon. Photons with a wavelength less than 1.12 μm possess more energy than is required to dislodge a valence electron in silicon. However, a single photon can dislodge only a single valence electron, and the difference between the band gap energy and the photon energy is absorbed as heat by the PV device. The graphical representation in Figure 9.3 is useful in delineating the behavior of a photovoltaic device exposed to solar irradiation.

The plot in Figure 9.3 was specifically generated for silicon, but the same features would hold for any semiconductor material (or band gap energy) specified. The solid curve represents the power spectral density at 5760 K (the temperature that corresponds to the spectral energy distribution of solar irradiation) normalized by its maximum value. The 1.12 μm wavelength is indicated in the figure. Wavelengths greater than 1.12 μm contain insufficient energy to dislodge a valence electron, while wavelengths less than 1.12 μm result in excess energy. If $E_{1.12}$ represents the energy required to dislodge a silicon valence electron, then Eqs. (9.1) and (9.2) can be converted to ratio form to express that portion of the energy required to dislodge a valence electron for a wavelength lower than 1.12 μm:

$$E_{1.12} = E\frac{\lambda}{\lambda_{1.12}} \tag{9-3}$$

and the excess energy term becomes

$$E - E_{1.12} = E - E\frac{\lambda}{\lambda_{1.12}} = E\left[1 - \frac{\lambda}{\lambda_{1.12}}\right] \tag{9-4}$$

Thus, the smaller the wavelength, the greater the excess energy of the photon. The generally low efficiencies of photovoltaic devices are attributable to this principle and to additional conversion inefficiencies associated with photon wavelengths less than those corresponding to the exact band gap energy.

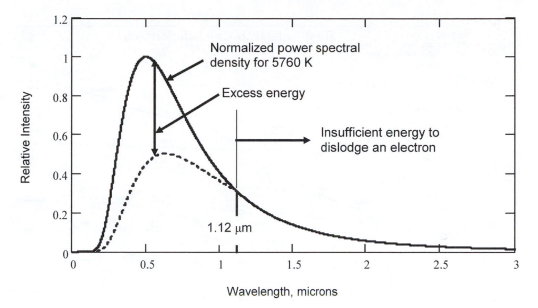

Figure 9.3 Solar irradiation and the band gap of silicon.

Table 9.2 lists theoretical and measured efficiencies for common PV cell materials. As they are the results of active and aggressive research programs, the measured efficiencies of some materials in the table are subject to change. NREL, for example, sponsors the Thin Film Partnership Program, focused on copper indium diselenide, cadmium telluride, and thin-film silicon, which has shown significant improvements in efficiency. The copper indium diselenide value in Table 9.2 was reported in 2007. The website is http://www.nrel.gov/pv/thin_film/.

TABLE 9.2 Theoretical and measured efficiencies of photovoltaic cells

Material	Theoretical (percent)	Measured (percent)
Si, silicon	24	18
CdTe, cadmium telluride	21	7
CdS, cadmium sulfide	16	7
$CuInSe_2$, copper indium diselenide	26	19
GaAs, gallium arsenide	24	11
GaP, gallium phosphide	17	1
InP, indium phosphide	23	3

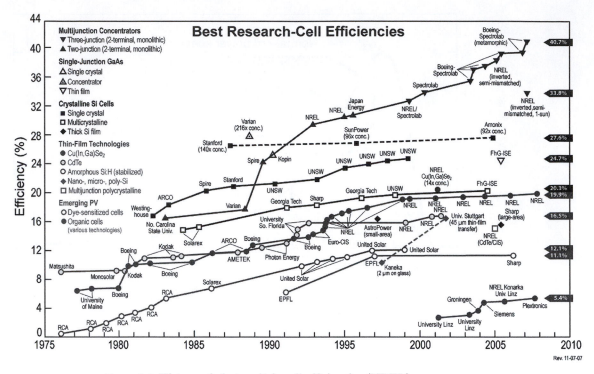

Figure 9.4 History of photovoltaic cell efficiencies (NREL).

A summary of the long-term gains in photovoltaic cell efficiencies was generated by NREL in June 2005 and is reproduced as Figure 9.4. The figure presents general trends in the efficiencies of various photovoltaic cells (and the research entities involved). Except for multijunction devices, the general trend is a slow but steady increase in cell efficiencies with time. One strategy used to overcome the relatively low conversion efficiencies of solar cells is to provide more than a single junction, that is, to fabricate two or more junctions, one above the other. Over the last decade, these multi-junction photovoltaic cells have shown impressive increases in conversion efficiencies.

Although the detailed physics of a PV cell are quite complex, simple models are available. Consider, as depicted in Figure 9.5, the current density flow at the *p-n* junction of a photovoltaic cell. The current density is the current, I, divided by the surface area, A, and is denoted as J.

At the *p-n* junction, a junction current density, J_j, is generated. The junction current density is the algebraic sum of the current density from the *p*-side to the *n*-side (the dark current or reverse saturation current), J_o, and the current density from the *n*-side to the *p*-side (the light-induced recombination current), J_r. In an illuminated

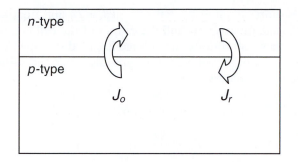

Figure 9.5 Current density at a *p-n* junction.

PV cell, the light-induced recombination current density is proportional to the reverse saturation current and is given by

$$J_r = J_o \cdot \exp\left(\frac{e_o V}{kT}\right) \tag{9-5}$$

where $e_o = 1.6 \times 10^{-19}$ J/V (the charge of one electron) and $k = 1.381 \times 10^{-23}$ J/K (Boltzmann's constant). The junction current density, J_j, is

$$J_j = J_r - J_o$$
$$= J_o\left(\exp\left(\frac{e_o V}{kT}\right) - 1\right) \tag{9-6}$$

The equivalent circuit for a photovoltaic cell is shown in Figure 9.6. In the equivalent circuit, the solar cell current output density, J_s, flows in parallel through either the junction or the load. The load current density, J_L, can be expressed as

$$J_L = J_s - J_j \tag{9-7}$$

Using Eq. (9-6), the load current density can be cast as

$$J_L = J_s - J_o\left[\exp\left(\frac{e_o V}{kT}\right) - 1\right] \tag{9-8}$$

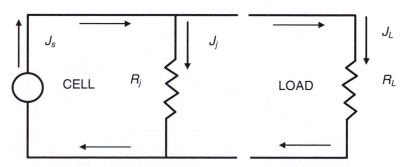

Figure 9.6 Photovoltaic cell equivalent circuit.

If the cell is short-circuited, $R_L = 0$, so that $J_s = J_L$, and Eq. (9-8) requires $V = 0$. If the circuit is open, then $J_L = 0$, and the output of the cell is directed through the junction. Under these conditions, the voltage is called the open circuit voltage, V_{oc}. The performance of a photovoltaic cell is typically described using the short circuit current and the open circuit voltage.

The open circuit voltage occurs when $J_L = 0$. If the voltage for $J_L = 0$ is denoted as V_{oc}, then Eq. (9-8) can be solved for V_{oc}, with the results

$$V_{oc} = \frac{kT}{e_o} \ln\left(\frac{J_s}{J_o} + 1\right) \tag{9-9}$$

Given the open circuit voltage and temperature, Eq. (9-9) can be solved for the ratio J_s/J_o. A typical value of V_{oc} is 0.6 V for a single solar cell. The J_s/J_o ratio is quite large. Dividing Eq. (9-8) by J_s yields an expression for the load current density to short circuit current density ratio:

$$\frac{J_L}{J_s} = 1 - \frac{J_o}{J_s}\left[\exp\left(\frac{e_o V}{kT}\right) - 1\right] \tag{9-10}$$

By using Eq. (9-10) and a specified value of V_{oc}, the current density versus voltage relationship can be determined. The power density delivered to the load is the load current density, J_L, times the voltage drop across the load, or

$$\text{Power} = J_L V A \tag{9-11}$$

where A is the photovoltaic cell area. Figure 9.7 presents, for $V_{oc} = 0.6$ V and $T = 300$ K, the variations of J_L/J_s and the normalized power as a function of the voltage. The power is normalized with the maximum power to the load, which occurs at $V \approx 0.5$ V. The general functional form of both the current density ratio and the

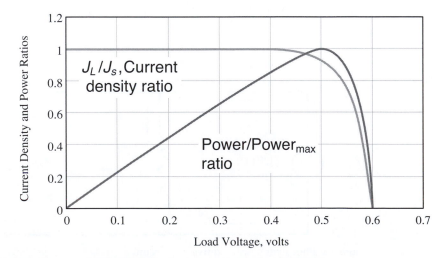

Figure 9.7 Current density ratio and power ratio versus voltage.

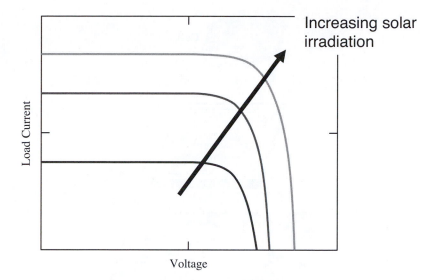

Figure 9.8 Effect of solar irradiation on solar cell performance.

power ratio are similar for most PV cells. The load current density decreases slowly as the voltage increases from zero, but as V_{oc} is approached, the current density rapidly decreases to zero at the open circuit condition. At the short circuit condition, the power delivered to the load is zero, and the power delivered to the load is also zero at the open circuit condition. Between these two extremes, the power reaches a maximum value, usually relatively close to the open circuit voltage.

Not surprisingly, the higher the solar irradiation, the higher the power output of a photovoltaic cell. Consider, as shown in Figure 9.8, the effect of irradiation on the performance of a solar cell. As the solar irradiation increases, the open circuit voltage increases slightly, but the load current is almost directly proportional to the solar irradiation. *The maximum power to the load is also almost directly proportional to the solar irradiation incident on the cell.*

With the fundamental physics of photovoltaic cells discussed, the next section examines actual components and systems.

9.3   PHOTOVOLTAIC COMPONENTS

No matter what kind of photovoltaic system is being considered, the fundamental element is the individual photovoltaic cell. Individual cells are assimilated into modules; modules are assembled into arrays, which are integrated into systems with a wide range of components. Figure 9.9 schematically illustrates the sequence. The web is a good source of information on all photovoltaic system components.

The basic building block is the individual solar cell. Although the web contains significant information on individual photovoltaic cells, much of the information is

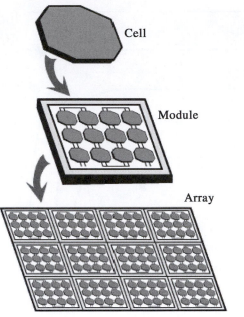

Cell

Module

Array

Figure 9.9 Cells, modules, and arrays (NASA).

qualitative, not quantitative, and does not provide the full specifications of a given photovoltaic cell. The website for Photowatt International, a French company, is one that does contain full technical specifications for the Photowatt PV cells. Figure 9.10 reproduces some of the information from the company site, www.photowatt.com, for Photowatt cell 125. The first portion of the figure reproduces a visual image of the cell as well as presenting information on the size and materials used and some construction details. The table in the middle segment of the figure delineates the open circuit voltage and short circuit current and the voltage, current, power, and efficiency at maximum power for different classes of the solar cell. The Photowatt *I-V* curve is illustrated as the last part of the figure. The *I-V* curve is similar to the theoretical one of Figure 9.7, implying that the simple model is realistic and useful. Most *I-V* plots from manufacturers also indicate the irradiation conditions for the characteristics reported. The general irradiation level used for photovoltaic cell testing is 1000 W/m^2; the Photowatt cell data conform to that level. The open circuit voltage is ~0.6 V, and the short circuit current is ~5 A. An examination of the numbers confirms that if appreciable power and voltage are to be attained, more than a single cell will be needed. Modules are constructed by placing photovoltaic cells in series and parallel arrangements.

Series and parallel configurations of solar cells follow the same rules as series and parallel DC circuits. For identical components placed in a series arrangement, the voltages add at constant current, and for identical components placed in parallel, the currents add at constant voltage. Consider the graphical representation of

High Efficiency Solar Cell Specifications	
Product	Multi-crystalline silicon solar cell
Size	125.5 mm X 125.5mm
Thickness	220 μm +/- 40μm
Front Face (-) Polarity minus.	Parallel straight lines grid pattern with 2 X 2 mm bus bars
Back Face (+) Polarity plus.	Back surface field (Aluminium) with 2 X 4 mm bus bars

Class	Voc	Isc	Voltage Vm @ Pmpp (V)	Current Im Pmpp (A)	Pmpp	Efficiency
Ah	0.608	5.65	0.493	5.13	2.53	16.1
Ag	0.607	5.57	0.491	5.06	2.49	15.8
Af	0.606	5.49	0.489	4.98	2.45	15.5
Ae	0.605	5.41	0.487	4.91	2.40	15.2
Ad	0.604	5.33	0.485	4.84	2.36	15,0
Ac	0.603	5.24	0.483	4.77	2.31	14.7
Ab	0.602	5.16	0.481	4.70	2.27	14.4
Aa	0.601	5.09	0.479	4.63	2.23	14.1

Solar irradiation = 1000 W/m^2

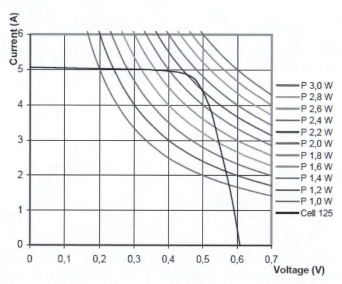

Figure 9.10 Typical photovoltaic cell specifications (Photowatt 125).

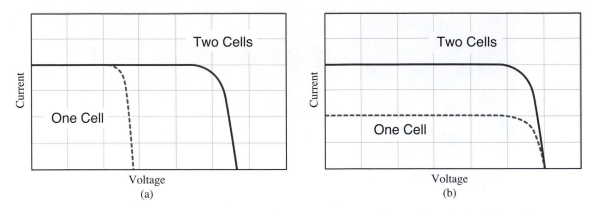

Figure 9.11 Behavior of cells in series and in parallel. (a) Two identical cells in series. (b) Two identical cells in parallel.

placing two identical cells in series and then in parallel in Figure 9.11. For cells in series, Figure 9.11(a), the voltages add at constant values of the current; hence, placing cells in series results in increased voltage. Cells in parallel, Figure 9.11(b), result in increased current. Consider the following example.

| Example 9.2 | Photowatt class Ah photovoltaic cells are to be arranged to provide an output of 12 V and a power of 120 W. Recommend an arrangement that meets the specifications. |

Solution From Figure 9.10, the voltage and current at maximum power are 0.493 V and 5.13 A, respectively. At these conditions each cell provides 2.53 W. The number of cells required for 120 W is

$$\text{Number of cells} = \frac{120 \text{ W}}{2.53 \frac{\text{W}}{\text{cell}}} = 47.4 \text{ cells}$$

To provide the correct voltage, 12 V, the number of cells in series must be

$$\text{Cells in series} = \frac{12 \text{ V}}{0.493 \frac{\text{V}}{\text{cell}}} = 24.3 \text{ cells}$$

Round the number of cells in series to 25. Two rows of 25 cells in parallel will require 50 cells with a total power of 126.5 W. The recommended configuration is sketched in Figure 9.12.

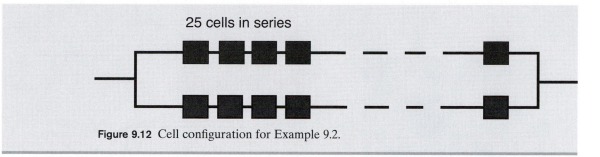

Figure 9.12 Cell configuration for Example 9.2.

Example 9.2 illustrates how individual solar cells can be arranged to meet specified voltage and power requirements. Such an arrangement is called a module, and this is the typical product marketed by photovoltaic manufacturers. Photovoltaic cells are arranged as described to meet certain specifications. However, to ensure proper module operation under realistic conditions, such as failure of an individual cell or shading of part of a module, in-series elements in arrays are usually provided with blocking diodes to prevent reverse current flow and other inefficiencies.

BP Solar, a large international manufacturer of photovoltaic cells and components, provides information on its website, www.bp.com, for a number of PV modules. Information on the BP 3160 photovoltaic module is provided in Figure 9.13. The BP 3160 measures 62.8 inches by 31.1 inches by 1.97 inches deep and weighs 33.1 pounds. The price of the BP 3160 is $719 as of 2008 (www.solarhome.org). An array of BP 3160 modules sufficient to supply 1 kW would cost about $5000. An image of the module appears in Figure 9.13 next to the *I-V* curve. The *I-V* information is parameterized by temperature but conforms to the expected shape. Additional performance details of the module are provided in the table. All of Figure 9.13 was taken from the BP Solar website.

The *I-V* curve and the tabular information are similar to that associated with an individual photovoltaic cell, which is not surprising since series and parallel arrangements of individual solar cells retain the same functional forms. NOTC (at the end of the table) is an abbreviation for normal operating cell temperature. The solar irradiation level corresponding to the performance data for the BP E3160 is 1000 W/m^2. Temperature coefficients are provided for current (α), voltage (β), and power (γ). These permit the effects on module performance of different temperatures to be estimated. The corrections for temperature are as follows:

$$I_{sc} = I_o[1 + \alpha \Delta T] \tag{9-12}$$

$$V_{oc} = V_o[1 + \beta \Delta T] \tag{9-13}$$

$$\text{Power} = \text{Power}_o[1 + \gamma \Delta T] \tag{9-14}$$

where the subscript "o" represents conditions at the reference temperature. An example is in order.

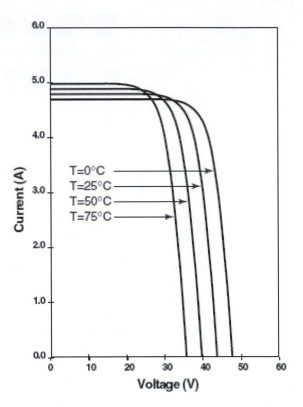

	BP 3160	BP 3150[2]
Maximum power (P_{max})[3]	160W	150W
Voltage at P_{max} (V_{mp})	35.1V	34.5V
Current at P_{max} (I_{mp})	4.55A	4.35A
Warranted minimum P_{max}	150W	140W
Short-circuit current (I_{sc})	4.8A	4.75A
Open-circuit voltage (V_{oc})	44.2V	43.5V
Temperature coefficient of I_{sc}	$(0.065\pm0.015)\%/°C$	
Temperature coefficient of voltage	$-(160\pm20)mV/°C$	
Temperature coefficient of power	$-(0.5\pm0.05)\%/°C$	
NOCT[5]	$47\pm2°C$	

Figure 9.13 Specifications for the BP 3160 photovoltaic module (BP Solar).

Example 9.3	Estimate the annual kWh production of one BP E160 module in Meridian, MS. Examine two cases: (1) constant tilt at the latitude angle and (2) two-axis tracking.

Solution   The solution is accomplished in Mathcad. The Mathcad worksheet is reproduced as Figure 9.14. The days in each month are defined. The *Solar Radiation Data Manual for Flat-Plate and Concentrating Collectors* (1994) provides the necessary

ORIGIN ≡ 1

Define month in terms of days.

The monthly-average conditions from NREL are used to obtain solar and weather data for the location, Meridian, MS, in this case.

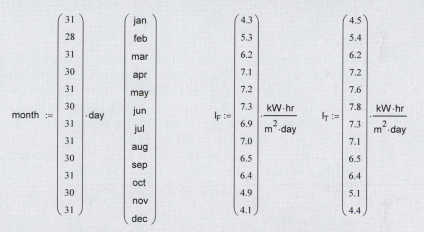

$$\text{month} := \begin{pmatrix} 31 \\ 28 \\ 31 \\ 30 \\ 31 \\ 30 \\ 31 \\ 31 \\ 30 \\ 31 \\ 30 \\ 31 \end{pmatrix} \begin{pmatrix} \text{jan} \\ \text{feb} \\ \text{mar} \\ \text{apr} \\ \text{may} \\ \text{jun} \\ \text{jul} \\ \text{aug} \\ \text{sep} \\ \text{oct} \\ \text{nov} \\ \text{dec} \end{pmatrix} \cdot \text{day} \qquad I_F := \begin{pmatrix} 4.3 \\ 5.3 \\ 6.2 \\ 7.1 \\ 7.2 \\ 7.3 \\ 6.9 \\ 7.0 \\ 6.5 \\ 6.4 \\ 4.9 \\ 4.1 \end{pmatrix} \cdot \frac{\text{kW} \cdot \text{hr}}{\text{m}^2 \cdot \text{day}} \qquad I_T := \begin{pmatrix} 4.5 \\ 5.4 \\ 6.2 \\ 7.2 \\ 7.6 \\ 7.8 \\ 7.3 \\ 7.1 \\ 6.5 \\ 6.4 \\ 5.1 \\ 4.4 \end{pmatrix} \cdot \frac{\text{kW} \cdot \text{hr}}{\text{m}^2 \cdot \text{day}}$$

Nominal power generation is 160 W at an irradiation of 1000 W/m² irradiation.

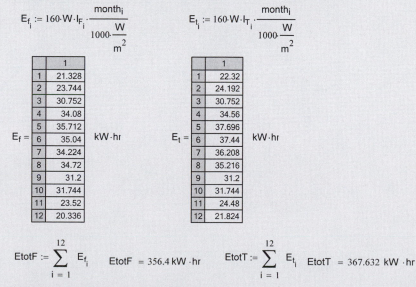

$$E_{f_i} := 160 \cdot W \cdot I_{F_i} \cdot \frac{\text{month}_i}{1000 \cdot \dfrac{W}{\text{m}^2}} \qquad\qquad E_{t_i} := 160 \cdot W \cdot I_{T_i} \cdot \frac{\text{month}_i}{1000 \cdot \dfrac{W}{\text{m}^2}}$$

$$E_f = \begin{array}{|c|c|} \hline & 1 \\ \hline 1 & 21.328 \\ 2 & 23.744 \\ 3 & 30.752 \\ 4 & 34.08 \\ 5 & 35.712 \\ 6 & 35.04 \\ 7 & 34.224 \\ 8 & 34.72 \\ 9 & 31.2 \\ 10 & 31.744 \\ 11 & 23.52 \\ 12 & 20.336 \\ \hline \end{array} \; \text{kW} \cdot \text{hr} \qquad E_t = \begin{array}{|c|c|} \hline & 1 \\ \hline 1 & 22.32 \\ 2 & 24.192 \\ 3 & 30.752 \\ 4 & 34.56 \\ 5 & 37.696 \\ 6 & 37.44 \\ 7 & 36.208 \\ 8 & 35.216 \\ 9 & 31.2 \\ 10 & 31.744 \\ 11 & 24.48 \\ 12 & 21.824 \\ \hline \end{array} \; \text{kW} \cdot \text{hr}$$

$$\text{EtotF} := \sum_{i=1}^{12} E_{f_i} \qquad \text{EtotF} = 356.4 \, \text{kW} \cdot \text{hr} \qquad\qquad \text{EtotT} := \sum_{i=1}^{12} E_{t_i} \qquad \text{EtotT} = 367.632 \, \text{kW} \cdot \text{hr}$$

Figure 9.14 Mathcad worksheet for Example 9.3.

solar data for the performance estimates. The data for Meridian, MS, from the manual are reproduced as Figure 7.12. For part 1 of the problem, the average monthly data for a surface tilted at the latitude angle is used for the solar irradiation incident on the module and is entered in the vector I_F. For part 2, the average monthly data for a surface with two-axis tracking (see Figure 7.3) is used for the solar irradiation incident on the module and is entered in the vector I_T. The irradiation is given in the units of kWh/m^2 day. The surface area of the module is computed from the dimensions (62.8 inches by 31.1 inches) given on the website for the BP 3160 module. The nominal power from the module is 160 W at an irradiation of $1000\ W/m^2$. The electrical generation per day for part 1 is computed as

$$\text{Elec } F_{day} = \frac{160\ W}{1000 \cdot \dfrac{W}{m^2}} \cdot I_F \qquad (9\text{-}15)$$

The above expression asserts that the electrical generation is directly proportional to the irradiation level. This assertion introduces some error into the calculations, but considering the uncertainties in the input values, the error is acceptable. The monthly electricity produced, E_t, is found by multiplying the daily production by the numbers of days in the month. The yearly total, EtotF, is obtained by adding the monthly values. The same procedure is followed for part 2, except that the average monthly data for a surface with two-axis tracking, I_T, is used. The yearly electricity produced by the BP 3160 module tilted at the latitude angle is 356 kWh, while the module with two-axis tracking produces 368 kWh. For the BP 3160 module placed at Meridian, MS, two-axis tracking does not significantly increase electricity production. The monthly generation is plotted in Figure 9.15 for both parts of the problem. Only a 3 percent

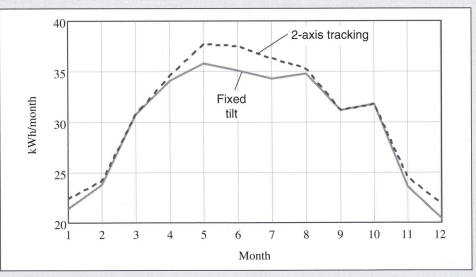

Figure 9.15 Monthly generation (kWh) for Example 9.3.

difference exists between fixed versus tracking setups for the conditions of this problem, which means that two-axis tracking is not economically attractive.

Modules, such as those discussed in Example 9.3, are assembled into arrays utilizing the same series/parallel principles as used for combining cells into modules. Figure 9.16 is an example of a PV array. The array is rated at 236 kW and was made by American Energy Technologies.

Figure 9.16 A 236-kW photovoltaic array (American Energy Technologies).

9.4 PHOTOVOLTAIC SYSTEMS

Photovoltaic systems come in a wide range of sizes and functions. The two principal classifications of photovoltaic systems are stand-alone and grid-connected (or utility interactive). Stand-alone photovoltaic systems are not connected to the electric grid. Functional uses for a hierarchy of increasing complexity of photovoltaic systems are as follows:

1. Direct-coupled
2. Systems with battery storage
3. Systems with backup (generator) power
4. Hybrid power systems
5. Systems connected to the electric grid
6. System for utility power production

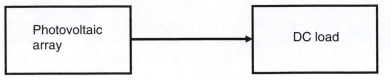

Figure 9.17 Direct-coupled photovoltaic system schematic.

The first four items in the list are generally stand-alone systems. The Florida Solar Energy Center (FSEC) presents schematics and some discussion on a variety of photovoltaic systems. Some of the schematics in this section are adaptations or modifications of the system schematics from the FSEC. The website is http://www.fsec.ucf.edu/en/consumer/solar_electricity/basics/types_of_pv.htm.

A direct-coupled photovoltaic system supplies DC power directly to the load. The schematic of a direct-coupled system is presented in Figure 9.17. Such systems are frequently used for specific, usually low-kW requirements—applications for which grid electricity is not available. With no grid connection, backup generation, or storage capability, stand-alone systems are completely dependent on solar irradiation for power generation and operation. Nonetheless, these are important "niche" applications for photovoltaics.

One of the disadvantages of direct-coupled PV systems is their inability to operate except when exposed to solar irradiation. A photovoltaic system with battery storage eliminates this disadvantage and permits operation during the night hours or when the solar irradiation is insufficient to meet the required electrical demand. Figure 9.18 schematically illustrates a PV system with battery storage. The charge

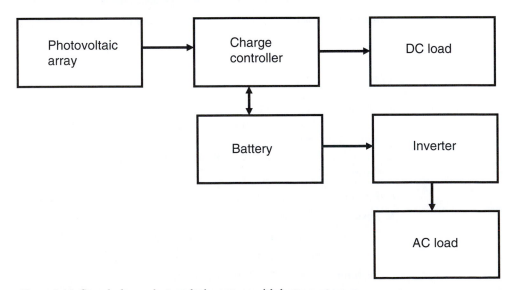

Figure 9.18 Stand-alone photovoltaic system with battery storage.

controller, a key component, directs the output of the PV array to the load and to battery recharging when electricity is being generated; when the array is not generating, it extracts electrical energy from the battery for the load. It also provides the capability to direct battery power to an inverter for AC loads.

Battery characteristics and requirements for photovoltaic system applications are important and merit review. Typical battery metrics include the voltage, the charge capacity, the cycle capability, and the depth of discharge. The first two metrics, voltage and charge capacity, are used to specify the energy storage/retrieval potential of the battery. The voltage across the battery terminals is specified in volts (V), and the charge capacity of a battery is rated in units of amp-hours (Ah). The product of the voltage and charge capacity yields the energy stored in the battery. For example, a 12-V battery with a charge capacity of 50 Ah stores 600 V A h = 600 W h = 0.6 kWh of energy.

Cycle capability refers to the number of charge/discharge cycles expected from the battery. The depth of discharge specifies how much (what percentage) of the stored energy at full capacity may be extracted without damage to the battery. Batteries for photovoltaic applications require high cycle capacity as well as high depth of discharge capability. Automotive batteries require a large amount of power draw over a short period of time (engine cranking time) with a low depth of discharge. Thus, automotive batteries are not suitable for solar storage applications.

An example of a deep cycle set is the 1275AH/12 V product marketed by ABS Alaskan (www.absak.com). The set provides 1275 amp-hours at 12 volts in a six-cell-battery pack; it is manufactured by East Penn Deka (www.eastpenn-deka.com) and sells for $2275 (as of 2008). Table 9.3 shows depth of discharge versus the number of expected cycles for the batteries contained in the set. The storage energy in a battery is relatively expensive, as the set stores 1.275 kWh for the $2275 cost. Indeed, the amortized cost just to store and retrieve 1 kW of electricity was stated at $0.19 kWh by Kreith and West. In many cases, this is more than 1 kWh on a grid would cost. Photovoltaic systems with battery backup require detailed economic analyses in order to confirm economic feasibility.

As Table 9.3 indicates, depth of discharge significantly affects battery life. Most solar photovoltaic systems require long life from batteries.

TABLE 9.3 Depth of discharge versus number of cycles (East Penn Deka)

Depth of Discharge (percent)	Cycles
10	5700
25	2100
50	1000
80	600
100	450

Example 9.4	How many East Penn Deka battery sets would be needed to store and retrieve 10 kWh per day for four days for a solar PV system?

Solution The battery system must store/retrieve

$$\text{energy storage} = 10\frac{\text{kWh}}{\text{day}} \cdot 4 \text{ day} = 40 \text{ kWh} = 40{,}000 \text{ W h}$$

If the depth of charge is specified to be 50 percent, then the battery storage capacity must be 80 kWh. A single battery set stores 1275 Ah $\cdot$ 12 V = 15,300 Ah V = 15.3 kWh. So the number of sets is

$$\text{number} = \frac{80 \text{ kW/set}}{15.3 \text{ kW}} = 5.2 \text{ sets}$$

Six battery pack sets would be required to meet the storage specifications. Five sets would likely be sufficient, but would not guarantee that the storage/retrieval requirement would be satisfied at a 50 percent depth of discharge. An alternative solution would be to purchase three sets and plan on replacing the three on a more frequent basis. The long-term cost of either option will be approximately the same.

PowerPod (www.powerpod.com) markets a range of stand-alone photovoltaic systems with battery backup. The specifications and cost of a PowerPod system suitable for a residence are given in Table 9.4.

A photovoltaic hybrid system results when a stand-alone system with battery storage is provided with a non-grid source of electricity (typically a generator driven by a prime mover). A schematic of a hybrid system is provided in Figure 9.19. The system schematic for a hybrid system is similar to that for a stand-alone system with battery storage, with the addition of a non-grid source of electricity and a rectifier. The rectifier is used to convert AC from an AC generator to DC for battery charging. When the photovoltaic output and/or the battery are insufficient to meet the loads (AC and DC), the "other" source of electricity is then used to make up the difference. Such a system provides great flexibility, but for grid-connected systems

TABLE 9.4 PowerPod system for residence

$8365 to $11,300 (2008)
24 V DC power with AC inverter
4 to 6 64-W PV arrays
2 battery sets with 600 Ah of storage
Charge controller
1500 to 2400 W inverter

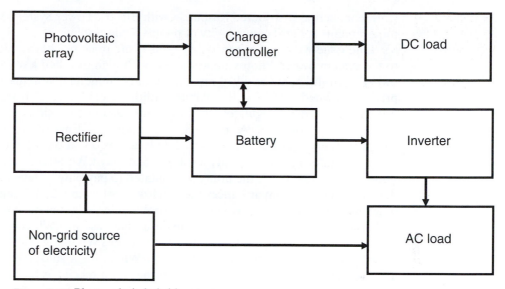

Figure 9.19 Photovoltaic hybrid system.

the utility rate structure can be economically constraining. If large tariffs or penalties are exacted when grid power is used, the cost per kWh of grid power can be excessive.

Photovoltaic systems can be used to supply electricity to the grid. A grid-connected photovoltaic system schematic is illustrated in Figure 9.20. In any photovoltaic system connected to a grid, the power from the solar arrays must be compatible with the voltage, frequency, and phase of the grid. IEEE Standard 1547

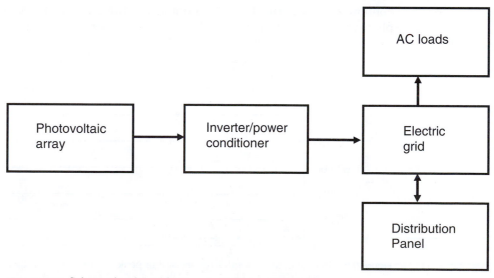

Figure 9.20 Schematic of a grid-connected photovoltaic system.

(Interconnecting Distributed Resources with Electric Power Systems) delineates the requirements for grid connectivity of photovoltaic systems.

One issue with grid-connected systems is the price per kWh paid by the utility to the system owner. Photovoltaic systems with high installed kW capacity may be owned by utilities, but low-kW residential and commercial systems are likely to be privately owned. PURPA, the Public Utilities Regulatory Policies Act of 1978, provides the federal regulatory requirements for most public utilities. Because of the complexity of PURPA, the U.S. Department of Energy provides a website, www.doe.energy.gov/purpa.htm, with details and explanations of many of PURPA's features. Under PURPA a utility must purchase any kWh offered by the owner of a photovoltaic system, but the utility is mandated to pay only its levelized cost of producing 1 kWh. In many instances the levelized cost is only 2 or 3 cents per kWh. A much more advantageous arrangement for the seller is "net metering." In net metering, power drawn from the grid results in an increase in kWh consumption, while power supplied to the grid results in a decrease in kWh usage. The net result is that the utility pays the same for a purchased kWh as for a sold kWh, a much better arrangement for a photovoltaic system owner than receiving the levelized cost for the purchase price of a kWh. Since photovoltaic generating is considered green power, many utilities are willing to participate in a net metering arrangement. The Energy Policy Act of 2005 requires utilities to consider net metering as well as other retail/wholesale, sale/purchase procedures; however, it does not require a utility to implement net metering, only to consider it. A formal agreement with the utility is needed by any owner contemplating selling power to the utility.

9.5 CLOSURE

Photovoltaic systems have become increasingly popular as the cost of energy continues to rise. The Energy Information Administration monitors the sale of photovoltaic systems. Figure 9.21 presents the yearly installed peak kW capacity

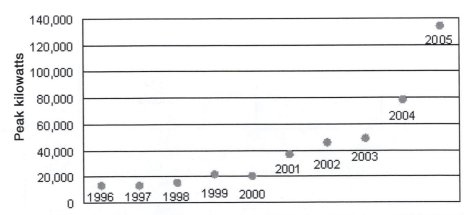

Figure 9.21 Photovoltaic cell sales (expressed in peak kW) from 1996 through 2005 (EIA).

for PV cells from 1996 through 2005. After remaining flat through the last of the decade of the 1990s, photovoltaic system sales dramatically increased with the new millennium. This behavior tracks the relatively constant energy prices (see Figures 1.11 and 1.15, for example) of the 1990s and the escalating energy prices of the last few years.

The Solar Energy Industries Association, SEIA, with the web address www.seia.org, provides an alphabetical listing of manufacturers of solar energy–related items, including photovoltaic cells.

REVIEW QUESTIONS

1. What two factors limit the amount of solar irradiation that can be converted to electrical energy in a photovoltaic cell?

2. Why do solar panels (modules) have many photovoltaic cells in series and in parallel arrangements?

3. For a house not connected to a utility electric grid, what components are required for a photovoltaic system?

4. If the short circuit current is 0.15 A and the open circuit voltage is 0.2 V, sketch the I versus V performance of a photovoltaic cell.

5. What is the band gap energy?

6. What is the conduction band?

7. What is a semiconductor? What is a dopant?

8. What is a p-n junction?

9. Why is $E = h\nu$ important in PV cell considerations?

10. What effect does increasing solar insolation have on the output of a PV cell?

11. How are batteries rated?

12. In the last few years, what has happened to the installed kW capacity of PV systems?

EXERCISES

1. Using a BP 5170 photovoltaic module, how many modules in what arrangement would be necessary to provide 144 V and 2 kW at rated conditions?

2. A company is marketing a photovoltaic cell that it claims has an output of 320 W/m^2 when tested at the standard 1000 W/m^2 irradiation condition. Do you believe this claim? Explain.

3. With a GEPV-030-M photovoltaic module (characteristics in the following table), how many modules in what arrangement would be required to provide 100 V and 900 W at the rated conditions?

Typical Performance Characteristics

		GEPV-050-M	GEPV-030-M
Peak power (W_p)	Watts	50	30
Maximum power voltage (V_{mp})	Volts	17.3	16.8
Maximum power current (I_{mp})	Amps	2.9	1.8
Open circuit voltage (V_{oc})	Volts	22.0	21.4
Short circuit current (I_{sc})	Amps	3.3	2.0
Short circuit temperature coefficient	mA/°C	+2	+1
Open circuit voltage coefficient	V/°C	−0.09	−0.09
Maximum power temperature coefficient	%/°C	−0.5	−0.5
Maximum series fuse	Amps	5	3

Standard test conditions: irradiance $= 1000$ w/m^2, cell temperature $= 77°F/25°C$, solar spectral irradiance per ASTM E892, rated power tolerance ± 10 percent.

4. How much energy could a 24-V, 200 amp-hour battery store?

REFERENCES

Duffie, J. A., and Beckman, W. A. 2006. *Solar Engineering of Thermal Processes*, 3rd ed. New York: Wiley.

Goswami, Y., Kreith, F., and Kreider, J. F. 2000. *Principles of Solar Engineering*, 2nd ed. New York: Taylor and Francis.

Kreith, F., and West, R. E. 1997. *CRC Handbook of Energy Efficiency*. Boca Raton, FL: CRC Press.

Patel, M. R. 2005. *Wind and Solar Power Systems*, 2nd ed. New York: Taylor & Francis.

Fuel Cells

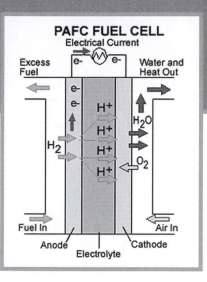

PAFC FUEL CELL

10.1 INTRODUCTION

In a conventional "fired" combustion system, a fuel (usually a hydrocarbon) reacts with an oxidizer (typically the oxygen in the air) to release the chemical energy stored in the fuel as heat. In a fuel cell, the chemical energy stored in the fuel is converted directly to useful work, electricity, without going through the process of combustion. For both combustion systems and fuel cells, the chemical equation describing the reaction is

$$\text{fuel} + \text{oxidizer} \rightarrow \text{products} + \text{chemical energy} \qquad (10\text{-}1)$$

If the thermodynamic states of the fuel, oxidizer, and products are the same for the two cases, then the heat released from the combustion process and the work from the fuel cell's electrochemical reaction must be equal.

A fuel cell is not a heat engine, so the Carnot bound on efficiency does not apply. One of the reasons fuel cells are of interest is that their efficiencies are higher than the efficiencies of heat engines.

10.2 FUEL CELL FUNDAMENTALS

Fuel cells are similar to batteries in that both produce direct current (DC) via an electrochemical process in the absence of direct combustion of a fuel source. However, a battery can deliver power only from stored energy (a finite amount), but a fuel cell can operate as long as a proper fuel source is available. A fuel cell consists of an anode, an electrolyte, and a cathode. Figure 10.1 presents a schematic of a typical hydrogen and oxygen fuel cell with a load. At the anode, a catalytic reaction splits the fuel into ions and electrons, and the ions pass from the anode, through an electrolyte, to the cathode.

The electrons pass through an external circuit to the load. At the anode, the hydrogen dissociates into two hydrogen ions, $2H^+$, and two electrons, $2e^-$. The

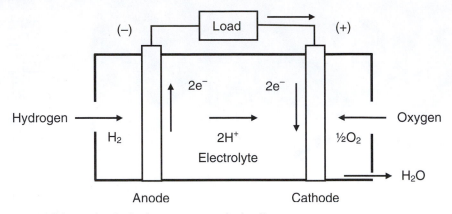

Figure 10.1 Schematic of a hydrogen-oxygen fuel cell.

hydrogen ions pass through the electrolyte to the cathode, where they combine with oxygen and the two electrons that have traversed the load to form water.

The fuel and the electrolyte determine the reactions at the anode and the cathode as well as the voltage produced by a single anode-cathode pair. Catalysts are generally needed at the anode and cathode to increase the reaction rates to acceptable levels. The catalysts required by some fuel cell types are precious metals, which increase the cost per kW for a fuel cell. The next section examines the reactions produced by different electrolytes. Although a number of fuel-oxidant-electrolyte combinations are available, the hydrogen-oxygen fuel cell is the most common.

Individual fuel cells produce between 0.5 and 0.9 V of DC electricity. Fuel cells are combined into "stacks" to produce usable voltage and power output. The same rules apply for combining fuel cells in series and parallel as for combining photovoltaic cells (see Chapter 9). A fuel cell stack is illustrated in Figure 10.2.

The repeating unit is a single fuel cell and contains the anode, electrolyte, cathode, fuel flow channel, and oxidant flow channel. The stack is constructed by assembling the repeating units in series, with the result that the voltages of the units add to each other while the transverse surface area determines the amperage.

A fuel cell system is composed of three major components: a fuel reformer (processor) that generates hydrogen-rich gas from fuel, a power section where the electrochemical process occurs, and a power conditioner (inverter) that converts the direct current (DC) electricity generated in the fuel cell to alternating current (AC) electricity. If a fuel cell is to be connected to a grid, then a power conditioner must synchronize the electrical output with the grid in accordance with IEEE Standard 1547. A simplified block diagram of a typical fuel cell system is illustrated in Figure 10.3.

Natural gas, which is mostly methane, is readily available and is one of the cleanest fuels (next to hydrogen) for powering fuel cells. Much research is being focused on natural gas–powered fuel cells, but the escalating price of natural gas has pushed fuel cell manufacturers and developers to consider other fuels, especially bio-derived fuels. However, fuel cells can also be powered by propane, diesel fuel, fuel oil, and other fuels.

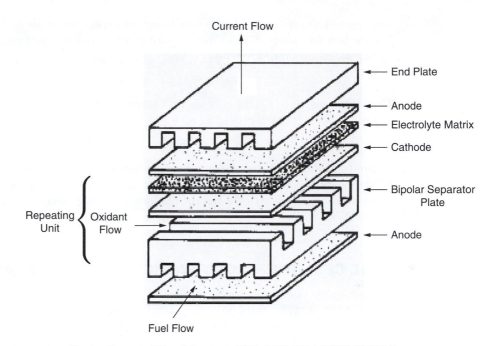

Figure 10.2 Fuel cell stack (Kinoshita et al. 1988, USDOE METC-88/6096).

For methane, the fuel processor utilizes steam reforming followed by a shift reaction. Reforming the fuel with steam concentrates the hydrogen and removes contaminants that could otherwise poison the catalyst in the cell. The chemical reaction for steam reforming is

$$CH_4 + H_2O \rightarrow CO + 3H_2 \tag{10-2}$$

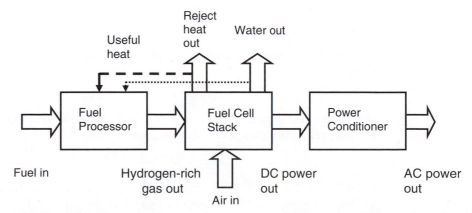

Figure 10.3 Fuel cell block diagram.

The steam reforming reaction is an endothermic reaction with $\Delta H = 206\ \text{kJ/mol}$. The carbon monoxide, CO, in the mixture can be used to produce more hydrogen by the water-gas "shift" reaction,

$$CO + H_2O \rightarrow CO_2 + H_2 \tag{10-3}$$

The water-gas shift reaction is an exothermic reaction with $\Delta H = -41\ \text{kJ/mol}$. Taken together the two processes result in

$$CH_4 + 2H_2O \rightarrow CO_2 + 4H_2 \tag{10-4}$$

The output of the fuel processor is thus carbon dioxide and hydrogen.

Hoogers (2002) contains a chapter on reforming techniques and procedures along with comparisons of different reactions and is a good choice for additional information.

10.3 FUEL CELL THERMODYNAMICS FUNDAMENTALS

The energy released in a chemical reaction is equal to the change in the enthalpy of formation, ΔH. ΔH is the sum of the enthalpies of formation for the products minus the sum of the enthalpies of formation of the reactants, or

$$\Delta H = \sum \Delta H_{\text{products}} - \sum \Delta H_{\text{reactants}} \tag{10-5}$$

ΔH is negative for an exothermic reaction and positive for an endothermic reaction. If all the energy in the chemical reaction were converted to electrical energy, then

$$W_e = \Delta H \tag{10-6}$$

where W_e is the electrical generation per mole of fuel. However, some of the chemical energy is converted to internal thermal energy. The minimum amount of internal thermal energy, if the heat transfer process were reversible, would be equal to

$$Q_{\text{te}} = \int T\,ds \tag{10-7}$$

The fuel cell is essentially an isothermal device and

$$Q_{\text{te}} = \int T\,ds = T\int ds = T\,\Delta S \tag{10-8}$$

The maximum amount of electrical energy produced by an isothermal fuel cell then becomes

$$W_e \le \Delta H - T\,\Delta S \tag{10-9}$$

where the equal sign is for a reversible reaction and the less-than sign is for a fuel cell with irreversibilities. The Gibbs free energy is defined as

$$G = H - TS \tag{10-10}$$

which for an isothermal process can be cast as

$$\Delta G = \Delta H - T\,\Delta S \tag{10-11}$$

Hence, for a fuel cell,

$$W_e \leq \Delta G \tag{10-12}$$

Since ΔH and ΔG are usually expressed per mole, the above equations are on a per mole basis and W_e is the electrical energy of a mole of electrons passing through the circuit. One kg mol of electrons contains 6.022×10^{26} electrons, each with a charge of 1.602×10^{-19} coulombs. The charge contained in a kg mol of electrons is thus 9.65×10^7 C/(kg mol electron), Faraday's constant (F_c).

For a given reaction, ΔG becomes

$$\Delta G = \sum \Delta G_{\text{products}} - \sum \Delta G_{\text{reactants}} \tag{10-13}$$

The general thermodynamic relationship from the First Law can be expressed as

$$V dp = dH - T ds \tag{10-14}$$

so that

$$dG = V dp \tag{10-15}$$

The thermal equation of state, in per mole form, is

$$V = \frac{R_g T}{p} \tag{10-16}$$

with R_g the universal gas constant, 8.314 kJ/(kg mol K) or 1545 (ft lbf)/(lbm mol R). Integrating Eq. (10-15) between a reference state, designated by the superscript o, at 1 atm pressure and a general state at p yields

$$\int_{G^o}^{G} dG = R_g T \int_{p^o}^{p} \frac{dp}{p} = G - G^o = R_g T \ln\left(\frac{p}{p^o}\right) \tag{10-17}$$

If the pressure is expressed in atmospheres, then $p^o = 1$ atm and

$$G = G^o + R_g T \ln p \tag{10-18}$$

The general chemical reaction for a fuel cell can be written as

$$a\,A + b\,B \rightarrow c\,C + d\,D \tag{10-19}$$

If the reactants and products are ideal gases with partial pressures of $p_A, p_B, p_C,$ and p_D, application of Eq. (10-19) to Eq. (10-18) yields

$$\Delta G = \Delta G^o + R_g T \ln\left(\frac{p_A^a p_B^b}{p_C^c p_D^d}\right) \tag{10-20}$$

If n moles of electrons are released by the reaction and the internal cell voltage is E_c, then for a fuel cell with reversible reactions, the electric power becomes

$$W_e = n F_c E_c \approx \Delta G \tag{10-21}$$

Since $E_c = \Delta G / n F_c$, Eq. (10-20) for the reaction of Eq. (10-19) takes the form (with the log term inverted)

$$E_g = E_g^o - \frac{R_g T}{n F_c} \ln\left(\frac{p_C^c p_D^d}{p_A^a p_B^b}\right) \tag{10-22}$$

which is a version (for ideal gases) of the Nernst equation. The Nernst equation provides the reversible, open circuit potential across a fuel cell and is one of the fundamental expressions in fuel cell thermodynamics.

The maximum conversion efficiency is defined as

$$\eta_{max} = \frac{W_e}{\Delta H^o} = \frac{\Delta G}{\Delta H^o} = 1 - \frac{T \Delta S}{\Delta H^o} \tag{10-23}$$

ΔH^o and ΔG^o must be known in order to apply the above equations. Thermodynamic properties are available from a number of sources, including a wide variety of software in thermodynamic and combustion textbooks. The JANAF Tables are widely referenced for properties, and the NIST website, webbook.nist.gov, lists a very inclusive set of thermodynamic properties. Care should be exercised to ensure that properties abstracted from different sources have unit compatibility. For example, results may be expressed in gm mol or kg mol and in J or kJ. A useful list of enthalpies of reaction at 25°C and 1 atm pressure for a number of possible fuel cell reactants and products is found in Culp (1991).

However, the chemical equation for the most common fuel cell reaction, the production of water from hydrogen and oxygen, is

$$H_2 + \tfrac{1}{2}O_2 \rightarrow 2e^- + 2H^+ + O \rightarrow H_2O \tag{10-24}$$

Values of ΔH^o and ΔG^o at 1 atm pressure and selected temperatures are provided in Table 10.1, which was adapted from Angrist (1982).

TABLE 10.1 ΔH^o and ΔG^o for H_2O at 1 atm pressure and selected temperatures

T (K)	ΔH^o (kJ/kg-mol)	ΔG^o (kJ/kg-mol)
298	-2.42×10^5	-2.29×10^5
400	-2.43×10^5	-2.24×10^5
500	-2.44×10^5	-2.19×10^5
1000	-2.48×10^5	-1.93×10^5
2000	-2.52×10^5	-1.35×10^5

Example 10.1

Find the output voltage and theoretical conversion efficiency of an oxygen-hydrogen fuel cell operating at 298 K. Oxygen is supplied from air at 1 atm, and all other products and reactants are at 1 atm.

Solution At 1 atm, $E_g^o = \Delta G^o / n F_c$. For the reaction of Eq. (10-24), 2 kg mol electrons are produced, so that $n = 2$ (kg mol electron)/(kg mol). From Table 10.1, $\Delta G^o = -229{,}000$ kJ/(kg mol). Using Faraday's constant, E_g^o can then be calculated:

$$E_g^o = \frac{\Delta G^o}{n F_c} = 229{,}000 \; \frac{kJ}{kg \; mol} \cdot \frac{kg \; mol}{2 kg \; mol \; electron} \cdot \frac{kg \; mol \; electron}{9.65 \cdot 10^7 C}$$

$$= 1.187 \; V$$

If air at 1 atm is used to supply the oxygen, the partial pressure of oxygen is 0.21 atm since oxygen is 20.95 percent by volume of air.

The reversible, open circuit voltage then becomes

$$E_g = E_g^o - \frac{R_g T}{n F_c} \ln\left(\frac{p_C^c \, p_D^d}{p_A^a \, p_B^b}\right) = E_g^o - \frac{R_g T}{n F_c} \ln\left(\frac{p_{H_2O}^1}{p_{H_2}^1 \, p_{O_2}^{1/2}}\right)$$

$$= 1.187\,V - 8.314\frac{kJ}{kg\,mol\,K} \cdot 298\,K \cdot \frac{kg\,mol}{2\,kg\,mol\,electron} \cdot \frac{kg\,mol\,electron}{9.65 \cdot 10^7 C}$$

$$\cdot \ln\left(\frac{1^1}{1^1 \cdot 0.21^{1/2}}\right)$$

$$= 1.177\,V$$

The efficiency is defined as $\eta_{max} = \Delta G / \Delta H^o$, with ΔG provided by Eq. (10-20). ΔG for the hydrogen-oxygen fuel cell at 298 K is

$$\Delta G = \Delta G^o + R_g T \ln\left(\frac{p_{H_2}^1 \, p_{O_2}^1}{p_{H_2O}^{1/2}}\right)$$

$$= 229,000\frac{kJ}{kg\,mol} + 8.314\frac{kJ}{kg\,mol\,K} \cdot 298\,K \cdot \ln\left(\frac{1^1 \cdot 0.21^{1/2}}{1^1}\right)$$

$$= 227,100\frac{kJ}{kg\,mol}$$

The efficiency can then be computed as

$$\eta_{max} = \frac{\Delta G}{\Delta H^o} = \frac{\Delta G}{LHV} = \frac{227,100\,(kJ/kg\,mol)}{242,000\,kJ/(kg\,mol)} = 0.938$$

The efficiency in the above expression is based on the product being water vapor when ΔH^o is the lower heating value, LHV. If the efficiency is based on the product being liquid water, then the higher heating value, HHV, is needed. The relationship between the two is

$$HHV = LHV + \Delta H_{fg} = 242,000 \ kJ/(kg\,mol) + 44,010 \ kJ/(kg\,mol)$$

where ΔH_{fg} is the heat of vaporization per kg mol of water. The efficiency based on the HHV is

$$\eta = \frac{\Delta G}{\Delta H^o} = \frac{227,100\,kJ/(kg\,mol)}{286,010\,kJ/(kg\,mol)} = 0.794$$

Even though the energy required for reforming the fuel is not considered in the efficiency calculations, this example illustrates why fuel cells are generating so much interest in a time of escalating energy costs. The conversion efficiency of fuel cells is significantly better than heat engine conversion efficiencies since the fuel cell is an isothermal device not constrained by the Carnot limit.

Example 10.2	Determine the effect of cell operating temperature on the fuel cell of Example 10.1. Present and discuss the appropriate results.

Solution   The effect of cell operating temperature on the output voltage and LHV and HHV efficiencies are the characteristics of interest. Example 10.1 accomplished these calculations for an operating temperature of 298 K. The same procedure will be followed in this example, except the operating temperatures will be varied. The results of the sequence of calculations are presented in Table 10.2.

TABLE 10.2 Summary of Results from Example 10.2

T (K)	$\Delta H°$ (kJ/kg mol)	$\Delta G°$ (kJ/kg mol)	E_g^o (V)	E_g (V)	ΔG (kJ/kg mol)	η_{LHV}	η_{HHV}
298	2.42×10^5	2.29×10^5	1.187	1.177	2.271×10^5	0.938	0.794
400	2.43×10^5	2.24×10^5	1.161	1.147	2.214×10^5	0.911	0.771
500	2.44×10^5	2.19×10^5	1.135	1.118	2.158×10^5	0.884	0.749
1000	2.46×10^5	1.93×10^5	1.000	0.966	1.865×10^5	0.758	0.643
2000	2.52×10^5	1.35×10^5	0.669	0.632	1.220×10^5	0.484	0.412

The cell output voltage and the LHV and HHV efficiencies are plotted as a function of temperature in Figure 10.4.

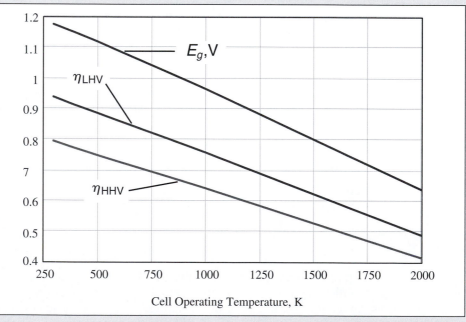

Figure 10.4 Cell voltage and efficiencies as a function of cell temperature.

Although this is a simple thermodynamic model, Table 10.2 and Figure 10.4 capture much of the behavior of a "real" fuel cell. As the cell operating temperature increases, the cell voltage decreases and both of the efficiencies (LHV and HHV) also decrease with cell operating temperature. Indeed, the effect of cell operating temperature on cell performance is quite pronounced.

The analysis presented in this section determines the salient features and limitations on the performance of a fuel cell. Fuel cell performance is limited by transport and chemical kinetics issues. Brouwer (2002) specifies the following 13 areas associated with losses in fuel cells:

1. Reactant transport to gas-electrolyte interface
2. Reactant dissolution in the electrolyte
3. Reactant transport through the electrolyte to the electrode surface
4. Pre-electrochemical chemical reactions
5. Absorption of species onto the electrode
6. Surface migration of absorbed species
7. Electrochemical reaction involving electrically charged species
8. Post-electrochemical reactions
9. Desorption of products
10. Post-electrochemical surface migration
11. Product transport away from the electrode
12. Evolution of products from the electrolyte
13. Transport of gaseous products from the electrolyte

These losses are generally grouped into three major categories:

1. Activation polarization
2. Ohmic polarization
3. Concentration polarization

The losses are irreversible and cause the cell voltage to be reduced from the ideal value. Activation polarization is the result of chemical kinetics—slow chemical reactions. Ohmic polarization losses are due to the resistance in the cell to the flow of electricity. Concentration polarization losses are caused by lower concentrations of reactants at the electrode surface than in the bulk flow.

Figure 10.5 illustrates the effects and regions of dominance of each of the three major categories of losses as a function of current density. The theoretical voltage is indicated, and the various current density regimes for which each of the effects dominate are illustrated. The three add together to produce the total loss. Generally, the overall losses due to activation polarization, ohmic polarization, and concentration polarization are important and cannot be ignored. References such as Brouwer (2002), Larminie and Dicks (2003), Li (2006,), and Hoogers (2002) treat in detail each of the loss categories.

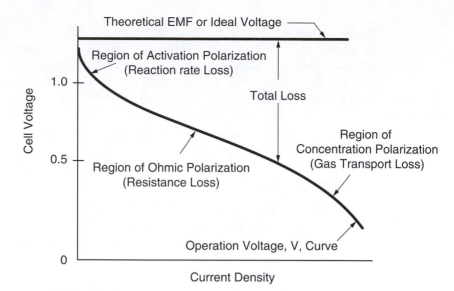

Figure 10.5 Fuel cell losses as a function of current density (Brouwer 2002).

In many thermal devices, economy of scale tends to demand systems with larger and larger capacities in order to attain near-maximum efficiencies. Fuel cells do not benefit economy of scale; indeed, making them larger tends to exacerbate concentration polarization. Hence, fuel cells are likely to remain relatively small in size. Large power requirements for installations will be met by assembling a number of fuel cells.

This section has examined the basic thermodynamics of fuel cells and briefly reviewed sources of losses in fuel cells. The examples were for a generic hydrogen-oxygen fuel cell. In actuality, fuel cell operating characteristics and electrochemistry details depend on the oxidant, the fuel, and the electrolyte. The most common fuel cell types are explored in the next section.

10.4 FUEL CELL TYPES

Most fuel cells are similar in structure to the schematic illustrated in Figure 10.1 but differ with respect to the electrolyte used. The five main types of fuel cells, distinguished by the electrolytes used, are as follows:

1. Alkaline (AFC)
2. Phosphoric acid (PAFC)
3. Molten carbonate (MCFC)
4. Solid oxide (SOFC)
5. Proton exchange membrane (PEMFC)

Each type will be examined in turn.

Alkaline fuel cells (AFCs) were the first type to be perfected and are used extensively in the space program to produce water and electricity on spacecraft. The electrolyte is potassium hydroxide in water. The catalyst at the anode and cathode is a non-precious metal, which is an advantage. Operating temperatures range from 23°C to 250°C (74°F to 482°F) with efficiencies as high as 60 percent. But a disadvantage is that alkaline fuel cells require pure hydrogen, which is difficult and expensive to produce. AFCs are not considered a major contender for most applications, although their niche application in the space program is well documented.

The electrochemical reactions of an alkaline fuel cell are presented in Figure 10.6. At the cathode water, electrons, and oxygen combine to form the hydroxyl ion. At the anode hydrogen and the hydroxyl ion (OH^-) are combined to form water, with the release of electrons (e^-). The hydroxyl ions pass through the electrolyte to the anode, and the electrons pass through the external circuit to the cathode. The reactions are as follows:

$$2\,H_2 + 4\,OH^- \rightarrow 4\,H_2O + 4\,e^- \qquad \text{Anode reaction}$$
$$O_2 + 2\,H_2O + 4\,e^- \rightarrow 4\,OH^- \qquad \text{Cathode reaction}$$

$$(10\text{-}25)$$

Phosphoric acid fuel cells (PAFCs) use liquid phosphoric acid as an electrolyte and are generally considered "first-generation" fuel cell technology. PAFCs use platinum as a catalyst and porous carbon electrodes for both the cathode and anode. The expensive catalyst is part of the reason for the high cost of PAFCs, $4000 to $4500 per installed kW. These fuel cells operate at about 200°C (400°F) and achieve 35 to 45 percent fuel-to-electricity efficiencies. UTC Power, a United Technologies Corporation company, markets the PureCell™ Model 200 fuel cell system and has

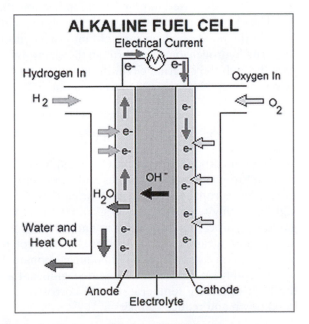

Figure 10.6 Alkaline fuel cell electrochemistry (EERE).

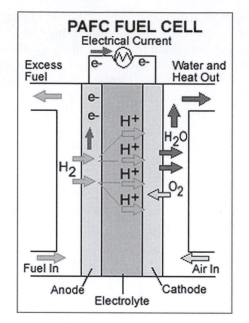

Figure 10.7 PAFC electrochemistry (EERE).

placed more than 275 systems in 19 countries. As of 2007 the PureCell (and its predecessor, the PC-25) had logged more than 8 million hours of operation. The PureCell Model 200 is discussed in more detail in Section 10.5.

The electrochemical reactions of phosphoric acid fuel cells are presented in Figure 10.7. At the anode, hydrogen is split into two hydrogen ions (H^+) and two electrons (e^-). The hydrogen ions pass through the electrolyte to the cathode, and the electrons pass through the external circuit to the cathode. At the cathode, the hydrogen ions, electrons, and oxygen combine to form water. The reactions are as follows:

$$H_2 \rightarrow 2H^+ + 2e^+ \qquad \text{Anode reaction}$$
$$\tfrac{1}{2}O_2 + 2H^+ + 2e^- \rightarrow H_2O \qquad \text{Cathode reaction} \tag{10-26}$$

Molten carbonate fuel cells (MCFCs) have the potential to reach 50 to 60 percent fuel-to-electricity efficiencies and can operate on hydrogen, carbon monoxide, natural gas, propane, landfill gas, marine diesel, and coal gasification products. The high operating temperature, 650°C (1200°F), of MCFCs makes direct operation on gaseous hydrocarbon fuels, such as natural gas, possible. Molten carbonate fuel cells utilize molten carbonate salt mixtures as electrolytes. The electrolyte typically consists of lithium carbonate and potassium carbonate. The high efficiencies and the ability to operate on a variety of fuels are significant advantages for MCFCs. However, the high operating temperature and corrosive electrolyte exacerbate component breakdown and reduce cell life.

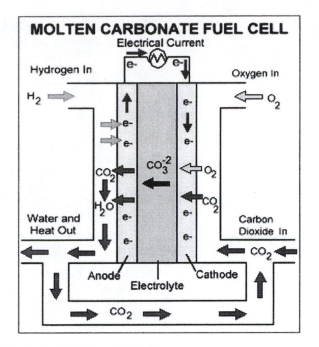

Figure 10.8 MCFC electrochemistry (EERE).

The electrochemical reactions of a molten carbonate fuel cell are presented in Figure 10.8. The MCFC reactions are as follows:

$$H_2 + CO_3^{2-} \rightarrow H_2O + CO_2 + 2e^- \qquad \text{Anode reactions}$$
$$CO + CO_3^{2-} \rightarrow 2CO_2 + 2e^- \qquad\qquad\qquad\qquad (10\text{-}27)$$
$$O_2 + 2CO_2 + 4e^- \rightarrow 2CO_3^{2-} \qquad \text{Cathode reaction}$$

An electrochemical reaction occurs at the anode between the hydrogen fuel and carbonate ions, CO_3^{2-}, from the electrolyte. This reaction produces water and carbon dioxide (CO_2) and releases electrons to the anode. At the cathode, oxygen and CO_2 are combined with electrons from the anode to produce carbonate ions, which enter the electrolyte.

Solid oxide fuel cells (SOFCs) operate at temperatures up to 1000°C (1800°F), thus permitting combined-cycle performance. As with other high-temperature fuel cell types, a fuel processor is not needed and the catalysts used are non-precious metals. Instead of a liquid electrolyte, solid oxide systems typically use a solid ceramic material. The solid ceramic construction provides a stable and reliable design, permits high temperatures, and enables flexibility in fuel choice. The SOFC is especially tolerant of sulfur compounds and carbon monoxide. Unlike other fuel cell types, solid oxide fuel cells can use CO as well as hydrogen as a fuel and are uniquely suited for fuels from coal gasification. SOFCs possess fuel-to-electricity efficiencies of 50 to 60 percent. The high operating temperatures also result in slow startup times and

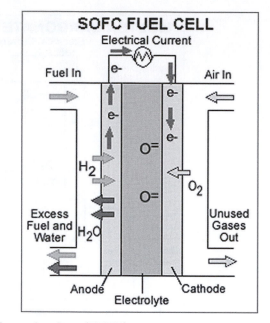

Figure 10.9 SOFC electrochemistry (EERE).

material durability problems. The electrochemical reactions of a solid oxide fuel cell are presented in Figure 10.9. The reactions occurring in a SOFC include

$$H_2 + O^{2-} \rightarrow H_2O + 2e^- \qquad \text{Anode reactions}$$
$$CO + O^{2-} \rightarrow CO_2 + 2e^- \qquad (10\text{-}28)$$
$$CH_4 + 4O^{2-} \rightarrow 2H_2O + CO_2 + 8e^-$$
$$O_2 + 4e^- \rightarrow 2O^{2-} \qquad \text{Cathode reaction}$$

In a SOFC, hydrogen or carbon monoxide, CO, in the fuel stream reacts with oxide ions, O^{2-}, from the electrolyte. These reactions produce water and CO_2 and provide electrons to the anode. The electrons pass through the load. At the cathode, oxygen molecules, O_2, from the air receive electrons and are converted to oxide ions. The oxide ions are returned to the electrolyte.

Polymer exchange membrane fuel cells (PEMFCs), also called proton exchange membrane fuel cells, contain a thin plastic polymer membrane through which hydrogen ions can pass. The membrane is coated on both sides with metal alloy particles (mostly platinum) that act as catalysts. The use of a precious-metal catalyst increases the system cost. Since the electrolyte in a PEMFC is a solid polymer, electrolyte loss is not an issue and does not affect stack life. The use of a solid electrolyte also eliminates the safety concerns and corrosive effects associated with liquid electrolytes. PEMFCs operate at relatively low temperatures (about 200°F). Polymer exchange membrane fuel cells offer high power density and low weight and volume compared with other fuel cell types. The low operating temperature results in fast startup and

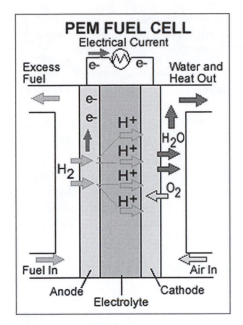

Figure 10.10 PEMFC electrochemistry (EERE).

enhanced component durability. PEMFCs are used for transportation and stationary power applications. The electrochemistry for the PEMFC is shown in Figure 10.10. The reactions that occur in the PEMFC are as follows.

$$H_2 \rightarrow 2H^+ + 2e^- \qquad \text{Anode reaction}$$
$$O_2 + 4H^+ + 4e^- \rightarrow 2H_2O \qquad \text{Cathode reaction}$$

(10-29)

Hydrogen ions and electrons are produced from the fuel at the anode. At the cathode, oxygen combines with electrons from the anode and hydrogen ions from the electrolyte to produce water. The solid electrolyte does not absorb the water; thus, the water is rejected from the back of the cathode into the oxidant gas stream.

All fuel cells have the same function: the electrochemical generation of electricity from a fuel and an oxidant. The five types of fuel cells discussed have different electrolytes, resulting in different operating temperatures, efficiencies, suitable fuels, fuel processor requirements, and availability. Table 10.3 gives an overview of fuel cell characteristics for the AFC, PAFC, SOFC, MCFC, and PEMFC. The information in the table is subject to change, as significant research on most fuel cell types is ongoing.

Virtually all fuel cell types have the advantages of quiet operation, low emissions, and high efficiency (compared with simple cycle thermal systems). The AFC and PAFC are relatively mature, proven technologies. All types of fuel cell are expensive compared on a $/kW basis with heat engines. The SOFC, MCFC, and PEMFC types are just emerging into commercialization (quite limited for some systems), and long-term reliability has not been established for these systems. The next section examines some of the currently available fuel cells.

TABLE 10.3 Overview of fuel cell characteristics
(adapted and modified from www.energy.ca.gov/distgen/)

	AFC	PAFC	SOFC	MCFC	PEMFC
Commercial status	Available	Available	Limited	Some available	Some available
Size range	2–100 kW	100–400 kW	1 kW–10 MW	250 kW–10 MW	3–250 kW
Operating temperature	480°F (250°C)	400°F (200°C)	1800°F (1000°C)	1200°F (650°C)	200°F (90°C)
Fuel	Pure hydrogen	Natural gas, landfill gas, digester gas, propane	Natural gas, hydrogen, landfill gas, fuel oil	Natural gas, hydrogen	Natural gas, hydrogen, propane, diesel
Efficiency	50–60%	40–45%	50–60%	50–60%	40–50%
Reforming		External	External/ internal	External/ internal	External
Catalyst	Platinum/ palladium	Platinum	Not required	Nickel/ nickel oxide	Platinum/ ruthenium
Environmental	Nearly zero emissions	Nearly zero emissions	Nearly zero emissions	Nearly zero emissions	Nearly zero emissions

  FUEL CELL AVAILABILITY

More than 275 PureCell™ Model 200 fuel cell systems from UTC Power, a United Technologies Corporation company, are in operation around the world. By the summer of 2007, the PureCell (and its predecessor, the PC-25) had generated more than 1 billion kWh. The PureCell Model 200 is a PAFC. Features of the PureCell are abstracted from the company literature (www.utcpower.com) in Figure 10.11. As indicated by the figure, the system output is 200 kW/235 kVA at 480 V, three-phase, 60-Hz power. Emissions are essentially negligible. The fuel cell dimensions are 121 inches by 114 inches by 212 inches, and the PureCell Model 200 weighs 40,000 lb. At a distance of 30 ft from the cell, the noise is only 60 dB—verification that the fuel cell is amazingly quiet and can be located inside a structure. Available system options are also listed. The lower portion of Figure 10.11 presents a schematic showing the fuel processor, fuel cell stack, and power conditioner.

The natural gas input is 2.1 MMBtu/h. The electrical efficiency is quoted as 39 percent. Heat recovery from the PureCell Model 200 is 900,000 Btu/h at 140°F. UTC Power is developing a next-generation phosphoric acid fuel cell stack that will deliver a 10-year operating life, 400 kW of power and more than 1.6 million Btu/h of thermal output for combined heat and power applications.

System Specifications

Power	Voltage and Frequency*	Cooling Module	Noise
200 kW/235 kVA	480 volts, 3-phase, 4-wire 60 Hz	Three fan air	60dBA @ 30 ft

Emissions (ppmV, 15% O_2 Dry)	Fuel Type	Fuel Flow	Power Module Dimensions	Cooling Module Dimensions
NOx - <1	Natural Gas	2050 scf/hr (avg.)	H: 121" (307 cm)	H: 50" (127 cm)
CO - <2	ADG	3500 scf/hr (avg.)	W: 114" (290 cm)	W: 49" (124 cm)
SOx, particulates, hydrocarbons - Negligible			L: 212" (538 cm)	L: 162" (411 cm)
			Weight	**Weight**
* Standard configuration			40,000 lb (18,144 kg)	1,700 lb (771 kg)

System Options

High Grade Heat Recovery
- 300,000 Btu/hr at rated power up to 250°F with remaining 450,000 Btu/hr up to 140°F.

Remote Data Acquisition and Control
- Key software application for monitoring, control, and maintenance activity.

400 Volt, 50 Hz or 480 Volt, 60 Hz Operation
- Power plant can be provided to operate at 400 Volt, 50 Hz or 480 Volt, 60 Hz.

Double Wall Heat Recovery Heat Exchanger
- In domestic hot water applications, ensures that potable water source cannot mix with customer's water stream.

Operating Fuel
- Power plant can be provided to operate on natural gas or waste methane from anaerobic digester gas.

Delete Cooling Module
- Cooling tower or other heat sink may be used in lieu of dry cooling module.

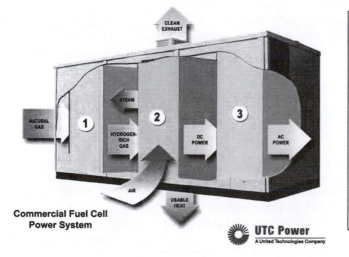

Commercial Fuel Cell Power System

UTC Power
A United Technologies Company

1 Fuel Processor (Reformer)
The Fuel Processor reforms the fuel (city-pressure natural gas or other compliant fuel) to a hydrogen rich gas to feed the fuel cell stack.

2 Fuel Cell Stack
Hydrogen rich gas and air are combined in an electrochemical process that produces Direct Current (DC) power, pure water and heat. The byproduct water is utilized in the operation of the power plant. The waste heat is available through an integral hot water heat exchanger for use in meeting other facility hot water requirements.

3 Power Conditioner
The DC power provided by the Fuel Cell Stack is conditioned to provide high quality Alternating Current (AC) output power.

Figure 10.11 Features of the PureCell Model 200 PAFC fuel cell (UTC).

TABLE 10.4 Companies developing/marketing fuel cells

Company	Website	Fuel Cell Type	Fuel Cell Power
Ballard	www.ballard.com	PEMFC	1.2–85 kW
Hydrogenics	www.hydrogenics.com	PEMFC	4–12 kW
Ida Tech	www.idatech.com	PEMFC	3 kW
Nuvera Fuel Cells	www.nuvera.com	PEMFC	5–125 kW
Plug Power	www.plugpower.com	PEMFC	5 kW
ReliOn	www.relion-inc.com	PEMFC	1.2–2 kW
Horizon Fuel Cells	www.horizonfuelcell.com	PEMFC	300 W
Fuel Cell Energy Corp.	www.fuelcellenergy.com	MCFC	300 kW–2.4 MW
GenCell Corp.	www.gencellcorp.com	MCFC	50 kW
UTC Fuel Cells	www.utcpower.com	PAFC	200 kW
Hydrogen, LLC	www.hydrogenllc.net	PAFC	400 kW
Fuel Cell Tech.	www.fuelcelltechnologies.ca	SOFC	1–50 kW
Rolls-Royce Fuel Cells	www.rolls-royce.com	SOFC	>1 MW
Siemens	www.siemens.com	SOFC	<3 MW
Acumentrics	www.acumentrics.com	SOFC	3 kW

A number of other companies are developing and/or marketing various sizes and types of fuel cells. Table 10.4 provides a list of such companies. Because the industry is quite fluid, other companies may appear and some listed in Table 10.4 may disappear. Fuel cell system specifications are also likely to change. The companies listed in Table 10.4 all have websites, and all the websites contain information on fuel cells. As is usual for the web, much of the information is generally "nontechnical," and detailed technical specifications may not be available except by contacting the company. Nonetheless, the table does provide an indication of what fuel cells are available in the marketplace and at least some indication of the power range and electrolyte types.

10.6 CLOSURE

Alkaline fuel cells have played an important role in the manned space program since its early days. But the relatively high cost per kW of fuel cells, in spite of their high efficiencies and other advantages, has limited their use as an alternative energy system. However, as the cost per kW decreases and as stable, pollution-free electric power becomes of more importance, fuel cells are becoming more attractive from economic and "green" perspectives. They may also play a greater role in providing emergency backup power.

REVIEW QUESTIONS

1. How is a fuel cell similar to a battery?

2. How does a fuel cell differ from a battery?

3. How does a fuel cell differ from an internal combustion engine?

4. Is the output of a fuel cell stack AC or DC?

5. What is the primary chemical reaction for a hydrogen-oxygen fuel cell?

6. What is the steam reforming process? Why is it needed? What are the chemical reactions required to transform methane into a hydrogen-rich fuel?

7. Name four fuel cell types based on the electrolyte.

8. Why are fuel cells not more widely used?

9. What is the power output of a PureCell Model 200 fuel cell?

10. What are the three primary components of a PureCell Model 200 fuel cell?

11. In addition to electricity, what are two other discharges or outputs of a PureCell Model 200 fuel cell?

12. Why are fuel cells composed of stacks?

13. What is the cost per kW of current fuel cells?

14. Are hydrogen and oxygen combusted in a fuel cell?

15. What fuel cell type (electrolyte) is best suited for use in automotive applications?

EXERCISES

1. Find the output voltage and theoretical conversion efficiency of an oxygen-hydrogen fuel cell operating at 298 K. Pure oxygen is supplied at 1 atm, and all other products and reactants are at 1 atm.

2. Determine the effect of cell operating temperature on the fuel cell of Exercise 1. Present and discuss the results.

3. Contrast the results of Exercises 1 and 2 using pure oxygen with the results from Examples 10.1 and 10.2 using oxygen from air.

4. A fuel cell produces 100 kW. If the fuel cell is run 8760 h/yr and the cost of electricity is \$0.035/kWh plus a \$10/kW demand charge per month, how much is the electrical power from the fuel cell worth per year?

5. If the electrical efficiency of the fuel cell in Exercise 4 is 60 percent and the maintenance cost is \$0.005/kWh, what price of natural gas (in \$/10^6 Btu) would make fuel cell electricity cheaper than grid-based electricity?

REFERENCES

Angrist, S. W. 1982. *Direct Energy Conversion*, 4th ed. Boston: Allyn and Bacon.

Appleby, A. J., and Foulkes, F. R. 1989. *Fuel Cell Handbook*. New York: Van Nostrand Reinhold.

Brouwer, J. 2002. "Fuel Cells," in *Distributed Generation: The Power Paradigm for the New Millennium*, edited by A. M. Borbely and J. Kreider. Boca Raton, FL: CRC Press.

Culp, A. W. 1991. *Principles of Energy Conversion*, 2nd ed. New York: McGraw-Hill.

Hoogers, G. 2002. *Fuel Cell Technology Handbook*. Boca Raton, FL: CRC Press.

Kinoshita, K., McLarnon, E. R., and Cairns, E. J. 1988. *Fuel Cells: A Handbook*. U. S. DOE, DOE/METC-88/6096.

Larminie, J., and Dicks, A. 2003. *Fuel Cell Systems Explained*, 2nd ed. New York: Wiley.

Li, X. 2006. *Principles of Fuel Cells*. New York: Taylor and Francis.

Rolle, K. C. 1999. *Thermodynamics and Heat Power*. Upper Saddle River, NJ: Prentice-Hall.

Weston, K. C. 1992. *Energy Conversion*. St. Paul, MN: West Publishing Co.

CHAPTER 11

Combined Heat and Power (CHP) Systems

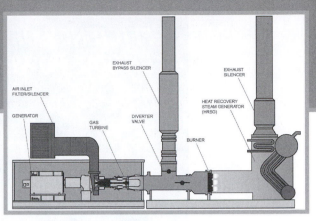

11.1 INTRODUCTION

Combined heat and power (CHP) systems are characterized by the decentralized generation of electricity by a prime mover and the capture and use of rejected thermal energy from the prime mover. The advantages for CHP systems are simple: increased overall system thermal efficiency, enhanced power quality, improved energy resiliency, and decreased emissions of greenhouse and other gases.

The electrical grid is based on large, centrally located power plants. The transmission of power from such plants to the end users is accomplished through an electrical grid that consists of high-voltage transmission systems and low-voltage distribution systems. The high-voltage transmission system carries electricity from the power plants and transmits it to substations, where the high-voltage electricity is transformed into low voltages and distributed to individual customers.

The overall thermal efficiency of generation, transmission, and distribution is relatively low. Figures 11.1 and 11.2 illustrate the losses inherent in the generation and delivery of electric power in both traditional and combined-cycle power plants. Termuehler (2001) presents a lucid history of power plant developments, and Kehlhofer et al. (1999) discuss in detail combined-cycle power plants. Traditional power plants convert about 30 percent of the available energy from the fuel into end-user electric power, and highly efficient, combined-cycle power plants convert about 50 percent of the available energy into end-user electric power. The majority of the energy content of the fuel is lost at the power plant through the discharge or rejection of waste heat. Additional losses are incurred in the transmission and distribution of power to the end user.

Inefficiencies and emission issues associated with conventional power plants provide the impetus for new and revisited paradigms in "on-site and near-site" or "decentralized" generation. New developments in prime movers and power electronic technologies, along with restructuring and deregulation of the electrical utility industry, have enhanced the economics of decentralized electricity generation.

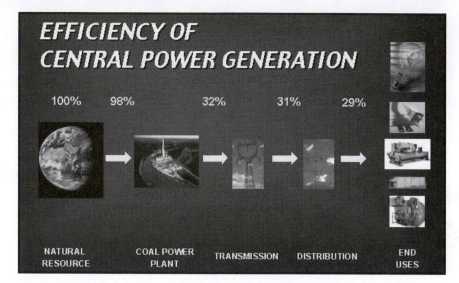

Figure 11.1 Efficiency of central power generation (www.bchp.org).

A technology consisting of power generation equipment coupled with thermally activated and/or thermal energy-consuming components has evolved under the rubric combined heat and power (CHP). *A successful CHP system requires both generated electricity/shaft power and thermal energy.* An operation that does not have the need for both will not benefit from CHP. CHP is especially beneficial for buildings and industrial processes, which typically use electric power and can have thermally activated HVAC system components or require thermal process energy.

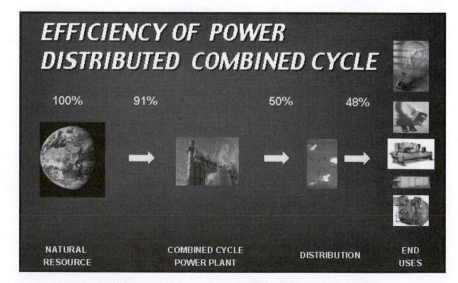

Figure 11.2 Efficiency of central power combined cycle (www.bchp.org).

CHP has become an emphasis for the U.S. Department of Energy (USDOE). The CHP program resides in the Energy Delivery and Energy Reliability (EDER) office of the USDOE. To enhance adoption and application of CHP technologies, the DOE has established eight regional CHP Application Centers. Detailed information, locations, key personnel, and web addresses are available at the following website: www.eere.energy.gov/de/chp/chp_applications/chp_application_centers.html. The Southeast CHP Regional Application Center website includes instructional modules on CHP and micro-CHP (< 200 kW). These modules contain additional information on CHP components, manufacturers of CHP components, and web/e-mail addresses pertinent to CHP topics. The United States Combined Heat and Power Association (www.uschpa.com) also provides useful information, as does the United States Environmental Protection Agency (www.epa.gov/chp/). The EPA's *Catalog of CHP Technologies*, available on the website, is a frequently referenced source of basic information on CHP components and metrics. Books on CHP topics include those by Boyce (2002), Horlock (1997), Kamm (1997), and Kolanowski (2000). Much of the information in the first sections of this chapter is from www.bchp.org; but visitors to that site are now directed to www.chpcentermw.org, the website of the Midwest CHP Applications Center.

The fundamentals of CHP systems are examined in the next section. CHP has the potential to reduce carbon and other emissions and to increase thermal energy efficiency dramatically. CHP produces both electric or shaft power and usable thermal energy on-site or near-site, converting as much as 80 percent of the fuel to usable energy. A higher thermal efficiency means less fuel is needed to meet the energy demands. Also, local power generation reduces grid demand and provides better power quality and resilient (emergency) capability. Figure 11.3 illustrates the increase

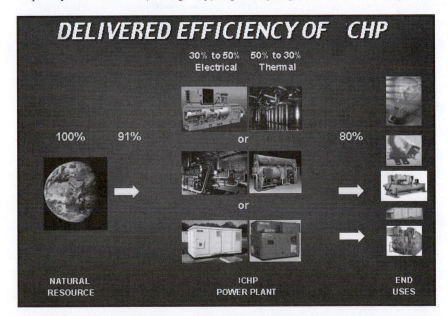

Figure 11.3 Efficiency of CHP systems (www.bchp.org).

in efficiency of CHP systems over the conventional, centralized power plant efficiencies (Figures 11.1 and 11.2).

11.2 CHP SYSTEM FUNDAMENTALS

A CHP system combines distributed power generation with thermally activated components to meet the cooling, heating, and power needs of buildings, or with heat recovery devices (such as heat recovery steam generators, HRSGs) to meet process energy needs. Specific types of distributed power generation and thermally activated technologies will be introduced and briefly discussed.

A number of technologies are commercially available for generating electric power or mechanical shaft power on-site or near the site where the power is used. The three major categories of prime movers for distributed generation (DG) are combustion (gas) turbines, internal combustion engines, and fuel cells. Combustion turbines were examined in Chapter 5, and fuel cells were discussed in Chapter 10. Characteristics of these prime movers for DG technologies are contrasted in Table 11.1.

TABLE 11.1 Comparison of DG technologies
(adapted from www.eren.doe.gov/der/chp/pdfs/chprev.pdf)

	Diesel Engine	Natural Gas Engine	Gas Turbine	Micro-turbine	Fuel Cells
Electric efficiency (LHV)	30–50%	25–45%	25–40% (simple) 40–60% (combined)	20–30%	40–70%
Power output (MW)	0.05–5	0.05–5	3–200	0.025–0.25	0.2–2
Footprint (ft²/kW)	0.22	0.22–0.31	0.02–0.61	0.15–1.5	0.6–4
CHP installed cost ($/kW)	800–1500	800–1500	700–900	500–1300	>3000
O&M cost ($/kW)	0.005–0.010	0.007–0.015	0.002–0.008	0.002–0.01	0.003–0.015
Availability (uptime)	90–95%	92–97%	90–98%	90–98%	>95%
Hours between overhauls	25,000–30,000	24,000–60,000	30,000–50,000	5,000–40,000	10,000–40,000
Startup time	10 s	10 s	10 min–1 h	60 s	3 h–2 days

TABLE 11.1 *(continued)*

	Diesel Engine	Natural Gas Engine	Gas Turbine	Micro-turbine	Fuel Cells
Fuel pressure (psi)	<5	1–45	125–500 (may require compressor)	40–100 (may require compressor)	0.5–45
Fuels	Diesel and residual oil	Natural gas, biogas, propane	Natural gas, biogas, propane, distillate oil	Natural gas, biogas, propane, distillate oil	Hydrogen, natural gas, propane
Noise	Moderate to high (requires building enclosure)	Moderate to high (requires building enclosure)	Moderate (enclosure supplied with unit)	Moderate (enclosure supplied with unit)	Low (no enclosure required)
NO_X emissions (lb/MW-hr)	3–33	2.2–28	0.3–4	0.4–2.2	<0.02
Uses for heat recovery	Hot water, LP steam, district heating	Hot water, LP steam, district heating	Direct heat, hot water, LP-HP steam, district heating	Direct heat, hot water, LP steam	Hot water, LP-HP steam
CHP output (Btu/kWh)	3400	1000–5000	3400–12000	4000–15,000	500–3700
Usable temperature for CHP (°F)	180–900	300–500	500–1100	400–650	140–700

The efficiency of electric power generation for combustion turbine systems operating in a simple-cycle mode (i.e., without heat recovery in the turbine exhaust) ranges from 25 to 40 percent. Combustion turbines produce high-quality thermal energy that can be used to generate steam or hot water for thermal applications, including heating and cooling, or for process energy. Industrial gas turbines represent a well-established technology for power generation. These turbines also represent the "high-capacity" end of power-generating equipment. Industrial turbines can provide from 1 MW to more than 200 MW of electric power. Microturbines, with <200 kW electrical generating capability, represent the "low-capacity" end of power-generating equipment but possess electrical efficiencies of 20–30 percent. The fuel source versatility of turbines allows their application in rural as well as urban areas.

A reciprocating engine, either four-cycle internal combustion or diesel, is frequently used as the prime mover in a CHP system. Internal combustion engines are a mature technology, and applications of reciprocating engines are well established and widespread. Engines can use natural gas, propane, or diesel fuel and are available in capacities ranging from 5 kW to 10 MW and possess electrical efficiencies of 25–50 percent. Reciprocating engines used for power generation have low capital cost, easy startup, proven reliability, good load-following characteristics, and significant heat recovery potential. Exhaust catalysts and better combustion design and control have significantly reduced IC engine emissions over the past few years. Thermal energy is captured from both the engine exhaust gases and the engine cooling system.

Fuel cells produce electric power by electrochemical reactions, generally between hydrogen and oxygen, without the combustion processes. Like a battery, a fuel cell produces direct current (DC). However, as discussed in Chapter 10, fuel cells come in complete packages in which the fuel cell stack is integrated with an inverter to convert the direct current to an alternating current (AC) and, if needed, a reformer to provide the hydrogen-rich fuel. Fuel cells have the highest electrical generating efficiency and lowest emissions of any CHP prime mover, but are the most expensive on an installed-kW basis of any CHP prime mover.

In addition to on-site generation of electricity, an equally important consideration is the capture and use of reject heat. Energy recovery from any of the prime movers in a CHP system is accomplished by heat exchangers. Heat exchangers used to recover reject heat from a CHP prime mover are classified according to the fluid pairs handled and the temperatures of the fluids entering the heat exchangers. The input streams can be gas-to-gas, gas-to-liquid, or liquid-to-liquid and are usually associated with high-, medium-, and low-temperature waste heat applications, respectively.

Gas-to-gas heat exchangers are often used as recuperators for preheating combustion air for internal combustion engines or combustion turbines. A cross-flow heat exchanger with flue gas flowing normal to a tube bundle containing air is called a convection recuperator. Such recuperators are used in low-temperature applications such as space heating, desiccant dehumidification systems, or direct-fired absorption chillers. Other types of heat exchangers used for gas-to-gas waste heat recovery include plate-fin, heat pipes, and rotary generators.

Gas-to-liquid heat exchangers include medium- to high-temperature heat recovery devices such as heat recovery steam generators (HRSGs), fluidized-bed heat exchangers, and heat pipes, as well as low- to medium-temperature heat exchangers used as economizers or fluid heaters. HRSGs, also called waste heat boilers, are used to generate steam from a prime mover, typically a gas turbine exhaust gas. HRSGs are very important in CHP applications and will be examined later. In a fluidized-bed heat exchanger, water or steam is heated by the exhaust gas (hot) stream that flows over a bed of finely divided solid particles. When the hot fluid reaches a critical velocity, the particles in the bed will float and act like a fluid, giving rise to an increase in the heat transfer coefficient. Fluidized-bed heat exchangers are often used for space or water heating, boiler feedwater heating, and process fluid heating. Gas-to-liquid heat pipes are similar to those used in gas-to-gas applications.

Economizers are cross-flow heat exchangers in which water flows through individually finned tubes with hot gases flowing normal to the tubes. Economizers are often used with the boiler flue gases to preheat the boiler feedwater or to heat water or other process liquids or to superheat steam. Fluid heaters are double-pipe heat exchangers that use hot gases to heat a process heat transfer fluid, which is circulated throughout the plant.

Liquid-to-liquid heat recovery exchangers are typically used in industrial applications that include prime movers that use hot oil or other liquid coolants. Shell-and-tube heat exchangers are generally specified for this type of heat recovery.

Heat recovery steam generators (HRSGs) are frequently employed in processes that require steam for thermally activated components or for process energy use and are classified as unfired, partially fired, or fully fired. Unfired HRSGs use only the hot exhaust gases for steam generation. A HRSG that uses a duct burner upstream to increase the exhaust gas temperature is considered partially fired. If the exhaust gas is used as preheated air to the combustion process, the HRSG is said to be fully fired.

Figure 11.4 presents a temperature-area diagram of the temperature states in a HRSG. The hot exhaust gas enters at a high temperature and is cooled as it traverses the HRSG. The energy extracted from the hot gas is used to heat, vaporize (evaporate), and superheat the water. The flow arrangement shown in the figure is counterflow. A HRSG generally includes sections identified as an economizer, an evaporator, and a superheater. The water is heated to saturation conditions by the low-temperature exhaust as it exits the HRSG. In the evaporator, the water is vaporized into saturated steam, and the saturated steam is superheated by the inlet high-temperature exhaust in the superheater. The pinch point, where the water first starts to vaporize, is the smallest temperature difference in the HRSG and is the limiting factor in its overall performance. The HRSG must produce the required pressure-temperature state of

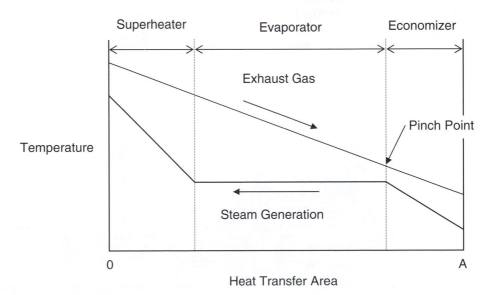

Figure 11.4 HRSG temperature-area diagram.

the steam. To avoid condensation in the HRSG, the exit temperature of the exhaust gas must be above the condensation temperature of vapors in the exhaust.

In the majority of installations, a flapper damper or "diverter" valve is employed to vary the flow across the heat exchanger to maintain a specific design temperature of the hot water or a specific steam generation rate. A schematic of a typical combustion turbine, diverter valve, and HRSG arrangement is presented in Figure 11.5. The turbine-generator is contained in a sound-insulated enclosure to which inlet air is ducted. The diverter valve leads to a stack for venting unneeded exhaust gas. A duct burner is illustrated upstream of the HRSG. The steam from the HRSG is then directed to thermally activated devices or for process energy use.

Thermally activated devices use thermal energy instead of electrical energy for providing heating, cooling, or humidity control for buildings. The two primary thermally activated devices used in CHP systems are absorption chillers and desiccant dehumidifiers.

Absorption chillers use heat as the primary source of energy for driving an absorption refrigeration cycle. These chillers require very little electrical power (0.02 kW/ton) compared to electric chillers, which need 0.47–0.88 kW/ton, depending on the type of electric chiller. Absorption chillers have fewer and smaller moving parts and are quieter during operation than electric chillers. These chillers are also environmentally friendly in that they use non-CFC refrigerants.

Commercially available absorption chillers can utilize one of the following sources of heat: steam, hot water, exhaust gases, or direct combustion. Absorption

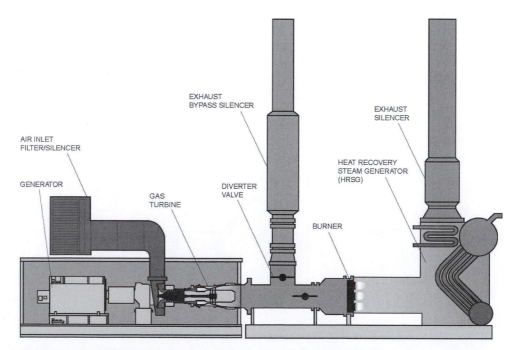

Figure 11.5 Typical CHP HRSG installation (www.solarturbines.com).

chillers, except for those that use direct combustion, are excellent candidates for providing cooling in a CHP system.

Two types of absorption chillers are commercially available: single-effect and multiple-effect. Compared to single-effect chillers, multiple-effect absorption chillers cost more (higher capital cost) but are more energy efficient and are thus less expensive to operate (lower energy cost). The overall economic attractiveness of each type of chiller depends on many factors, including the cost of capital and the cost of energy. There are two separate aspects of space conditioning for comfort cooling:

- Lowering the temperature of the air (sensible cooling)
- Reducing humidity in the air (latent cooling)

The humidity level should remain below 60 percent relative humidity (RH) to prevent growth of mold, bacteria, and other harmful microorganisms in buildings and to prevent adverse health effects. Conventional air-conditioning systems reduce the air temperature below the air dewpoint temperature so that moisture in the incoming air condenses on the outside of a cooling coil over which the air passes. The cooled and dehumidified air is then sent to the space being conditioned. Reducing humidity in the air by cooling often requires lowering the air temperature below a comfortable level and may necessitate reheating of the cooled dehumidified air if comfort is to be maintained.

Desiccant dehumidifiers reduce humidity in the air by using solid or liquid desiccants to attract and hold moisture. Desiccant dehumidifiers operate independently of chillers. In CHP systems, thermal energy from the prime mover is used for regenerating desiccant material in these dehumidifiers. By reducing the moisture content of the air, desiccant dehumidifiers satisfy the latent cooling load and, thus, reduce the load of the chillers to only the sensible cooling (reducing the temperature). Desiccant components in effect permit the sensible and latent cooling functions to be separated and independently controlled.

11.3 CHP SYSTEM ECONOMICS AND OPERATION

CHP systems are attractive for a number of reasons, including economics, thermal efficiency, limited greenhouse gas emissions, power quality, and resiliency. However, for CHP to be viable for a given scenario, the following three attributes must be present:

1. A use for the electricity generated
2. A use for the thermal energy recovered
3. A sufficient "spark spread"

The spark spread refers to the difference, generally expressed in $/million Btu, between the costs of electricity and the prime mover fuel, typically natural gas. Example 11.1 illustrates how the spark spread can be calculated.

Example 11.1

What is the spark spread if electricity costs \$0.12/kWh and natural gas is \$0.90/ccf?

Solution

As indicated in the problem statement, the energy charge for electricity is generally assessed as a cost per kWh, and the cost of natural gas is expressed as the cost per cubic foot. The abbreviation ccf stands for 100 cubic feet. Another common billing unit for natural gas is mcf, for 1000 cubic feet. The energy content of natural gas is nominally 100,000 Btu/ccf or 1,000,000 Btu/mcf. Thus,

$$\text{Cost of natual gas} = \frac{\$0.90}{ccf} \frac{ccf}{100,000 \text{ Btu}} \frac{10}{10} = \frac{\$9.00}{1,000,000 \text{ Btu}} = \$9.00/10^6 \text{ Btu}$$

Recall from Chapter 1 that 1 kWh = 3412 Btu. Then

$$\text{Cost of electricity} = \frac{\$0.12}{kWh} \frac{kWh}{3412 \text{ Btu}} \frac{10^6}{10^6} = \frac{\$35.17}{1,000,000 \text{ Btu}} = \$35.17/10^6 \text{ Btu}$$

So the spark spread becomes

$$\text{Spark spread} = \$35.16 / 10^6 \text{ Btu} - \$9.00/10^6 \text{ Btu} = \$26.16/10^6 \text{ Btu}$$

Hence, electricity is \$26.16 per million Btu more expensive than natural gas. A spark spread of this magnitude is very favorable for CHP if the first two conditions are met.

However, the determination of the economic feasibility of a CHP system is more involved than just calculating the spark spread. A first-order economic analysis of the feasibility of a CHP system requires consideration of the operating strategy of a candidate facility. Consider the five most common operating strategies (indicated as A through E) depicted in Figure 11.6. In the figure, the axes are the thermal energy and the electrical energy. The three dotted lines represent the characteristics of the prime mover. The larger the slope, the more electricity generated for a given amount of thermal energy recovered. Because of variations in the electricity and thermal energy needs, position A, which illustrates a perfect match between the thermal and electrical requirements, is virtually never met. Positions $B–E$, however, represent different but realistic strategies. For position B, the electrical energy requirement is met, but the thermal energy recovered is greater than required. Unless a neighboring facility can purchase the excess thermal energy, it must be rejected to the atmosphere, thus negating the economic benefit of some of the energy harvested. Position C corresponds to meeting the thermal requirement but not meeting the electrical requirement. Electricity would have to be purchased from the grid. Condition C is a common operating strategy for a CHP system. In many instances, connection to the grid provides power redundancy for a facility. For position D, the electrical load is satis-

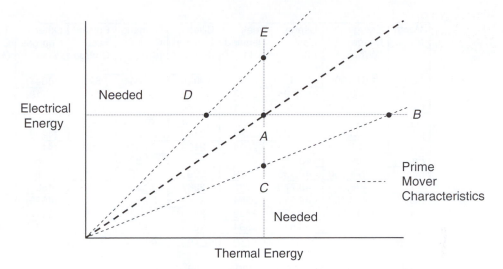

Figure 11.6 Operating strategies for a CHP system.

fied, but the thermal needs are not met, so additional thermal energy must be provided. Condition D can be a useful operating strategy. Position E corresponds to meeting the thermal needs, but generating excess electricity, which must be sold. Unless the excess power can be sold at a price sufficient to pay the cost of generation, option E is not economically attractive and is infrequently specified.

Figure 11.7 presents the utility bills for a year and the utility rate structures for a manufacturing facility. Consider the electric bill first. Commercial and industrial electrical rate schedules usually involve energy and demand charges and can include items such as power factor charges and time-of-day or time-of-year tariffs. The electricity bill illustrated in Figure 11.7 contains energy, demand, and power factor charges as well as industrial credits ("Other Charges"). The electrical energy is expressed in kWh and is billed at some specified rate. The demand is the maximum sustained (usually for a 15- or 30-minute period) power draw from the grid per billing period. Once set, the maximum demand is not reset until the start of the next billing period. The rate schedule shown in the figure is relatively simple; rate schedules in many instances are more complex. Generally, natural gas is billed only as an energy charge. The usual billing units are ccf or mcf (as defined previously).

Plotting the electricity and gas usages on a monthly basis often reveals important trends in energy use and is the initial step in analyzing energy usage patterns. Figures 11.8 and 11.9 contain graphical representations of the energy usages from Figure 11.7.

The trends indicated in Figures 11.8 and 11.9 are typical of manufacturing facilities with large electricity and natural gas usages for process energy. The natural gas usage is essentially constant throughout the year with no dominant use in the winter

Month	Energy Usage (kWh)	Usage Cost ($)	Peak Demand (kW)	Billed Demand (kW)	Demand Cost ($)	Power Factor Charge ($)	Other Charges ($)	TSI* Cost ($)	Total Cost ($)
Aug	1,518,000	50,428	2259	2338	22,344	825	(4641)	1155	70,111
Sept	1,550,400	51,504	2283	2352	22,596	721	(4719)	1173	71,275
Oct	1,331,400	44,229	2244	2354	22,187	1153	(4252)	1064	64,382
Nov	1,331,700	44,239	2130	2235	20,992	1102	(4177)	1046	63,202
Dec	1,400,400	46,521	2085	2130	20,521	466	(4258)	1063	64,314
Jan	1,337,700	44,438	2094	2126	20,615	336	(4122)	1032	62,300
Feb	1,299,900	43,183	2115	2151	20,835	375	(4056)	1017	61,353
Mar	1,274,400	42,336	2079	2091	20,458	121	(3964)	994	59,945
Apr	1,498,200	49,770	2289	2289	22,659	0	(4568)	1137	68,998
May	1,490,400	49,511	2364	2364	23,445	0	(4598)	1145	69,503
June	1,617,300	53,727	2439	2439	24,231	0	(4917)	1220	74,260
July	1,400,100	46,511	2394	2394	23,759	0	(4424)	1105	66,951
Total	17,049,900	566,398	26,775	27,262	264,642	5099	(52.696)	13,151	796,593

*TSI Cost: Tax, service charge, and industrial credit

Electric Rate Schedule:
Usage Rate $ 0.03322/kWh
Demand Rate 9.15/kW; first 1000 kW of demand
10.48 for all additional kW
Service Charge 50.0/month
Tax Rate 1.5 percent

(a)

Month	Gas Usage (ccf)	Usage Cost ($)	Gas Cost ($/ccf)	TSI Cost ($)	Total Cost ($)
Aug	191,467	48,699	0.254	742	49,441
Sept	163,219	36,252	0.222	552	36,804
Oct	177,653	47,262	0.266	720	47,982
Nov	193,294	51,088	0.264	778	51,866
Dec	193,791	53,686	0.277	818	54,503
Jan	181,893	43,480	0.239	662	44,142
Feb	168,669	40,632	0.241	619	41,251
Mar	192,318	43,080	0.224	656	43,736
Apr	187,298	47,046	0.251	716	47,763
May	193,764	60,660	0.313	924	61,584
June	190,402	54,399	0.286	828	55,227
July	180,018	52,418	0.291	798	53,216
Total	2,213,786	578,702	0.261	8,813	587,515

Tax Rate 1.5 percent

(b)

Figure 11.7 Sample utility bills (electricity and gas). (a) Electricity bill and rate structure. (b) Natural gas bill and rates.

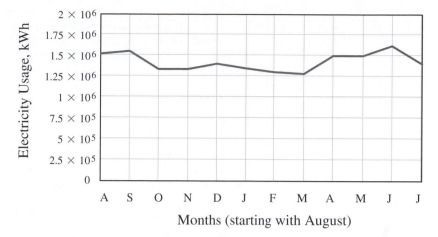

Figure 11.8 Electricity usage from Figure 11.7.

months. This behavior means that only a small percentage of the natural gas is used for space heating; thus, most of the natural gas is used for the manufacturing process. Likewise, the near-constant electricity usage, with about a 20 percent increase in the summer, implies that most of the electricity is also used for the process and not for space conditioning.

Another revealing metric is the usage factor for electricity. The usage factor is the percent of time, for a given billing period, that the peak demand (kW) must be used in order to account for the total energy usage (kWh). The energy usage (kWh)

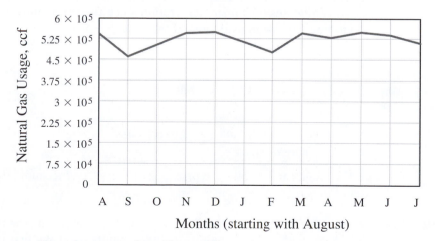

Figure 11.9 Natural gas usage from Figure 11.7.

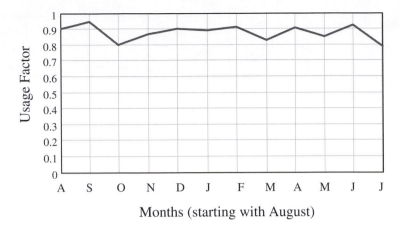

Figure 11.10 Usage factor per month.

divided by the product of the peak demand (kW) and the number of hours in the billing period yields the usage factor:

$$\text{Usage factor} = \frac{\text{kWh usage}}{\text{peak kW} \cdot \text{days in billing period} \cdot 24 \text{ hours / day}} \quad (11\text{-}1)$$

For the usage and demand represented by Figure 11.7, the monthly usage factor is presented in Figure 11.10.

As demonstrated in Figure 11.10, the usage factor varies between 0.8 and 0.95 and indicates that the facility runs most of the equipment most of the time. In a manufacturing scenario, the process must be run three shifts for five to seven days a week to achieve such a high usage factor. Facilities with high electrical usage factors are good candidates for on-site generation since the prime mover/generator would be operating near peak load most of the time.

Example 11.2

Based on the three CHP attributes,

1. A use for the electricity generated
2. A use for the thermal energy recovered
3. A sufficient "spark spread"

is the facility whose bills are provided in Figure 11.7 a candidate for further investigation for a CHP system?

Solution Figures 11.7–11.10 contain information useful in analyzing the three requirements. The kWh and ccf usage by the facility certainly meet the first two conditions. The spark spread is calculated next. The average cost is $0.265/ccf, so the average cost per million Btu becomes

$$\text{Cost of natual gas} = \frac{\$0.265}{\text{ccf}} \frac{\text{ccf}}{100,000 \text{ Btu}} \frac{10}{10} = \frac{\$2.65}{1,000,000 \text{ Btu}} = \$2.65/10^6 \text{ Btu}$$

The average cost per kWh of electricity, including demand, service charges, and rebates, from Figure 11.7 is

$$\text{Cost per kWh} = \frac{\$796,593}{17,049,900} = \$0.047/\text{kWh}$$

Then

$$\text{cost of electricity} = \frac{\$0.047}{\text{kWh}} \frac{\text{kWh}}{3412 \text{ Btu}} \frac{10^6}{10^6} = \frac{\$14.58}{1,000,000 \text{ Btu}} = \$14.58/10^6 \text{ Btu}$$

So the spark spread becomes

$$\text{spark spread} = \$14.58/10^6 \text{ Btu} - \$2.65/10^6 \text{ Btu} = \$11.93/10^6 \text{ Btu}$$

On a per million Btu basis, the cost of electricity is nearly $12 more than the cost of natural gas. This magnitude of spark spread is usually taken as a positive indicator for CHP. Hence, all three of the attributes are satisfied, so that this facility is a candidate for further investigation. This short analysis does not confirm that a CHP system is appropriate, only that further investigation is warranted.

If the "short" analysis of Example 11.2 indicates only that further investigation is appropriate, what is needed for the next level of feasibility study? The next section explores a more involved economic assessment of CHP suitability.

11.4 ECONOMIC ASSESSMENT OF CHP SUITABILITY

The economics and feasibility of a CHP system can be assessed on a first-order basis by considering the yearly utility bills. This technique uses the energy consumption information in conjunction with industry-accepted efficiencies for both conventional and CHP components to estimate the savings, if any, from meeting the energy needs using CHP. Acceptable values of component efficiencies and electrical and thermal energy production are assigned as follows:

Conventional:
  Boiler $\eta_{\text{boiler}} = 0.85$
  Central power plant $\eta_{\text{cpp}} = 0.35$
  Central chiller $\text{COP}_{\text{cc}} = 2.5$

CHP:

Electricity (gas turbine)	$\eta_e = 0.30$
Recovered thermal energy	$\eta_r = 0.35-0.50$
Absorption chiller	$COP_{abc} = 0.7-0.8$

The first-order assessment procedure is best described by means of a detailed example.

Example 11.3

Perform a first-order assessment for CHP suitability using the utility bills presented in Figure 11.7.

Solution

The first step in the procedure is to establish the operating mode (conditions A–E of Figure 11.6) congruent with the energy usages. Insight into the appropriate operating mode can be gained by computing the ratio of the electrical energy (expressed in Btu) to the gas energy:

$$\text{Electrical energy} = \text{kWh uasge} \cdot 3412 \frac{\text{Btu}}{\text{kWh}}$$

$$= 17,049,900 \cdot \text{kWh} \cdot 3412 \frac{\text{Btu}}{\text{kWh}}$$

$$= 58,170 \cdot 10^6 \text{ Btu}$$

and

$$\text{Natural gas energy} = \text{ccf usage} \cdot 100,000 \frac{\text{Btu}}{\text{ccf}}$$

$$= 2,213,786 \cdot \text{ccf} \cdot 100,000 \frac{\text{Btu}}{\text{ccf}}$$

$$= 221,400 \cdot 10^6 \text{ Btu}$$

The ratio of electrical energy to gas energy is then

$$\text{Energy ratio} = \frac{\text{electrical energy}}{\text{natural gas energy}} = \frac{58,170 \cdot 10^6 \text{ Btu}}{221,400 \cdot 10^6 \text{ Btu}} = 0.263$$

Since the electrical energy required is much less than the gas energy required, a logical choice of the operating mode is case D of Figure 11.6. Mode D provides all the electricity required, but it does not meet the thermal requirements. The natural gas required for a gas turbine to generate the required kWh value can be calculated by dividing the electrical energy required by the gas turbine electrical generation efficiency, η_e, or

$$\text{Natural gas required} = \frac{\text{electrical energy required}}{\eta_e}$$

$$= \frac{58,170 \cdot 10^6 \text{ Btu}}{0.30} \frac{\text{ccf}}{100,000 \text{ Btu}} = 1.939 \cdot 10^6 \text{ ccf}$$

Thus, a gas turbine operating at an electrical efficiency of 30 percent will require $1.939 \cdot 10^6$ ccf to produce the required 17,049,900 kWh. The thermal energy recovered from the turbine is the natural gas required times the recovery efficiency, η_r:

$$\text{Thermal energy recovered} = \text{natural gas required} \cdot \eta_r$$

$$= 1.939 \cdot 10^6 \text{ ccf} \cdot 0.34 = 660,000 \text{ ccf}$$

However, to obtain the input energy content of the thermal energy recovered from the boiler, the boiler efficiency, η_{boiler}, is needed:

$$\text{Thermal energy equivalent} = \frac{\text{thermal energy recovered}}{\eta_{\text{boiler}}}$$

$$= \frac{660,000 \text{ ccf}}{0.85} = 776,000 \text{ ccf}$$

The recovered thermal energy is equivalent to 776,000 ccf of natural gas. Since this amount is considerably less than the required thermal energy of 2,213,786 ccf, the thermal energy recovered does not exceed the thermal energy required. Operating mode D is confirmed as appropriate for the utility usage of Figure 11.7. The total natural gas usage required would be the natural gas currently purchased to meet the thermal needs, plus the additional natural gas required for the CHP system, minus the thermal energy equivalent recovered from the turbine exhaust. For this example, the natural gas required is thus

$$\text{Natural gas required} = \text{currently purchased} + \text{CHP required} - \text{equivalent recovered}$$

$$= 2,213,786 \text{ ccf} + 1,939,000 \text{ ccf} - 776,000 \text{ ccf}$$

$$= 3.377 \cdot 10^6 \text{ ccf}$$

At an average cost of \$0.265/ccf and a tax rate of 1.5 percent, the cost of natural gas to run the facility with a CHP system in mode D becomes

$$\text{NG cost} = 3.377 \cdot 10^6 \text{ ccf} \cdot \frac{\$0.265}{\text{ccf}} \cdot 1.015 = \$908,400$$

From Table 11.1 a reasonable operation and maintenance (O&M) charge for a gas turbine is 0.5 cents/kWh. The O&M cost is thus estimated as

$$\text{OM cost} = 17,049,900 \text{ kWh} \cdot \frac{\$0.005}{\text{kWh}} = \$85,250$$

The total cost for operating the CHP system is the sum of the natural gas cost and the O&M cost, or

$$\text{Total cost} = \$908{,}000 + \$85{,}250 = \$993{,}250$$

The cost of operating the current system, electricity and natural gas, is $1,384,000. The savings for switching to a CHP system are thus

$$\text{CHP savings} = \$1.384 \cdot 10^6 - \$0.993 \cdot 10^6 = \$390{,}000$$

The operation of a CHP system with a gas turbine would save $390,000 per year over the existing, conventional system. For operating mode D, all of the electricity requirements are met, and the facility could be disconnected from the grid. However, with no grid reliance, the facility would be without power during CHP system inspections or downtimes unless backup sources were provided. In many instances, the choice would be made to maintain a grid connection to ensure continued operation if the CHP system goes down. The cost analysis was done under the assumption of no grid connection charges.

If the turbine is to take the full load of the facility, a turbine with an electrical output of at least 2439 kW (the maximum demand from Figure 11.7) is needed. A nominal 2.5-MW turbine would meet the existing needs but would provide little capability for expansion. A ballpark cost per kW output for a turbine is $800, from Table 11.1. A 2.5-MW gas turbine is estimated to cost $2,000,000. The simple payback period, based on turbine cost only (no heat exchanger cost), becomes

$$\text{Simple payback} = \frac{\$2{,}000{,}000}{\$390{,}000/\text{year}} = 5.1 \text{ years}$$

This is as far as this analysis can go without additional information on the facility, including management policies. In addition to economics, other factors that would enter into a decision to convert to a CHP system include, but are not limited to, power quality, CO_2 credits, company "green" policies, and energy security concerns.

The preceding example illustrates a first-order approach to assessing the economic feasibility of a CHP system. For systems operating in mode C, a similar procedure would be used, except that electricity would have to be purchased from the grid. The cost of that power would be determined by the electrical rate schedule.

Example 11.3 contained a detailed explanation of the procedure used in the first-order assessment procedure, and the results were obtained "by hand." A more flexible approach is to use a computational system such as Mathcad to do the calculations. Example 11.4 is structured about the Mathcad approach to CHP economic assessment.

<table>
<tr><td>**Example 11.4**</td><td>Determine the cost savings and simple payback period for the facility discussed in Example 11.3 if 50 percent of the exhaust energy is recovered. This represents essentially the maximum that could reasonably be expected to be recovered. Use Mathcad.</td></tr>
</table>

Solution   The same procedure would be followed as in Example 11.3 except that the recovered energy, η_r, would be 0.50 instead of 0.34. The Mathcad worksheet is provided in Figure 11.11. The variable names in the worksheet are similar to the names in Example 11.3 except NG is used in place of "natural gas" and EE is the electrical energy. If more thermal energy is recovered, the savings increase to $488,600 and the simple payback period falls to 4.1 years.

$$kWh := kW \cdot hr \qquad ccf := 100 \cdot ft^3$$

$$\eta_e := 0.30 \qquad \eta_r := 0.50 \qquad \eta_{boiler} := 0.85$$

$$EEy := 17049900 kWh \qquad Demandy := 26775 kW \qquad NGy := 2213786 ccf$$

$$UFy := \frac{EEy \cdot 12}{Demandy \cdot 8760 \, hr} \qquad UFy = 0.872$$

At a generation efficiency of 30 percent, the natural gas required to generate the electricity is:

$$NGelecy := \frac{EEy \cdot ccf}{\eta_e \cdot 100000 BTU} \qquad NGelecy = 1.939 \times 10^6 \, ccf \qquad NGccfy := \frac{0.265}{ccf}$$

$$Erecy := NGelecy \cdot \frac{\eta_r}{\eta_{boiler}} \qquad Erecy = 1.141 \times 10^6 \, ccf$$

This is mode D of the narrative for which the electricity is supplied. The natural gas required is the old plus the new (electricity generation) minus the recovered.

$$NGreqy := NGy + NGelecy - Erecy \qquad NGreqy = 3.012 \times 10^6 \, ccf$$

$$NGcosty := NGreqy \cdot NGccfy \cdot 1.015 \qquad NGcosty = 8.102 \times 10^5$$

$$OMcosty := EEy \cdot \frac{0.005}{kWh} \qquad OMcosty = 8.525 \times 10^4$$

$$CHPcosty := NGcosty + OMcosty \qquad CHPcosty = 8.955 \times 10^5$$

$$ConvCost := 796593 + 587515 \qquad ConvCost = 1.384 \times 10^6$$

$$CHPsavingy := ConvCost - CHPcosty \qquad CHPsavingy = 4.886 \times 10^5$$

$$TurbineCost := \frac{800}{kW} \cdot 2500 kW \qquad TurbineCost = 2 \times 10^6$$

$$Paybacky := \frac{TurbineCost}{CHPsavingy} \qquad Paybacky = 4.093$$

Figure 11.11 Mathcad solution for Example 11.4.

11.5 CHP SYSTEM EXAMPLE

The Mississippi Baptist Medical Center (MBMC) in Jackson, MS, provides a good example of a CHP system with a long history of economic success and reliability. Figure 11.12 is a photograph of the MBMC campus. The MBMC is a 624-bed, full-service hospital with a medical staff of 500 and 3,000 employees. In 1990, in an effort to reduce utility costs, the hospital investigated the possibility of installing a CHP system, at the time called a co-generation system. The energy profile in 1990 was as follows:

1. High electricity requirement
2. High steam requirement
3. Significant price differential per Btu (spark spread) between electricity and gas
4. Centralized physical plant
5. Low daily variations in energy requirements

Thus, the MBMC possessed all the attributes necessary to make CHP economically feasible. The results of a detailed engineering study resulted in the following specifications and cost estimates for a CHP system:

1. Projected savings of $800,000/year
2. Initial system cost of $4.2 million
3. Simple payback period of 6.3 years
4. Provide >70 percent of electricity requirement

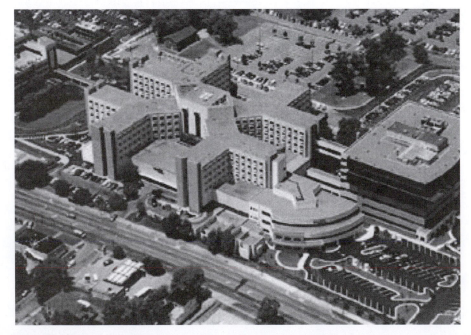

Figure 11.12 MBMC campus.

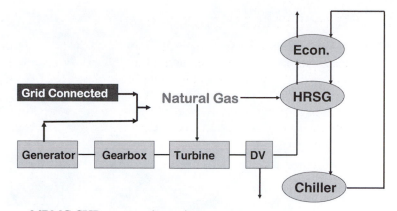

Figure 11.13 MBMC CHP system schematic.

5. Provide 95 percent of steam required

6. Provide 75 percent of cooling load (via absorption chillers)

A system schematic is presented in Figure 11.13 and is similar to the layout discussed in conjunction with Figure 11.5. Since the system was planned to supply about 70 percent of the electricity, the facility is normally grid connected. The turbine exhaust is directed to a HRSG that provides steam for two absorption chillers and sterilization.

A diverter valve (DV), controlled by the HRSG, directs the exhaust gas to the HRSG or out of the bypass stack to maintain the required steam pressure. The HRSG contains a duct burner that can be operated in two modes: (1) with a heat addition up to 5.8 MMBtu to supplement the turbine exhaust or (2) with direct fire up to 41.5 MMBtu when the turbine is off-line. The economizer utilizes the remaining waste heat to preheat boiler feed water and to add water treatment chemicals to the feed water.

Table 11.2 contains descriptive details of MBMC CHP system components. Figure 11.14 is a cutaway illustration of a Solar Turbines Centaur H combustion turbine.

TABLE 11.2 MBMC CHP system component information

Component	Description	Performance
Turbine	Solar Centaur H	ISO 5,600 hp/4.3 MW/13,800 V Natural gas fired
Controller	Allen Bradley PLC 5/20 microprocessor	
Diverter valve		
HRSG	ABCO	30,000 lb/h steam at 125 psi
Absorption chillers	York Paraflow double effect Trane double effect	1250 tons 750 tons
Switch gear	Powell Metal-clad 4-bay	Synchronizing controls 2 sec to secondary grid

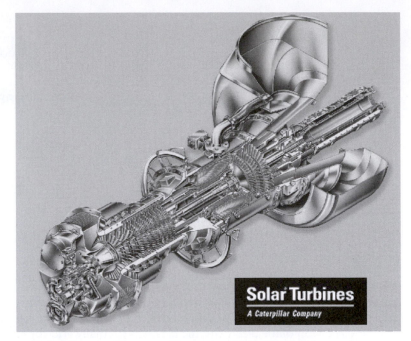

Figure 11.14 Solar Centaur H cutaway illustration (Solar Turbines).

The system went online in 1992 and has been in use since then. Table 11.3 provides a summary of savings and usage for several years of CHP operation.

The results in Table 11.3 indicate the expected savings during the initial years of operation. The system paid for itself in the projected 6 years. In subsequent years, the savings from the system were substantial. However, one unexpected benefit was the experience of MBMC during Hurricane Katrina. After landfall, Katrina's eye passed close to Jackson and severely damaged the city's electrical grid. The MBMC's experience during Katrina is illustrated in Figure 11.15.

TABLE 11.3 Yearly CHP operation summary for several years

Year	Electricity Savings	Natural Gas Cost	Maintenance Cost	Savings
1994	$1,250,000	$402,000	$159,000	$686,000
1995	1,240,000	432,000	159,000	648,000
1996	1,400,000	468,000	163,000	770,000
Average cost avoidance				$701,000
2001	1,415,770	708,549	174,000	533,225
2002	1,587,074	776,950	174,000	636,123
2003	1,823,494	743,852	174,000	905,643
2004	2,022,926	972,401	174,000	876,524
Average cost avoidance				$737,879

About one hour after Katrina impacted the area, the main power grid (MPG) failed. The switch gear shifted to the alternative electrical feed, and city water pressure was lost. The Jackson Fire Department placed a pumper truck near the MBMC CHP system and provided water for the duration of the emergency. The MPG was restored but proved to be unstable. Five hours into the emergency a decision was made by the MBMC to shed sufficient load to enable it to disconnect totally from the grid and to use the CHP system as the only source of electricity. The hospital elevators were switched to the backup diesel generator, and MRI use was discontinued. MBMC operated for 52 hours on the CHP system and was literally an island of light and comfort during the emergency. Approximately 57 hours after Hurricane Katrina hit, the MPG was restored and normal operation was initiated. During this unforeseen event, the CHP system demonstrated its "islanding" capability and its power resiliency.

Because of supply problems in the Gulf of Mexico due to Katrina, the price of natural gas spiked to about $15/10^6$ Btu in the months after Katrina. At this rate,

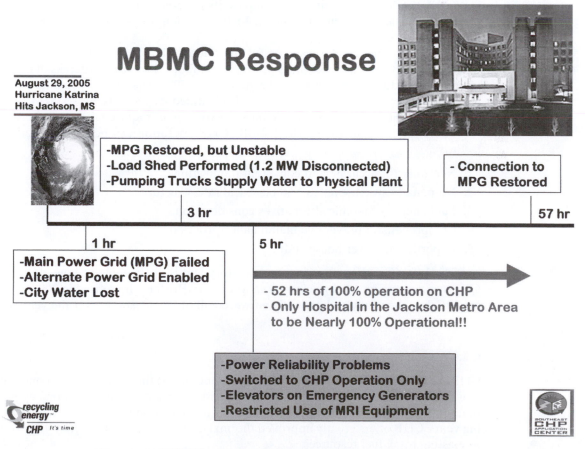

Figure 11.15 MBMC's response to Hurricane Katrina.

TABLE 11.4 Maintenance details and costs

Turbine gearbox overhaul	3.5 years
Oil change	As needed (years between)
Filter replacement	3 months
Routine inspections	8 weeks
Intermediate inspections	6 months
Annual inspection	
200 off-line hours/year for inspections	
Monthly maintenance contract	$14,500

electrical power purchased from the grid was cheaper than using the CHP system, so the MBMC used economics to determine if the grid power or CHP-generated power should be used during periods of excessive natural gas costs. Once natural gas prices decreased from the post-Katrina high, normal CHP operation was resumed. However, the MBMC continues to exercise its economic option when natural gas prices exceed an economic maximum. In this fashion, the CHP system provides fuel/energy diversity.

Two keys to the long-term successful operation of the MBMC CHP system have been attention to maintenance and staff expertise. Table 11.4 provides details of maintenance procedures and cost as of the summer of 2007. The monthly maintenance fee includes replacement components and scheduled inspections. Personnel under contract with Solar Turbines provide on-site expertise as required.

The lessons learned from the MBMC CHP operation can be summarized as follows:

1. Necessity for accurate and consistent monitoring
2. No-penalty switch-over to grid electrical rate structure
3. Fuel/energy economic alternatives consideration
4. Comprehensive preventive maintenance program
5. Expertise of power house staff

The MBMC CHP system is a success story and is an example of how effective CHP systems can be. (*Acknowledgment:* Mr. Harry Brister, who retired in 2007 from MBMC, has cheerfully and patiently answered all of my questions over the years.)

11.6 CLOSURE

CHP is a new version of an old paradigm (cogeneration) that offers many economic as well as operational advantages. The web provides considerable information beyond that presented here. Indications are that CHP will increase in importance in the coming years. CHP offers greatly improved thermal efficiencies and is an alternative use of existing fossil fuel resources.

REVIEW QUESTIONS

1. Describe a CHP system.

2. What is an advantage of a CHP system over conventional central power plants?

3. Contrast the prime movers for distributed generation systems.

4. In addition to electricity, what is the other important output for a CHP system?

5. What is the pinch point?

6. What is a HRSG?

7. What is a desiccant?

8. Why is an absorption chiller a natural component for a CHP system?

9. What is demand? What are the units of demand?

10. What are the three general requirements needed for a CHP system to be economical?

11. Why is mode *E* of Figure 11.6 not a good operating strategy for a CHP system?

12. What is the spark spread? Why is it important to CHP system economics?

13. What is the electrical usage factor?

EXERCISES

1. The exhaust of a 1-MW combustion turbine operating with a 940° C turbine inlet temperature and a compressor ratio of 10 transfers heat to operate an absorption chiller with a COP of 0.80. The exhaust gases leave the chiller at 160° C. The compressor inlet conditions are 101.3 kPa and 20° C, and the compressor and turbine efficiencies are 0.86 and 0.91, respectively. How many tons (12,000 Btu/h) of refrigeration are produced by the chiller?

2. The exhaust of a 2-MW gas turbine operating with a 960° C turbine inlet temperature and a compressor ratio of 9 transfers heat without loss to operate an absorption chiller with a COP of 0.8. The exhaust gases leave the chiller at a temperature of 130° C. The compressor inlet conditions are 105 kPa and 25° C, and the compressor and turbine isentropic efficiencies are 0.82 and 0.89, respectively. How many tons (12,000 BTU/h) of refrigeration are produced by the chiller?

3. A food-processing facility has monthly electricity and gas consumption and costs as shown. Because of the complexity of the electrical rate schedule, use an average value per kWh (often called the "blended" cost of electricity) for electrical cost estimates. Acceptable values for component efficiencies are presented in the chapter. A gas turbine is specified as the prime mover.

(a) Plot the monthly electricity and gas usages.

(b) Compute and plot the electrical load factor.

(c) What are the costs for 10^6 BTU of electricity and gas? What is the spark spread?

(d) Does the operation meet the three criteria for cogeneration system candidacy?

(e) Assess the potential for cogeneration for this facility. This facility operates in mode B, so all the recovered energy cannot be used.

(f) Assess the potential for cogeneration for this facility if the electrical usage remains the same, but the natural gas usage and costs are doubled. This places the operation in mode D.

Electricity usage:

Month	Energy Usage (kWh)	Usage Cost ($)	Peak Demand (kW)	Billed Demand (kW)	Demand Cost ($)	TSI Cost ($)	Total Cost ($)
Apr	1,899,750	72,479	3,292	3,292	36,351	−11,238	97,592
May	2,033,500	77,555	3,419	3,419	37,829	−12,106	103,278
June	2,163,250	82,479	3,598	3,598	40,004	−13,228	109,254
July	2,368,500	90,268	4,006	4,006	44,782	−14,607	120,442
Aug	2,503,750	95,400	4,053	4,053	45,295	−15,253	125,443
Sept	2,479,250	94,471	4,105	4,105	45,960	−14,910	125,521
Oct	2,170,500	82,754	3,891	3,891	43,413	−13,909	112,258
Nov	1,957,250	74,661	3,754	3,754	41,849	−12,321	104,189
Dec	2,100,750	80,107	3,241	3,241	35,808	−12,032	103,882
Jan	1,896,750	71,981	3,216	3,216	35,539	−11,328	96,192
Feb	1,823,750	69,211	3,156	3,156	34,820	−10,799	93,232
March	1,782,750	67,655	3,068	3,068	33,776	−10,553	90,877
Total	25,179,750	959,024	42,799	42,799	475,431	−152,291	1,282,164

Rate schedule: GSA E-2
Usage rate $0.03795/kWh
Demand rate $10.31/kW for the first 1000 kW
         $11.74/kW for all additional kW

Tax rate   1.5%
TSI   Tax, service charges, and incentives

Natural gas usage:

Month	Gas Usage (ccf)	Usage Cost ($)	Gas Cost ($/ccf)	TSI Cost ($)	Total Cost ($)
Apr	76,516	55,170	0.721	827	55,997
May	72,818	65,391	0.898	980	66,372
June	66,398	48,862	0.736	732	49,595
July	66,110	52,763	0.798	791	53,554
Aug	52,602	40,262	0.765	603	40,866
Sept	58,558	45,767	0.782	686	46,453
Oct	60,410	39,352	0.651	590	39,942
Nov	65,191	43,141	0.662	647	43,789
Dec	80,040	51,927	0.649	778	52,706
Jan	79,876	76,870	0.962	1,153	78,023
Feb	80,101	79,913	0.998	1,198	81,112
March	74,742	66,657	0.892	999	67,657
Total	833,362	666,081	0.799	9,991	676,072

Tax rate   1.5%
TSI   Tax, service charges, and incentives

4. Using the utility bills of Exercise 3, answer the following questions.
   (a) What is the cost of 10^6 Btu of electricity?
   (b) What would be the electrical output rating of an internal combustion engine–generator set sufficient to make the facility grid-independent?
   (c) What is the cost of 10^6 Btu of gas energy?
   (d) Does air conditioning account for significant electrical energy usage in this facility? Explain.
   (e) Estimate the amount of natural gas used for heating.

REFERENCES

Boyce, M. P. 2002. *Handbook for Cogeneration and Combined Power Cycles*. New York: ASME Press.

Horlock, J. H. 1997. *Cogeneration: Combined Heat and Power*. Malabar, FL: Krieger.

Kamm, F. 1997. *Heat and Power Thermodynamics*. Clifton Park, NY: Delman.

Kehlhofer, R., Bachman, R., Nielsen, H., and Warner, J. 1999. *Combined Cycle Gas and Steam Turbine Power Plants*. Tulsa, OK: PennWell.

Kolanowski, B. F. 2000. *Small-Scale Cogeneration Handbook*. Upper Saddle River, NJ: Prentice-Hall.

Kreith, F. 2000. *CRC Handbook of Thermal Engineering*. Boca Raton, FL: CRC Press.

Kreith, F., and West, R. E. 1997. *CRC Handbook of Energy Efficiency*. Boca Raton, FL: CRC Press.

Termuehler, H. 2001. *100 Years of Power Plant Development*. New York: ASME Press.

Biomass

Hybrid Poplars | Switchgrass Reed Canary Grass | Willows Hybrid Poplars Silver Maple Black Locust

Switchgrass Poplar Tropical Grasses Sycamore Sweetgum Sorghum Black Locust Miscanthus

Hybrid Poplars Eucalyptus

Eucalyptus

Miscanthus Switchgrass Hybrid Poplars Silver Maple Reed Canarygrass Black Locust Sorghum

Eucalyptus

12.1 | INTRODUCTION

Biomass: Any material of recent biological origin. Examples include crops and forest products and residue associated with crops and forest products, as well as animal fat and food processing waste.

In the pre-twentieth-century era (see Figure 1.3 in Chapter 1), wood, a biomass, was the primary fuel used. However, biomass as a solid has many disadvantages over liquids such as petroleum or gases such as natural gas. The goal of many biomass processes is to convert solid fuel into more useful forms: gaseous or liquid fuels. Biomass has received much attention in the media and by political leaders, both in relation to alternative fuel sources and as a remedy for the United States' dependence on imported fossil fuels. Biomass is attractive as at least part of a strategy for reducing dependence on foreign oil imports because it is domestic, secure, and abundant. Recent books addressing biomass fuel issues include Drapcho et al. (2008) and Speight (2008). This chapter explores the salient features of biomass and biomass conversion processes.

Part of the economic impetus for biomass fuels have been the subsidies enacted to support the production and use of biomass-based fuels such as ethanol and biodiesel, both of which are discussed in this chapter. For example, the Volumetric Ethanol Excise Tax Credit (VEETC) is $0.51 per gallon of pure ethanol (minimum 190 proof) blended with gasoline. The excise tax credit may be claimed on a quarterly basis and is available to those companies that have produced, sold, or used ethanol in their trade or business. Filing for eligibility is accomplished through the Internal Revenue Service (IRS). The VEETC, as of late 2008, is scheduled to expire on December 31, 2010.

The Biodiesel Mixture Excise Tax Credit (BMETC) is $1.00 per gallon of pure agricultural-based biodiesel blended with conventional diesel fuel, or $0.50 per gallon of biodiesel made from other sources (waste grease, for example). Only those businesses that are registered with the IRS and produce, sell, or use a qualified biodiesel mixture are eligible. To qualify for the BMETC, the biodiesel must be certified as meeting the requirements of ASTM D6751. The Biodiesel Income Tax Credit (BITC) is $1.00 per gallon for pure agricultural-based biodiesel (B100, or 100 percent biodiesel) or $0.50 per gallon of B100 from other sources. The B100 must be registered as a fuel with the United States Environmental Protection Agency (EPA) for it to qualify. The BMETC and the BITC expire on December 31, 2009 but are likely to be extended by the U.S. Congress. These and other subsidies along with increased interest in biomass-based fuels have altered agricultural crop patterns and prices in recent years. For example, near Mississippi State University the staple grain crop for more than 50 years has been soybeans, but in 2007 soybean acreage was dramatically displaced by grain corn to be used as a biomass source for ethanol production.

Biomass raw materials (feedstocks) exhibit considerable variation in chemical composition and cell structure. These features determine the products that can be obtained from biomass feedstocks as well as the processes and difficulties in obtaining useful fuel and other chemicals from biomass. Wright et al. (2006) provide a comprehensive survey of biomass data from diverse sources. The statistics and information they present characterize the biomass industry from feedstocks to end use.

Biomass chemical compositions are broadly classed as cellulosic biomass or starch/sugar biomass. However, both biomass classes are based on photosynthesis using energy supplied by the sun. The basic photosynthesis process utilizes carbon dioxide (CO_2) and water (H_2O) to produce simple sugars in the presence of solar energy, chlorophyll, and other plant ingredients. A representative photosynthesis chemical reaction for the production of glucose ($C_6H_{12}O_6$) is

$$6CO_2 + 6H_2O \rightarrow C_6H_{12}O_6 + 6O_2 - 480\frac{kJ}{mol} \qquad (12\text{-}1)$$

The detailed chemical path to glucose, as represented by Eq. (12-1), is quite complicated, but the end products are glucose and oxygen. The energy comes from the sun, but as indicated in Table 1.2 and as confirmed by Decher (1994), while the local conversion efficiency may be as high as a few percentage points, the global efficiency of photosynthesis may be as low as 0.3 percent. The plants use the sugars of reactions, such as those represented by Eq. (12-1), to produce the myriad chemical substances needed for growth and life processes. One of the primary selling points of biomass is the potential use of carbon dioxide, a greenhouse gas, to produce fuel (energy) and oxygen—a win-win situation.

Indeed, one of the great attractions of biomass is the belief that biomass may be CO_2 neutral. The central concept behind a CO_2-neutral biomass fuel is that the CO_2 generated by combustion is balanced by the CO_2 incorporated by photosynthesis during the growth process. The implication is that the CO_2 produced by harvesting and combustion of the biomass is completely offset by growth of the biomass. But, as Schobert (2002) and others point out, this balance is difficult to achieve for a number of reasons—the most important being the loss of CO_2 to the soil and the generation of CO_2 by the fuels required to cultivate, harvest, and process biomass. Additionally, the time scale required for biomass growth can be on the order of years to decades, depending on the particular biomass. Another environmental concern with burning biomass is that the combustion process produces emissions that may contain a diverse variety of pollutants, including carbon monoxide, nitrogen oxides, and ash/soot particulate matter.

12.2   BIOMASS AVAILABILITY

The U.S. Department of Energy's Energy Efficiency and Renewable Energy (EERE) website, www.eere.energy.gov, contains a wide variety of useful information relating to biomass and is a good source of additional, readily available, public-domain information on topics presented in this chapter. Following EERE, a delineation of biomass resources follows.

- Agricultural crops—commodity products such as corn, soybeans, and wheat that yield sugar and oils.
- Agricultural crop residues—residues, such as stalks and leaves, of agricultural crops that are not harvested for commercial use. Corn stover consists of the stalks, leaves, husks, and cobs of corn, for example.
- Herbaceous energy crops—perennials that are harvested annually after reaching maturity. These are grasses such as switchgrass, miscanthus, bamboo, and fescue.
- Woody energy crops—hardwood trees harvested within five to seven years of planting. Examples are poplar, willow, maple, cottonwood, and sweetgum.
- Forestry residues—biomass not harvested or removed during logging, and materials removed during forest management operations.
- Aquatic crops—aquatic biomass such as kelp, seaweed, and algae.
- Industrial crops—crops developed to produce specific chemicals or feedstocks. Kenaf is an example.
- Animal waste—waste resulting from farm and processing operations.
- Municipal waste—waste from residential, commercial, and industrial sectors, which can contain significant biomass with energy content.

Figure 12.1, adapted from www.geonomics.energy.gov, provides a breakdown of the sources of biomass. The sources include forest lands and agricultural lands as well as secondary sources such as municipal solid waste, animal manure, and residues from food processing. Wright et al. (2006) present a quantitative compilation of biomass feedstock resources.

Much of the United States is well suited for the production of herbaceous and woody energy crops in addition to agricultural crops. Figure 12.2 illustrates by region energy crops that are particularly suited to the region. Thus most of the United States, except for the mountains and desert Southwest, could produce biomass feedstock. The question of availability of biomass is partially answered by a 2005 study jointly sponsored by the U.S. Departments of Agriculture and Energy (Perlack et al. 2005). The authors conclude that land resources are sufficient to sustain a biomass budget of 1 billion tons/yr—enough biomass to displace 30 percent of the fossil fuels used for transportation. Moreover, Perlack et al. suggest that an increase to more than

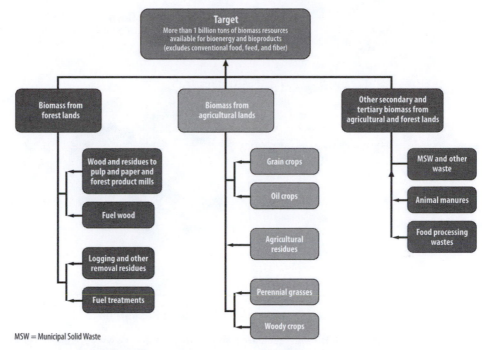

Figure 12.1 Biomass sources (www.geonomics.energy.gov).

Figure 12.2 Biomass energy crop suitability by region (www.genomics.energy.gov).

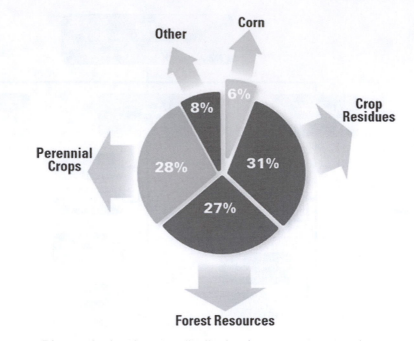

Figure 12.3 Biomass feedstock source distribution (www.eere.energy.gov).

1.3 billion tons/yr is possible with relatively modest changes in agricultural and forestry practices. They project a total of 1.366 billion tons/yr, with 998 billion tons/yr from agricultural resources and 368 billion tons/yr from forestry resources.

The results of an assessment by Perlack et al. (2005) of the sources for biomass in the United States are summarized in Figure 12.3. In the assessment, crop residues and forest resources account for 58 percent of the total biomass feedstock, while crops and corn account for 34 percent of the biomass feedstock.

12.3 BIOMASS FUNDAMENTALS

The primary products of interest from biomass processing are as follows.

1. Liquid fuels

   (a) Ethanol—ethyl alcohol for use as a fuel
   (b) Methanol—methyl alcohol produced as a by-product of gaseous fuels
   (c) Biodiesel—chemically modified vegetable oils suitable for diesel fuel
   (d) Vegetable oil—oils contained in grains and seed
   (e) Pyrolysis oil—a liquid fuel with a heating value of 17–20 MJ/kg

2. Gaseous fuels

   (a) Biogas—a mixture of CH_4 and CO_2 (methane and carbon dioxide), 55–70 percent methane by volume

   (b) Producer gas—a flammable gas mixture containing CO, H_2, CH_4, N_2, CO_2, and higher hydrates

   (c) Synthesis gas—a mixture of CO and H_2 (carbon monoxide and hydrogen)

Each of the possible biomass fuels will be examined in a later section of this chapter.

The U.S. Department of Energy (DOE) has classified biomass feedstock to fuels or other products in terms of platform. The two primary platforms for biomass and their relationships to feedstocks, processes, and fuels/products are illustrated in Figure 12.4. The primary platforms of interest for this discussion are the sugar (biochemical) and the syngas (thermochemical) platforms. Biomass feedstocks are used in both platforms, with the primary outputs being fuels and chemicals. However, for both platforms fuels can be used in combined heat and power (CHP) applications. CHP systems are discussed in Chapter 11.

Biomass and biomass for fuel conversion are intimately involved with chemistry. To establish a uniform basis, the following definitions are introduced.

Chemical Formulae and Molecular Masses

Acetone, C_3H_6O: molecular mass = 58.09 g/g mol

Ethane, C_2H_6: molecular mass = 58.09 g/g mol

Ethanol, C_2H_5OH (ethyl alcohol): molecular mass = 46.06844 g/g mol

Ethylene, C_2H_4: molecular mass = 28.05 g/g mol

Glycerol (a clear, colorless, viscous, sweet-tasting liquid belonging to the alcohol family of organic compounds), $C_3H_5(OH)_3$: molecular mass = 92.09382 g/g mol

Isopropanol, C_3H_8O: molecular mass = 60.10 g/g mol

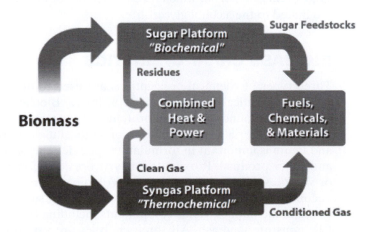

Figure 12.4 DOE biomass platforms (www.nrel.gov/biomass/biorefinery).

Methane, CH_4: molecular mass = 16.0425 g/g mol

Methanol, CH_3OH (methyl alcohol): molecular mass = 32.04 g/g mol

Phenol, C_6H_5OH (hydroxybenzene): molecular mass = 94.11 g/g mol

Sucrose, $C_{12}H_{22}O_{11}$: molecular mass = 844.33 g/g mol

Definitions

Acetate: Man-made textile fiber produced from the plant substance cellulose.

Ash: The powdery residue of matter that remains after burning.

Carbohydrates: Compounds that include sugars, starches, cellulose, and a wide variety of other cellular products. The name carbohydrates arose because many of these compounds have the formula $C_n (H_2O)_m$ (where m and n are integers)— apparently making them "hydrates of carbon," although this description was later found to be inaccurate.

Cellulose: $(C_6H_{10}O_5)_x$, the chief constituent of the cell walls of plants.

Ester: Any of a class of organic compounds that react with water to produce alcohols and organic or inorganic acids.

Hemicellulose: Complex carbohydrates that surround the cellulose fibers of a plant cell.

Hydrolysis: The breaking of hydrogen bonds in long-chained organic molecules.

Lignin: Complex oxygen-containing organic substance that, with cellulose, forms the chief constituent of wood.

Protein: Any group of nitrogenous organic compounds of high molecular weight that are synthesized by plants and animals and are required for life processes.

Starch: $(C_6H_{10}O_5)_n$, a white, granular, organic chemical that is produced by all green plants.

Sugar: Any of numerous sweet, colorless, water-soluble compounds present in the sap of seed plants and the milk of mammals and that make up the simplest group of carbohydrates.

Triglycerides: Esters in which three molecules of one or more different fatty acids are linked to the alcohol glycerol.

12.4   BIOMASS CHARACTERISTICS

Three types of chemical descriptions are used to describe the attributes of biomass: biochemical, proximate, and ultimate. Biochemical analysis describes an organic compound in terms of proteins, oils, sugars, and fiber (lignocellulose). For cellulosic materials, the breakdown of lignocellulose into cellulose, hemicellulose, and lignin is important in formulating specific processes for extracting useful fuel and other products. Proximate analysis breaks a compound into its constituents of volatiles, ash, and fixed carbon. Volatile material is the fraction that can be decomposed by heat (~400° C) in non-oxidizing surroundings. Ultimate analysis, as the name implies, yields components including the elements (C, H, O, N, S, and Cl), moisture, and ash. All three analyses are useful in understanding biomass processes. Table 12.1 illustrates biochemical analysis of some common biomass feedstocks. The feedstocks are divided into those that are food based and those that are lignocellulose based. Table 12.2 shows proximate analysis of feedstocks,

TABLE 12.1 Biochemical composition of biomass feedstocks (rounded to integer percentages) (adapted from Goswami et al. 2000)

Food-based	Protein	Oil	Starch	Sugar	Fiber
Corn (grain)	10	5	72	<1	12
Wheat (grain)	14	<1	80	<1	5
Artichoke	<1	<1	<1	75	25
Sugar cane	<1	<1	<1	50	50
Sorghum	<1	<1	<1	50	50

Cellulosic	Cellulose	Hemicellulose	Lignin	Other
Bagasse	35	25	20	20
Corn stover	53	15	16	16
Corn cobs	32	44	13	11
Wheat straw	38	36	16	10
Woody crops	50	23	22	5
Herbaceous crops	45	30	25	10
Waste paper	76	13	11	0

TABLE 12.2 Proximate analysis of common biomass feedstocks (adapted from Goswami et al. 2000)

Feedstock	HHV (MJ/kg)	Volatile[*] (%)	Ash[*] (%)	Fixed Carbon[*] (%)
Alfalfa	18.45	73	7	20
Black locust	19.71	81	1	18
Black oak	18.65	86	1	13
Cedar	20.56	87	~0	13
Corn cobs	18.77	80	1	19
Corn stover	17.65	75	6	19
Corn grain	17.20	87	1	12
Douglas fir	20.37	87	~0	13
Pine, ponderosa	20.02	83	~0	17
Poplar	19.38	82	1	16
Redwood	20.72	80	~0	20
Rice, hulls	16.14	65	18	17
Rice, straw	14.56	62	24	13
Sudan grass	17.39	73	9	19
Sugarcane bagasse	17.33	74	11	15
Switchgrass	18.64	81	4	15
Wheat straw	17.51	71	9	20

[*] Rounded to integer percentages

TABLE 12.3 Ultimate analysis of common biomass feedstocks (rounded to integer percentages) (adapted from Goswami et al. 2000)

Feedstock	C %	H %	O %	N %	S %	Cl %	Ash %
Black locust	51	6	42	1	—	—	1
Black oak	49	6	43	—	—	—	1
Corn cobs	47	6	45	—	—	—	1
Corn stover	44	6	43	1	—	1	6
Corn grain	44	6	47	1	—	—	1
Douglas fir	51	6	43	—	—	—	—
Pine, ponderosa	49	6	44	—	—	—	—
Poplar	48	6	44	—	—	—	1
Redwood	51	6	43	—	—	—	—
Rice, hulls	41	4	36	—	—	—	18
Rice, straw	35	4	36	1	—	—	25
Sugarcane bagasse	45	5	40	—	—	—	10
Switchgrass	47	6	42	1	—	—	—
Wheat straw	43	5	39	1	—	—	11

and Table 12.3 presents results of ultimate analyses for most of the feedstocks listed in Table 12.2.

A wide range of liquid and gaseous fuels can be obtained from biomass. Biomass-derived fuels may be used directly in some internal combustion engines, may be used for process energy, or may be further processed into more desirable fuels or chemical feedstocks. This next section of this chapter examines the characteristics of a number of biomass-based fuels and products.

12.5 BIOMASS-BASED FUELS AND PRODUCTS

Ethanol

Ethanol (C_2H_5OH), also called ethyl alcohol or grain alcohol, is the alcohol in "alcoholic " beverages and can be used as a fuel in modified internal combustion engines. The nominal heating value of ethanol as a fuel is typically quoted as 101,000 Btu/gal, about three-fourths the value of gasoline (on a volumetric basis). Thus a motor vehicle that has a gas mileage of 24 miles/gal of gasoline would likely have an "ethanol mileage" of 18 miles/gal.

Ethanol is used to improve the octane rating and emission characteristics of gasoline and is commonly used in the form of a 10 percent blend in "gasohol" called E10

(10 percent ethanol and 90 percent gasoline). The highest blend, E85, contains 85 percent ethanol and is used in flex fuel vehicles. Most (about 90 percent) of the ethanol in the United States is produced by fermentation of corn or other biomass feedstocks. The remaining amount, called synthetic ethanol, is produced from ethylene, a petroleum by-product, and is used mostly in industrial applications. Ethylene is one of the most highly produced organic chemicals and is important as an industrial feedstock. However, the extensive production of ethanol from petroleum-based ethylene will not eliminate our dependence on imported fossil fuels.

The feedstock for fermentation to produce ethanol must be high in sugars. Sugar cane, sugar beets, and sorghum all contain high levels of suitable sugars. Sugar beets are the basis for the fuel ethanol industry in Brazil. A variety of carbohydrates (starches, hemicellulose, and cellulose) can also serve as feedstock if the carbohydrates can be broken down into sugars capable of undergoing fermentation. Corn, which is high in starches, has been the basis of the United States ethanol industry for more than two decades. Cellulose and hemicellulose can be converted to fermentable sugar compounds, but the process is more difficult than for starches.

Ethanol produced from grain, such as corn, starts as a water-based slurry. After the starches are converted to sugars, the sugars are fermented into ethanol. In the fermentation process, the enzymes produced by yeast or other microorganisms decompose the sugars into ethanol and CO_2. For the production of ethanol from glucose, $C_6H_{12}O_6$, the chemical reaction is

$$C_6H_{12}O_6 \rightarrow 2\,C_2H_5OH + 2\,CO_2 \tag{12-2}$$

After fermentation, distillation is necessary to separate the ethanol from the water. The primary steps are illustrated in Figure 12.5.

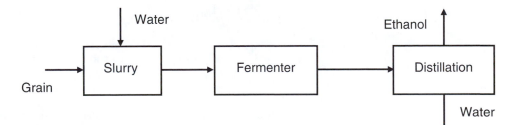

Figure 12.5 Principal steps in ethanol production from grain.

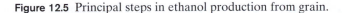

Example 12.1	How much ethanol can be produced from a ton of sucrose?

Solution   Fermentation of sucrose, $C_{12}H_{22}O_{11}$, with water results in ethanol and carbon dioxide; the chemical equation is

$$C_{12}H_{22}O_{11} + H_2O \rightarrow 4C_2H_5OH + 4CO_2$$

The molecular masses (expressed in kg/kg-mol) of C, H, O, and OH are as follows:

$$C_{mm} = 12.01 \frac{kg}{kg \cdot mol} \quad H_{mm} = 1.008 \frac{kg}{kg \, mol} \quad O_{mm} = 16.00 \frac{kg}{kg \, mol}$$

$$OH_{mm} = O_{mm} + H_{mm} = 17.008 \frac{kg}{kg \, mol}$$

Sucrose (SUC) has a molecular mass of

$$SUC_{mm} = 12 \cdot C_{mm} + 22 \cdot H_{mm} + 11 \cdot O_{mm} = 844.332 \frac{kg}{kg \, mol}$$

For water (WAT), ethanol (ETH), and carbon dioxide (CO_2), the molecular masses are

$$WAT_{mm} = 2 \cdot H_{mm} + O_{mm} = 18.016 \frac{kg}{kg \, mol}$$

$$ETH_{mm} = 2 \cdot C_{mm} + 5 \cdot H_{mm} + OH_{mm} = 46.068 \frac{kg}{kg \, mol}$$

$$CO_{2mm} = C_{mm} + 2 \cdot O_{mm} = 44.01 \frac{kg}{kg \, mol}$$

For 1 ton of sucrose, the water required for the reaction, and the ethanol and carbon dioxide formed, are

$$WAT = \frac{WAT_{mm}}{SUC_{mm}} \cdot 1 \, ton = 0.053 \, ton = 105.3 \, lb$$

$$ETH = \frac{4 \cdot ETH_{mm}}{SUC_{mm}} \cdot 1 \, ton = 0.538 \, ton = 1077 \, lb$$

$$CO_2 = \frac{4 \cdot CO_{2mm}}{SUC_{mm}} \cdot 1 \, ton = 0.514 \, ton = 1029 \, lb$$

Although grain is the leading feedstock for ethanol production in the United States, the significant replacement of gasoline by grain-based ethanol would be very difficult. Availability is the primary limitation. If all the corn in the United States were used to produce ethanol, the ethanol produced would replace only about 10 percent of the gasoline currently used (Schobert 2002). Hence, for close to 100 percent replacement of gasoline by ethanol, the production of corn would have to increase tenfold. Given the amount of cultivatable land in the United States, such an increase is not possible. Moreover, the widespread utilization of corn grain for ethanol would mean diverting a primary food stock to a nonfood use. Because of these limitations, ethanol from corn cannot be viewed as a solution to foreign oil dependence. However, the use of other biomass feedstocks for ethanol production, especially cellulosic feedstock, has significant long-term potential. Indeed, the U.S. Department of Energy has announced a biofuels initiative with the goal of reducing the cost of cellulosic-based ethanol. Because of this potential, further examination of cellulosic ethanol is warranted.

The conversion of cellulosic biomass to ethanol involves the three primary polymers (cellulose, hemicellulose, and lignin) that make up the cell walls of a plant. Depending on the plant, a cell wall typically consists of 35 to 50 percent cellulose, 20 to 25 percent hemicellulose, and 10 to 25 percent lignin. Cellulose is the most abundant biomass on Earth (www.genomics.energy.gov). From a biochemical standpoint, cellulose is a linear polymer of glucose residues. Hemicellulose is a branched sugar polymer formed mostly of five-carbon sugars (pentose) with some six-carbon sugars (hexose). Lignin is a complex, cross-linked polymer that is covalently bonded to hemicellulose. These three polymers impart strength to mature cell walls, and this strength is part of the reason cellulosic biomass is difficult to break down into its component sugar compounds. The process of breaking down the cellulosic biomass to extract the sugars is called hydrolysis. The long-term, conventional process is acid hydrolysis, but the biochemical hydrolysis process is thought to have the potential to convert cellulosic biomass in quantities sufficient to replace a significant amount of gasoline with ethanol.

The biochemical hydrolysis of cellulosic biomass is a complex process requiring a detailed understanding of the roles of various enzymes in degrading the cell structure into sugars. The website, www.genomics.energy.gov, and Himmel (2008) are excellent sources of information and details on cellulosic biomass. Fundamentally, the enzymatic biochemical conversion of cellulosic biomass consists of three steps: (1) pretreatment to make the complex polymers more liable to enzymatic breakdown, (2) the use of appropriate enzymes to hydrolyze the plant cell wall polymers (cellulose, hemicellulose, and lignin) into sugars, and (3) fermentation to convert the sugars to ethanol. Figure 12.6, taken from the aforementioned website, provides a

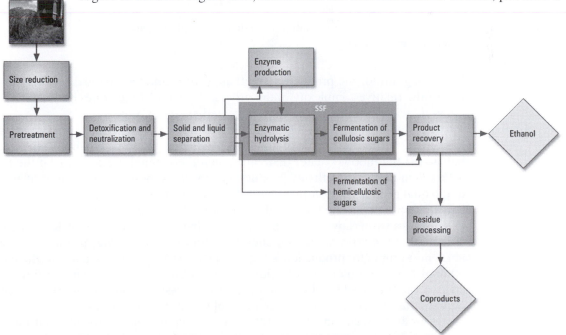

Figure 12.6 Enzymatic biochemical process to produce ethanol from cellulosic biomass (www.genomics.energy.gov).

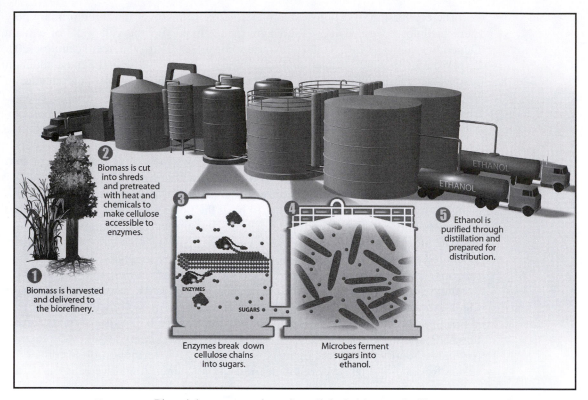

② Biomass is cut into shreds and pretreated with heat and chemicals to make cellulose accessible to enzymes.

① Biomass is harvested and delivered to the biorefinery.

ENZYMES

SUGARS

⑤ Ethanol is purified through distillation and prepared for distribution.

Enzymes break down cellulose chains into sugars.

Microbes ferment sugars into ethanol.

Figure 12.7 Pictorial representation of a cellulosic biomass facility (www.genomics.energy.gov).

block diagram for the production of ethanol from cellulosic biomass and is indicative of the relative complexity of the process. Figure 12.7 illustrates how a cellulosic ethanol facility, with the three steps indicated, might appear.

An issue of concern with grain-based ethanol is the ratio of the energy per gallon of ethanol to the energy required to produce a gallon of ethanol—the energy balance. If the ratio is greater than unity, then the ethanol saves energy; if the ratio is less than unity, then ethanol production consumes more energy than is available in the final product. Not surprisingly, different various groups, agencies, and individuals line up on opposite sides of the question. The U.S. Department of Energy, for example, is a strong advocate of the energy-saving potential of ethanol. Kreith (2007) accepts that the ratio of ethanol energy content (heating value) to the nonrenewable energy input for production is 1.25, but he is a strong advocate of cellulosic-based ethanol, with a ratio of about 5. Tidwell and Weir (1986) concur with Kreith, especially if the fuel used for distillation is part of a waste stream. Schobert (2002) recognizes that the energy balance for ethanol might be negative. Pimentel in a series of paper, articles, and interviews (2003, for example) has maintained that ethanol takes more energy to produce than it provides. In comparison with conventional gasoline, ethanol produces less in the way of emissions (NO_x, SO_x, soot, and CO_2).

Methanol

Methanol (CH_3OH), also called wood alcohol, can be used as a fuel in modified internal combustion engines. The heating value of methanol as a fuel is nominally 76,000 Btu/gal, only about 58 percent of the value of gasoline (on a volumetric basis). Thus, a motor vehicle that has a gas mileage of 24 miles/gal would likely have a methanol mileage of about 14 miles/gal.

Most methanol produced in the United States uses natural gas as the feedstock. The production of methanol from natural gas involves three chemical processes. First, natural gas (primarily methane, CH_4) reacts with steam to produce synthesis gas (carbon monoxide and hydrogen) according to

$$CH_4 + H_2O \rightarrow CO + 3H_2 \tag{12-3}$$

In the second step, the synthesis gas is combined with carbon dioxide to alter the ratio of hydrogen and carbon monoxide in the fashion

$$H_2 + CO_2 \rightarrow H_2O + CO \tag{12-4}$$

For the third step of the process, carbon monoxide and water, in the presence of a catalyst, are used to produce methanol:

$$CO + 2H_2O \rightarrow CH_3OH \tag{12-5}$$

While production of methanol from natural gas only increases the demand for natural gas, an important observation in relation to the three-step process is that only step 1, the production of synthesis gas, involves natural gas. Synthesis gas can also be produced from a variety of carbonaceous biomass feedstocks, as well as coal. Thus, large-scale production of methanol does not have to be dependent on natural gas. Because of its importance, synthesis gas is covered as a separate topic in this section.

Methanol has many advantages over conventional gasoline: the methanol-air flame temperature is lower than that for gasoline, so that NO_x production would be decreased; methanol produces no SO_x, and methanol has limited soot and CO_2 emissions. As for drawbacks, in addition to the lower energy density, methanol presents severe health hazards if ingested or absorbed, and it is soluble in water. Also, since methanol burns with a nearly colorless flame, accidental ignitions can be hard to see.

Biodiesel/Vegetable Oil

Biodiesel is a fuel composed of fatty acid alkyl esters and is made from feedstocks such as vegetable oils and animal fats. This is not a particularly useful definition for non–chemical engineers, so additional discussion is needed. A plethora of oils can be extracted from seeds, nuts, and other plant parts. Additionally, recycled cooking oil and animal fats are sources of biodiesel feedstocks. Examples include corn oil, linseed oil, olive oil, soybean oil, and sunflower oil. Although these and similar plant oils can be used to fuel diesel engines, the direct use of plant oils in diesel engines shortens operating life and leads to engine maintenance issues. Additionally, plant oils have a higher viscosity (by a factor of as much as 20) than diesel fuel and have a lower cetane number, which is a performance metric. Schobert (2002) states that the expected 10,000 hours between engine overhauls can be reduced to as little as 600 hours for sunflower oil operation or 100 hours for operation on linseed oil. If, however, plant oil is reacted with methanol, the resulting methyl ester product is a generic biodiesel. The generic chemical reaction is

$$\text{triglyceride} \; + \; \text{methanol} \; \rightarrow \; \text{methyl ester} \; + \; \text{glycerol} \tag{12-6}$$
$$\text{(plant oil)}$$

A specific example for a slightly simplified composition for olive oil, from Schobert, is

$$(C_{17}H_{33}CO_2)_3\,C_3H_5 + 3CH_3OH \rightarrow 3C_{17}H_{33}CO_2CH_3 + C_3H_5(OH)_3 \tag{12-7}$$

where $C_{17}H_{33}CO_2$ is the fatty acid component of olive oil. This chemical procedure is called transesterification. The resulting methyl ester is a biodiesel fuel that possesses characteristics much closer to those of conventional diesel fuel than the original plant oil. As Van Gerpen et al. (2004) point out, 100 lb of oil and 10 lb of methanol input to a transesterification reaction will yield about 100 lb of biodiesel and 10 lb of glycerol.

Example 12.2	Verify that the transesterification reaction of olive oil agrees with the assertion of Van Gerpen et al. about biodiesel yields.

Solution Equation (12-7) represents the reaction. The molecular masses for C, H, O, and OH are given in the solution to Example 12.1.

The fatty acid component, $(C_{17}H_{33}CO_2)_3$, of olive oil has a molecular mass of

$$FA_{mm} = 3 \cdot (17 \cdot C_{mm} + 33 \cdot H_{mm} + C_{mm} + 2 \cdot O_{mm}) = 844.332 \; \frac{kg}{kg\,mol}$$

The molecular mass of olive oil is thus

$$PO_{mm} = FA_{mm} + 3 \cdot C_{mm} + 5 \cdot H_{mm} = 885.402 \; \frac{kg}{kg\,mol}$$

For methanol (MET), biodiesel (BD), and glycerol (GLY), the molecular masses are

$$MET_{mm} = 3 \cdot (C_{mm} + 3 \cdot H_{mm} + OH_{mm}) = 96.126 \; \frac{kg}{kg\,mol}$$

$$BD_{mm} = FA_{mm} + 3 \cdot (C_{mm} + 3 \cdot H_{mm}) = 889.434 \; \frac{kg}{kg\,mol}$$

$$GLY_{mm} = 3 \cdot C_{mm} + 5 \cdot H_{mm} + 3 \cdot OH_{mm} = 92.094 \; \frac{kg}{kg\,mol}$$

For 100 lb of olive oil feedstock, the methanol required and the biodiesel and glycerol formed are thus

$$MET = \frac{MET_{mm}}{PO_{mm}} \cdot 100 \text{ lb} = 10.85 \text{ lb}$$

$$BD = \frac{BD_{mm}}{PO_{mm}} \cdot 100 \text{ lb} = 100.45 \text{ lb}$$

$$GLY = \frac{GLY_{mm}}{PO_{mm}} \cdot 100 \text{ lb} = 10.40 \text{ lb}$$

Thus, the assertion is substantially correct. Moreover, since most plant oils have similar (large) molecular masses and since the transesterification process involves changes of a few atoms, the generalization of Van Gerpen et al. is likely to be an appropriate rule of thumb for most biodiesel transesterification reactions.

Biodiesel has a viscosity comparable to that of conventional diesel fuel and a cetane number of about 50. Moreover, operating experience with biodiesel suggests that biodiesel provides for a much longer time between overhauls compared with the use of plant oils. It also burns cleaner than conventional diesel fuel and significantly reduces emissions of carbon monoxide, unburned and aromatic hydrocarbons, and sulfates, as well as particulate matter. Biodiesel is compatible with most emission control technologies. Blends of up to 20 percent biodiesel with conventional fuel can be used without any engine or infrastructure (storage and distribution) modifications. Similar to the practice for ethanol, biodiesel blends are labeled with a B followed by the percentage of biodiesel; B20, a commonly used blend, consists of 20 percent biodiesel. B5, B10, and B20 are the usual biodiesel blends. EERE (Tyson et al. 2004) estimates that sufficient biomass feedstock exists to supply 1.9 billion barrels of biodiesel, about 5 percent of the on-road of use of conventional diesel fuel. Biodiesel that is classified as fuel grade must be produced in compliance with ASTM D6751. Unprocessed vegetable oil does not meet biodiesel specifications and is not a legal motor fuel (www.eere.energy.gov/RE/bio_fuels.html).

The three basic processes for producing biodiesel are

1. Base-catalyzed transesterification
2. Direct acid-catalyzed transesterification
3. Conversion of the oil to its fatty acids and then to biodiesel

Base-catalyzed transesterification is the most prevalent process and possesses the following advantages:

1. The process occurs at low temperature and pressure.
2. The conversion efficiency is high—98 percent.
3. The process involves minimal side reactions.
4. The conversion to biodiesel is direct, with no intermediate compounds.
5. No special materials are needed in the reactor.

TABLE 12.4 Characteristics of biodiesel fuels

Specific gravity	0.87–0.89
Cetane number	46–70
Higher heating value	17,000–18,000 Btu/lb

Since biodiesel can be made from a wide variety of plant oils and animal fats, the physical properties of biodiesel have some variation. Generally accepted values (from www.eere.energy.gov) are presented in Table 12.4.

Pyrolysis Liquids

Pyrolysis is defined as the thermal decomposition of organic compounds in the absence of oxygen. Charcoal is a simple example of pyrolysis that results when wood is heated to about 250° C and the moisture and volatile products are driven off by the heat. The resulting product has a very high carbon content. Pyrolysis is an endothermic reaction that leads to the degradation of solid fuels into a variety of gases, organic vapors, and liquids. In contrast, combustion is an exothermic reaction that involves the rapid oxidation of fuel with a large release of heat.

The product stream from pyrolysis depends on the rate and duration of heating. In the production of oil using fast pyrolysis, cellulosic biomass feedstock is ground to a fine power that is heated for a short period of time (<0.5 sec) to 400–600° C and then rapidly quenched. Rapid quenching is necessary to prevent high-molecular-weight liquids from decomposing into lower-molecular-weight gases. Liquids from fast pyrolysis processes have a low viscosity and are dark brown in appearance. Pyrolysis liquids are highly oxygenated, complex mixtures of hydrocarbons that contain significant amounts of water. Such liquids are highly unstable and highly corrosive (low pH). Because of the corrosive nature of pyrolysis liquids, storage is difficult, and the high oxygen and water content makes them incompatible with conventional hydrocarbon fuels. Hence, conversion of pyrolysis liquids into more conventional hydrocarbon fuels is desirable, although pyrolysis liquids can be used in place of heavy fuel oils such as Bunker C.

Biogas

Biogas is a mixture of CH_4 and CO_2, typically 50–80 percent methane and 20–50 percent carbon dioxide by volume, with traces of hydrogen, carbon monoxide, and nitrogen. The International Energy Agency offers a compact examination of biogas production and utilization as IEA Bioenergy: T37:2005-01 (www.berr.gov.uk/files/file39295.pps). Biogas is the final product of anaerobic digestion of organic waste, such as animal manure and food processing waste. Anaerobic digestion is the decomposition of such wastes into gaseous fuels by the action of bacteria in an oxygen-free environment. Biogas differs from natural gas in that natural gas is composed of more than 70 percent methane, with the remaining constituents being other hydrocarbons. Biogas is also called swamp gas, landfill gas, or digester gas. Although the biological processes that result in biogas are relatively complex, anaerobic digestion systems are relatively simple. Biogas is produced in stages by different types of bacteria that break down the complex organic compounds. The following four stages, with the associated bacteria function, form the generally accepted path from organic waste to biogas: (1) hydrolytic bacteria break down the waste into sugars and amino acids; (2) fermentative bacteria convert the sugars and amino acids to organic acids; (3) acidogenic bacteria convert the organic acids to hydrogen, carbon dioxide, and acetate; and (4) methanogenic (methane-forming) bacteria produce biogas from the acetic acid, hydrogen, and carbon dioxide.

Anaerobic digestion takes place in an airtight chamber called a digester. Digesters may operate in batch mode or in continuous mode. Continuous digesters produce a steady stream of biogas and are typically associated with large-scale operation. The digestion process is temperature sensitive; a temperature of at least 68° F is required, but temperatures as high as 150° F are possible. The higher the temperature, the shorter the process time and smaller the digester volume, but high-temperature digesters are more difficult to operate and maintain and require close monitoring. Boyle (2004) states that a typical anaerobic digester can produce 200–400 m³ of biogas, with a 50–75 percent methane content, per metric ton of feedstock. A typical anaerobic digester consists of a pre-mixing tank, a digester vessel, a system that uses the biogas, and a means of spreading or distributing the digested effluent. Figure 12.8 presents a schematic of a typical batch anaerobic digester. In many instances the "system" is an engine-generator set. The dotted line from the system to the digester represents an energy stream directed from the system to the digester to maintain digester temperature.

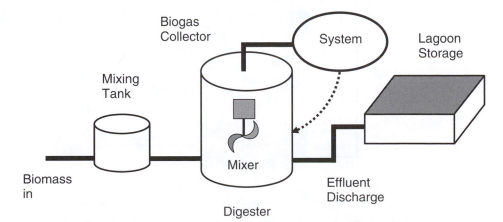

Figure 12.8 Batch anaerobic digester schematic.

The Alberta (Canada) Agriculture and Rural Development Agency is heavily involved in promoting anaerobic digesters for agricultural waste. Figure 12.9, taken from the agency's website, is an example of an anaerobic digester system that utilizes manure and provides the potential for biogas use on-site in a co-generation arrangement as well as for upgrading to natural gas for pipeline connection. Additionally, concentrated nutrients and water from the effluents are provided. The carbon dioxide and water must be removed, leaving mostly methane, if biogas is to be upgraded to natural gas and fed to the pipeline.

The U.S. Environmental Protection Agency Landfill Methane Outreach Program (LMOP) provides assistance for landfill issues; details are available at www.epa.gov/lmop/index. The Tillamook Digester Facility, www.potb.org/methane-energy.htm, located in Tillamook, Oregon, makes use of manure from the extensive dairy farms that support the famous Tillamook cheese industry in the area. The facility has two 400,000-gals digester cells that produce biogas, which is used to run two grid-connected 200-kW engine-driven generators.

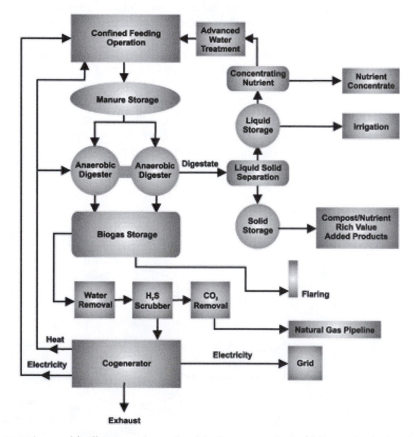

Figure 12.9 Anaerobic digester schematic with diverse products (Alberta Agriculture and Rural Development Agency).

Producer Gas

Producer gas is a low- to medium-Btu flammable gas mixture containing CO, H_2, CH_4, N_2, CO_2, and higher hydrates and is made from gasification of wood or coal. Producer gas can be burned directly or used as a feedstock for liquid fuels. The conversion of biomass to producer gas is often called thermal gasification. Thermal gasification is the partial oxidation of a solid fuel, such as a solid biomass, at elevated temperatures into a flammable gas mixture called producer gas. As Boyle (2004) points out, all gasification processes involve the interaction of steam and oxygen with a solid feedstock. The producer gas contains varying percentages of H_2, CO, CH_4, and CO_2 as well as smaller quantities of higher hydrates; the specific chemical constituents are determined by the temperature to which the biomass is heated, the composition of the biomass feedstock, and the gasifier layout. Table 12.5 shows the varying composition of producer gas.

Brown (2004) presents a compact, lucid description of thermochemical biomass technologies, including gasification. He describes thermal gasification as a multi-step process involving heating and drying, pyrolysis, solid-gas reactions that consume char, and gas-phase reactions that determine the final composition of the producer gas. At about 100° C, moisture is driven out by heating and drying. Other authors, including Goswami et al. (2000), view thermal gasification as a combination of combustion and pyrolysis. (The distinctions between combustion and pyrolysis were discussed previously in the coverage of pyrolysis liquids.) Since pyrolysis is an endothermic process, energy is required for pyrolysis to proceed, and combustion (by means of the addition of oxygen or air to the gasifier) is used only to the extent necessary to produce energy for pyrolysis. At about 400° C, the biomass begins to break down into gases, vapors, and liquids. Only char, a porous solid containing carbon and ash, remains. Some idea of the percentage composition of products from pyrolysis can be obtained from the proximate analysis of the biomass feedstock (see Table 12.2). The volatile content is approximately equal to the pyrolysis yield, and the carbon and ash percentages are roughly equal to the char. At about 700° C, the char begins to react with any oxygen as well as carbon dioxide and water vapor to produce additional flammable gases. Gasification involves a number of different chemical reactions. Table 12.6 presents a summary of the important reactions associated with biomass gasification.

The arrows in the reactions column of Table 12.6 indicate that the gasification reactions can proceed in either direction depending on the fuel, oxygen, and steam added to the gasifier as well as the temperature and pressure. Low temperatures and

TABLE 12.5 Producer gas composition variation (*percent by volume*) (adapted from Goswami et al. 2000)

H_2	CO	CO_2	CH_4	N_2	Heating Value (MJ/m^3)
9–31	14–48	20–0	7–21	50–0	5.4–17.4

TABLE 12.6 Gasification reactions
(adapted from Goswami et al. 2000 and Tester et al. 2005)

	Chemical Reaction	Comment
Carbon-water	$C + H_2O \leftrightarrow H_2 + CO$	Endothermic $\rightarrow$
Water-gas shift	$CO + H_2O \leftrightarrow H_2 + CO_2$	
Hydrogenation	$C + 2H_2 \leftrightarrow CH_4$	
Carbon-oxygen	$2C + O_2 \leftrightarrow 2CO$	
Boudouard	$C + CO_2 \leftrightarrow 2CO$	
Methanation	$CO + 3H_2 \leftrightarrow H_2O + CH_4$	Exothermic $\rightarrow$

high pressures favor the formation of CH_4 while high temperatures and low pressures favor the formation of H_2 and CO.

In addition to producer gas, tars (viscous, liquid high-molecular-weight compounds) and particulate matter are likely to be in the gasifier output streams. Thus, most gasifiers include processes for gas cleaning. The thermodynamic efficiency of gasifiers varies widely depending on the type and operating conditions. Gasifiers that can harvest more than 90 percent of the chemical energy of the biomass feedstock are available but are expensive. Most gasifiers harvest between 70 and 80 percent of the available energy in the feedstock.

Synthesis Gas

Synthesis gas, or syngas, is a mixture of CO and H_2 and can be produced by a gasification process using oxygen rather than air. The oxygen requirement for syngas production is about one-third of the oxygen required for complete combustion. Synthesis gas is so named because, in addition to being a useful, high-energy fuel, it can be used as the feedstock for the synthesis of many useful chemicals. The chemistry of the production of syngas from natural gas, coal, and heavy oil fractions has been well known for more than a century. Reyes et al. (2003) provide a historical perspective on the production of syngas from non-biomass sources and review recent technological advances in the production of syngas from traditional feedstocks. In recent years, the production of syngas from biomass resources has received much attention as at least a partial solution to the energy needs of many countries, including the United States. Spath and Dayton (2003) assess the potential for biomass-derived syngas in the United States, and Cobb (2007) provides a survey of biomass gasification facilities in the United States.

Syngas is used via the Fischer-Tropsch process to produce a wide range of hydrocarbon products, typically in liquid form. Documents related to the history and development of the Fischer-Tropsch process are archived in electronic form at www.fischer-tropsch.org. This website is unusual, but its existence points out the importance of the process in hydrocarbon chemical engineering. Fischer-Tropsch synthesis takes place at low temperature (200–240° C) or at high temperature (300–350° C) with iron or cobalt serving as catalyst in reactors that operate at pressures ranging from 10 to

TABLE 12.7 Fischer-Tropsch synthesis examples (adapted from Spath and Dayton 2003).

Reaction	Product
$CO + 2H_2 \rightarrow -CH_2- + H_2O$	Methyl species
$CO + 3H_2 \rightarrow CH_4 + H_2O$	Methane
$nCO + (2n + 1)H_2 \rightarrow C_nH_{2n+2} + nH_2O$	Paraffins
$nCO + 2nH_2 \rightarrow C_nH_{2n} + nH_2O$	Olefins
$nCO + 2nH_2 \rightarrow C_nH_{2n+1}OH + (n - 1)H_2O$	Alcohols

40 atmospheres (Spath and Dayton 2003). The most common hydrocarbons synthesized using the Fischer-Tropsch process are listed in Table 12.7. In the first reaction in Table 12.7, $-CH_2-$ indicates methyl species that can serve as initiators for chain growth via the sequential addition of $-CH_2-$. Depending on the ratio of carbon monoxide to hydrogen, a wide range of hydrocarbons can result from the Fischer-Tropsch synthesis. Paraffins are alkane hydrocarbons and have the general chemical formula) C_nH_{2n+2}. For $20 < n < 40$ paraffins are solid and waxy, as suggested by the name. Octane, C_8H_{18}, is an example of a paraffin. Olefins are alkene hydrocarbons; an example is propene, C_3H_6. Two characteristics of such syntheses are (1) the unavoidable production of a wide range of hydrocarbon products and (2) the liberation of significant amounts of heat from the exothermic synthesis reactions. The distribution of the products is determined by temperature, pressure, the CO/H_2 ratio, and catalyst (type and composition). Products generated by the Fischer-Tropsch process must generally undergo upgrading procedures that result in product separation, concentration, and tailoring. The upgrading procedures generally involve further chemical reactions.

The extensive Fischer-Tropsch literature should be consulted for additional details. For biomass syngas, Spath and Dayton (2003) is a comprehensive reference.

Example 12.3

For 1 kg of CO, calculate the hydrogen required and the resulting products for methane, ethanol, octane, and propene for Fischer-Tropsch synthesis.

Solution The molecular masses for CO, H_2, CH_4, H_2O, ethanol, octane, and propene are

$$CO_{mm} = 28.01\frac{kg}{kg\ mol} \qquad H2_{mm} = 2.016\frac{kg}{kg\ mol}$$

$$CH4_{mm} = 16.042\frac{kg}{kg\ mol} \qquad H2O_{mm} = 18.016\frac{kg}{kg\ mol}$$

$$ETHANOL_{mm} = 46.068\frac{kg}{kg\ mol} \qquad OCTANE_{mm} = 114.224\frac{kg}{kg\ mol}$$

$$PROPENE_{mm} = 42.078\frac{kg}{kg\ mol}$$

For methane, the reaction is

$$CO + 3H_2 \rightarrow CH_4 + H_2O$$

For 1 kg of CO, the hydrogen required and the methane and water formed become

$$H_2 = \frac{3 \cdot H2_{mm}}{CO_{mm}} \cdot 1\ kg = 0.216\ kg \qquad CH_4 = \frac{CH4_{mm}}{CO_{mm}} \cdot 1\ kg = 0.573\ kg$$

$$H_2O = \frac{H2O_{mm}}{CO_{mm}} \cdot 1\ kg = 0.643\ kg$$

For ethanol (C_2H_5OH), an alcohol, the general alcohol reaction from Table 12.7 with $n = 2$ can be used:

$$nCO + 2nH_2 \rightarrow C_nH_{2n+1}OH + (n - 1)H_2O$$

For 1 kg of CO, the hydrogen required and the ethanol and water formed become

$$H_2 = \frac{4 \cdot H2_{mm}}{2 \cdot CO_{mm}} \cdot 1\ kg = 0.144\ kg$$

$$ETHANOL = \frac{ETHANOL_{mm}}{2 \cdot CO_{mm}} \cdot 1\ kg = 0.822\ kg$$

$$H_2O = \frac{H2O_{mm}}{2 \cdot CO_{mm}} \cdot 1\ kg = 0.322\ kg$$

For methanol (CH_3OH), also an alcohol, the general alcohol reaction with $n = 1$ is required. No water is produced. For 1 kg of CO, the hydrogen required and the methanol and water formed become

$$H_2 = \frac{4 \cdot H2_{mm}}{CO_{mm}} \cdot 1\ kg = 0.144\ kg$$

$$METHANOL = \frac{METHANOL_{mm}}{CO_{mm}} \cdot 1\ kg = 1.144\ kg$$

$$H_2O = 0.0\ kg$$

For octane (C_8H_{18}), the general paraffin reaction with $n = 1$ is used.

$$nCO + (2n + 1)H_2 \rightarrow C_nH_{2n+2} + nH_2O$$

For 1 kg of CO, the hydrogen required and the octane and water formed become

$$H_2 = \frac{17 \cdot H2_{mm}}{8 \cdot CO_{mm}} \cdot 1\ kg = 0.153\ kg \qquad OCTANE = \frac{OCTANE_{mm}}{8 \cdot CO_{mm}} \cdot 1\ kg = 0.51\ kg$$

$$H_2O = \frac{8 \cdot H2O_{mm}}{8 \cdot CO_{mm}} \cdot 1\ kg = 0.643\ kg$$

For propene (C_3H_6), the general olefin reaction with $n = 3$ is used.

$$nCO + 2nH_2 \rightarrow C_nH_{2n} + nH_2O$$

For 1 kg of CO, the hydrogen required and the propene and water formed become

$$H_2 = \frac{6 \cdot H2_{mm}}{3 \cdot CO_{mm}} \cdot 1 \text{ kg} = 0.144 \text{ kg}$$

$$PROPENE = \frac{PROPENE_{mm}}{3 \cdot CO_{mm}} \cdot 1 \text{ kg} = 0.501 \text{ kg}$$

$$H_2O = \frac{3 \cdot H2O_{mm}}{3 \cdot CO_{mm}} \cdot 1 \text{ kg} = 0.643 \text{ kg}$$

A summary of the reactants and products for each of the reactions is presented in Table 12.8. The variety of possible Fischer-Tropsch products is illustrated by these results. The ratio of carbon monoxide to hydrogen is the same for some reactions, but the products are different. The products are different for reactions with different carbon monoxide to hydrogen ratios—hence the first characteristic of Fischer-Tropsch synthesis, relating to the unavoidable production of a wide range of hydrocarbon products.

TABLE 12.8 Summary of Fischer-Tropsch synthesis reactions for Example 12.3

Reaction	CO (kg)	H$_2$ (kg)	H$_2$O (kg)	Product (kg)
Methane (CH$_4$)	1	0.216	0.643	0.573
Ethanol (C$_2$H$_5$OH)	1	0.144	0.322	0.822
Methanol (CH$_3$OH)	1	0.144	0.0	1.144
Octane (C$_8$H$_{18}$)	1	0.153	0.643	0.510
Propene (C$_3$H$_6$)	1	0.144	0.643	0.501

Although not strictly a process with biomass feedstock, a proposal by Uhrig et al. (2007) utilizes CO_2 from fossil fuel–fired power plants in conjunction with hydrogen from non–fossil fuel sources and the Frischer-Tropsch process to produce synthetic fuels for motor vehicles. In this process the feedstocks are carbon dioxide and water; electrolysis is used to produce hydrogen from the water. The "reverse" water shift reaction (Table 12.6) is used for production of carbon monoxide from carbon dioxide and water, followed by the Fischer-Tropsch reaction to produce CH_2— the fundamental building block for hydrocarbon synthetic fuels.

12.6 MUNICIPAL SOLID WASTE (MSW)

Because of its high content of organic materials, municipal solid waste (MSW) is often classified as a biomass feedstock source. MSW is trash or garbage that is usually thrown away by households, and does not include industrial, hazardous, or construction waste. The Environmental Protection Agency (EPA) provides a yearly compilation of municipal solid waste information that addresses MSW generation, composition, and disposal. Additionally, historical data on MSW are presented so that trends can be ascertained. Most of the first portion of this section was abstracted from the 2006 EPA compilation *Municipal Solid Waste Generation, Recycling, and Disposal in the United States: Facts and Figures for 2006,* available at www.epa.gov.

Figure 12.10 presents the United States' total MSW (in millions of tons) generated per year and the per capita generation (pounds per person per day) from 1960 through 2006. In the forty-six years represented in Figure 12.10, the total MSW generation rate increased by 185 percent while the population increased by 67 percent! On a per capita basis, the increase was 72 percent. About the only positive comment that can be made about the information in the figure is that since 2000 the per capita generation rate has stabilized at about 4.6 lb/person day. Figure 12.10 illustrates in a rather dramatic fashion the "disposal" philosophy of the economy of the United States.

Two questions are germane: (1) What is the composition of MSW? (2) What happens to MSW? Figure 12.11 is a pie chart showing the composition of MSW for 2006. The EPA defines MSW as composed of yard trimmings, wood, plastics, metals, glass, paper, food scraps, rubber/leather/textiles, and other. Paper, at 33.9 percent (by weight), is by far the largest category, followed by food scraps, yard trimmings, and plastics, each at nearly 12 percent. The remaining categories each represent between 3 and 8 percent. Plastics, glass, and metals are non-biomass components in MSW. Later in this section, the biomass versus non-biomass composition of MSW is examined in more detail.

Table 12.9, with results expressed in millions of tons, addresses the second question: what happens to MSW? The EPA compiles information on MSW recovered for

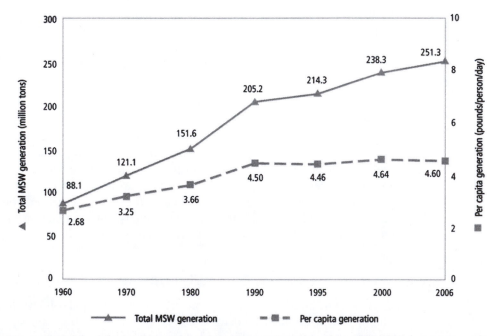

Figure 12.10 Municipal solid waste total and per capita generation, 1960–2006 (www.epa.gov).

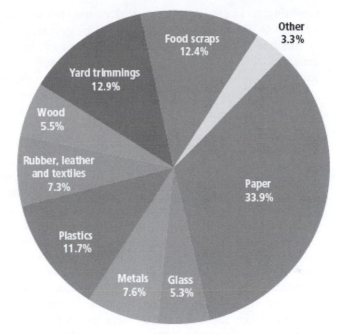

Figure 12.11 Municipal solid waste composition in 2006 (www.epa.gov).

TABLE 12.9 Generation, recovery, composting, combustion with energy recovery, and discards of municipal solid waste (millions of tons), 1960–2006 (www.epa.gov)

Activity	1960	1970	1980	1990	2000	2002	2004	2005	2006
Generation	88.1	121.1	151.6	205.2	238.3	239.4	249.2	248.2	251.3
Recovery for recycling	5.6	8.0	14.5	29.0	52.8	53.8	57.5	58.6	61.0
Recovery for compositing[a]	Negligible	Negligible	Negligible	4.2	16.5	16.7	20.5	20.6	20.8
Total materials recovery	5.6	8.0	14.5	33.2	69.3	70.6	77.9	79.1	81.8
Combustion with energy recovery[b]	0.0	0.4	2.7	29.7	33.7	33.4	34.4	33.4	31.4
Discards to landfill, other disposal[c]	82.5	112.7	134.4	142.3	135.3	135.5	136.9	135.6	138.2

[a] Composting of yard trimmings, food scraps, and other MSW organic material. Does not include backyard composting.

[b] Includes combustion of MSW in mass burn or refuse-derived fuel form, and combustion with energy recovery of source-separated materials in MSW (e.g., wood pallets and tire-derived fuel).

[c] Discards after recovery minus combustion with energy recovery. Discards include combustion without energy recovery. Details may not add to totals due to rounding.

recycling, recovered for composting, combusted with energy recovery, and discarded to landfills. The good news is that while the generation rate has remained nearly constant since 2000, the amount recovered for recycling, the total composted, and the amount of materials recovered have all increased. The MSW combusted in facilities producing energy has actually decreased slightly since 1990, reflecting the increasing cost of such facilities, the more stringent emissions requirements, and the increase in difficulties with permitting practices.

While landfilling has long been the usual solution for disposal of MSW, landfilling is associated with difficulties such as availability, siting, costs, transportation, effluent generation, and capacity. The number of available landfills in the United States decreased from 7924 in 1988 to 1967 in 2000 to 1754 in 2006 (EPA 2006). One way to reduce the landfill volume is incineration—burning MSW at high temperatures. Incineration decreases the volume of MSW to 10–12 percent of the original for "typical" MSW, allowing existing landfills to last longer.

This discussion focuses on facilities that combust MSW and produce useful energy—waste-to-energy (WTE) facilities. Such facilities use the energy from the MSW to make steam that is used in district heating and cooling systems or to power a generator to produce electricity for the grid. In 2007 the Integrated Waste Energy Association (IWEA) identified 89 WTEs in the United States. Many of the developed countries make more extensive use of MSW than the United States. Most of the countries of western Europe, for example, have extensive WTE facilities and programs. The web is an excellent source of detailed information about WTE facilities. The primary purpose of a WTE facility is to reduce the volume of solid waste with minimum adverse impact on the environment. The energy captured is secondary, but is important especially as a selling point for local governments with both solid waste disposal and energy concerns.

WTE facilities are broadly classed as mass-burn systems or refuse-derived fuel (RDF) systems. Mass-burn WTE facilities combust, with excess air, the MSW as received and usually have only a single combustion chamber. A moving, sloping grate is used to ensure that the MSW is well mixed with air; the grate also provides a path for ash to enter an ash collection system. For RDF systems, the MSW is processed to a more homogeneous mix compared with the as-received MSW for the mass-burn system. Mass-burn WTEs are the most common, with the MSW being used in same fashion as other biomass fuels in direct combustion technologies.

The mechanical processing for RDF can include grinding/shredding as well as pelletization and removal of metals and other bulky items. Refuse-derived fuels can be used as the primary fuel or co-fired with fossil fuels in conventional utility boilers. Although the number varies according to the local composition of municipal solid waste, the IWEA states on its website, www.wte.org, that on average each ton of MSW used in a modern WTE facility produces about 550 kWh that can be supplied to the grid.

The Pinellas County (Florida) Utilities WTE facility is an example of a facility that burns MSW and uses the energy to generate electricity, which is supplied to the grid. Information is available at www.pinellascounty.org/utilities/wte. Figure 12.12 is a schematic illustrating the primary features of the Pinellas County WTE. MSW

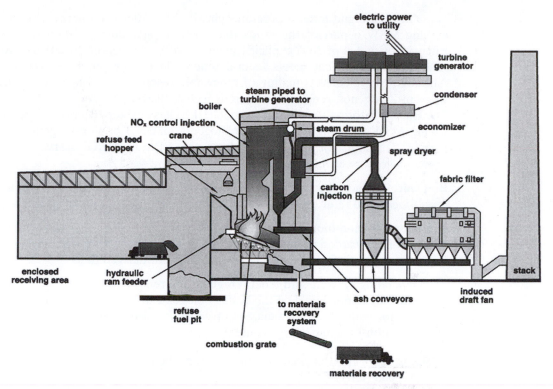

Figure 12.12 WTE facility schematic (www.wheelabratortechnologies.com).

arrives at the facility by truck and is dumped into a receiving pit. A crane feeds the MSW into the combustion section of the boiler; ash falls through the combustion grate and is routed for post-combustion treatment. Steam is produced in the boiler and directed to a turbine-generator that is grid connected.

The facility processes about one million tons annually, although the capacity is 3150 tons of MSW per day. The nominal output is 75 MW, of which 60 MW is routed to the electrical grid and 15 MW is used internally. The Pinellas County WTE uses modern pollution control technology to ensure that emissions are within EPA standards. The metals in the ash resulting from the combustion of the MSW are recovered and sold for recycling. The remaining ash, after being size separated, is used for landfill cover, interior berm fill, and roadways. WTE facilities are quite complex, and only a brief overview has been presented here; however, Kreith and West (1997) present a detailed discussion of most aspects of waste-to-energy technology. Many local governments with WTE facilities have websites with statistics and operational details. A web search using "WTE" as the key word yields many hits. Many of these websites contain technical information and relate operational experiences; however, a non-negligible number of websites report details of litigation involving solid waste disposal or counterclaims by individuals or citizen interest groups. WTE facilities thus generate some controversy.

Many state and federal programs classify MSW-derived energy as renewable, making MSW a part of the renewable portfolio standard program. The federal Energy Policy Act of 2005 explicitly includes MSW-derived electricity as a form of renewable energy that meets federal renewable energy purchase requirements. Although MSW consists mainly of renewable resources such as food, paper, and wood products, non-renewable materials such as tires and plastics are also included. The Energy Information Administration (EIA) has developed a procedure for classifying MSW as being composed of biogenic material or non-biogenic material. Formal definitions of these terms from the EPA (EPA Glossary available at www.epa.gov)are as follows:

Biogenic: Produced by biological processes of living organisms. [*Note*: EIA uses the term *biogenic* to refer only to organic nonfossil material of biological origin.]

Non-biomass (non-biogenic) waste: Material of non-biological origin that is a by-product or a discarded product. "Non-biomass waste" includes municipal solid waste from non-biogenic sources, such as plastics, and tire-derived fuels.

Depending on state and federal regulations, renewable-energy credit for using MSW may be apportioned on the basis of biogenic and non-biogenic contributions. Table 12.10, available at www.eia.gov, presents the history of MSW in terms of biogenic and non-biogenic energy content.

TABLE 12.10 MSW biogenic and non-biogenic heat content from 1989 to 2005 (EIA)

Year	Heat Content (million Btu/ton)	Total MSW Energy	
		Shares of Biogenic	**Non-biogenic**
1989	10.08	0.67	0.33
1990	10.21	0.66	0.34
1991	10.40	0.65	0.35
1992	10.61	0.64	0.36
1993	10.94	0.64	0.36
1994	11.15	0.63	0.37
1995	11.11	0.62	0.38
1996	10.94	0.61	0.39
1997	11.17	0.60	0.40
1998	11.06	0.60	0.40
1999	10.95	0.60	0.40
2000	11.33	0.58	0.42
2001	11.21	0.57	0.43
2002	11.19	0.56	0.44
2003	11.17	0.55	0.45
2004	11.45	0.55	0.45
2005	11.73	0.56	0.44

Note: Years in bold are EPA data collection years.

The trend indicated in Table 12.10 is a decrease in MSW biogenic content and an increase in non-biogenic content through the years shown. Because the non-biogenic material generally has a higher heating value than the biogenic portion, the energy content per ton has also increased over time.

Example 12.4	Using the IWEAs (www.wte.org) estimate that each ton of MSW processed in a WTE facility provides 550 kWh to the grid, estimate the electrical efficiency using the EIA estimate of the energy content per ton of MSW (Table 12.10). If 50 percent of the MSW were used in the WTEs, estimate the base-load power that could be supplied to the electrical grid.

Solution From Table 12.10, the EIA estimated that the energy content of an "average" ton of MSW was 11.73×10^6 Btu in 2005. If 550 kWh can be supplied to the grid, then the electrical efficiency becomes

$$\eta = \frac{550 \, \dfrac{kWh}{ton} \cdot 3412 \, \dfrac{Btu}{kWh}}{11.73 \cdot 10^6 \, \dfrac{Btu}{ton}} = 0.16$$

Thus, a WTE can harvest and supply to the grid about 16 percent of the energy content of average MSW. If this amount seems small, remember that the MSW volume has been reduced by 88–90 percent in the WTE and that much of the remaining ash might go toward other uses.

The electricity that might be generated from the use of 50 percent of the MSW in the United States in 2006 can be estimated using information in Table 12.9. Consider

$$E_{generated} = 0.5 \cdot 251.3 \cdot 10^6 \, ton \cdot 550 \cdot \frac{kWh}{ton} = 6.91 \cdot 10^{10} \, kWh$$

The only viable strategy for WTE usage is as a base-load supply to the electrical grid. For a base-load situation, the total kWh generated would be constant with respect to time. Since a year contains 8760 hr, the power supplied is estimated as

$$Power = \frac{6.91 \cdot 10^{10} \, kWh}{8760 \, hr} = 7890 \, MW$$

Used in a uniform manner as a base load to the grid, 50 percent of the MSW could supply nearly 8000 MW.

The estimates made in this example contain significant uncertainties since they are based on "average" energy content, expected electrical harvesting, and yearly data. Nonetheless, the electrical efficiency, the kWh generated, and the average power available are reasonable first-order estimates.

CLOSURE

As is true for many alternative energy sources, the energy content and potential of biomass are staggering, but the technical and economic challenges are also imposing. Table 12.11 provides a convenient summary of biomass technologies, processes, feedstocks, and end-point uses. Biomass is certainly going to be an increasingly important source of energy in the United States.

The outlook for the widespread use of biomass feedstock for fuels and chemical applications is summarized in a quote from Tester et al. (2005):

> As a raw material we can think of biomass as a nearly "universal feedstock" for producing energy and a plethora of energy intensive fuels and chemicals. Because of this versatility and the renewability of biomass, is it then the preferred "solution" for a sustainable planet? ... A variety of challenges must first be surmounted to push biomass beyond its current 5% contribution to energy needs in the developed world.

TABLE 12.11   Biomass summary
(developed from the Oregon Department of Energy Biomass Program)

Technology	Conversion Process Type	Major Biomass Feedstock	Energy or Fuel Produced
Direct combustion	Thermochemical	Wood	Heat
		Agricultural waste	Steam
		Municipal solid waste	Electricity
Gasification	Thermochemical	Wood	Producer gas
		Agricultural waste	
		Municipal solid waste	
Pyrolysis	Thermochemical	Wood	Synthetic fuel oil
		Agricultural waste	(biocrude)
		Municipal solid waste	Charcoal
Anaerobic digestion	Biochemical (anaerobic)	Animal manure	Methane
		Agricultural waste	
		Landfills/wastewater	
Ethanol production	Biochemical (aerobic)	Sugar or starch crops	Ethanol
		Wood waste	
		Pulp sludge	
Biodiesel production	Chemical	Seed oils	Biodiesel
		Waste vegetable oil	
		Animal fats	
Methanol production	Thermochemical	Wood	Methanol
		Agricultural waste	
		Municipal solid waste	

REVIEW QUESTIONS

1. What is the most common source of biomass?

2. What is the most common biomass feedstock?

3. What does "CO_2 neutral" mean?

4. True or false? All biomass fuels and products are CO_2 neutral. Explain.

5. Can biomass supply a significant portion of the United States' energy requirements? Explain.

6. What is an energy concern with ethanol production from biomass?

7. What is anaerobic digestion? What is the primary useful product of anaerobic digestion?

8. What is gasification?

9. How does anaerobic digestion differ from fermentation? Answer in terms of process, feedstock, and product.

10. How does the energy content (on a volume basis) differ for gasoline, ethanol, and methanol?

11. What biochemical compound is required for fermentation to produce ethanol?

12. Could the United States produce enough ethanol from food crops to displace gasoline? Explain your answer.

13. If a motor vehicle gets 30 miles per gallon on conventional gasoline, what gas mileage would be expected from pure ethanol? From methanol?

14. Why aren't vegetable oils (from grain and seeds) used directly in a diesel engine?

15. What is the relationship between biodiesel and transesterification?

16. What is an environmental problem with biomass?

17. What is producer gas? With what process is it associated?

18. What is the municipal solid waste (MSW) problem?

19. What are two benefits resulting from waste-to-energy conversion?

20. In the United States, what is the largest component of MSW?

21. What is the best heating value (Btu/lbm) that can be expected from MSW?

22. What is the typical volume reduction of waste for an MSW operation?

23. What is the environmental downside of MSW treatments?

24. Why is the production of ethanol from cellulosic sources more difficult than the production of ethanol from grain sources?

25. What is an important advantage of cellulosic ethanol over grain-source ethanol?

EXERCISES

1. How much ethanol can be produced from a ton of glucose, $C_6H_{12}O_6$?

2. How much methanol can be made from a ton of natural gas (methane) according to the following sequence of reactions?

$$CH_4 + H_2O \rightarrow CO + 3H_2$$
$$H_2 + CO_2 \rightarrow H_2O + CO$$
$$CO + 2H_2O \rightarrow CH_3OH$$

3. How much biodiesel can be made from a ton of soybean oil? The average molecular weight of soybean oil methyl esters is 292.2 kg/kg mol (from www.biodiesel.org/pdf_files/fuelfactsheets/Weight&Formula.PDF).

4. For 1 kg of CO, calculate the hydrogen required and the resulting products for ethylene (C_2H_4), isopropanol (C_3H_8O), and pentane (C_5H_{12}) for Fischer-Tropsch synthesis.

REFERENCES

Boyle, G., ed. 2004. *Renewable Energy.* Oxford, UK: Oxford University Press.

Brown, R. C., 2004, "Thermochemical Technologies for Biomass Energy," *Proceedings of the 2004 IEEE Power Engineering Society General Meeting*, Vol. 2, pp. 1650–1652.

Chynoweth, D. P., and Isaacson, R. E., eds. 1987. *Anaerobic Digestion of Biomass.* New York: Springer-Verlag.

Cobb, J. T. 2007. "Production of Synthesis Gas by Biomass Gasification," in *Proceedings of the 2007 Spring National AIChE Meeting, Houston, TX.*

Decher, R. 1994. *Energy Conversion.* New York: Oxford University Press.

Drapcho, C., Nghiem, J., and Walker, T. 2008. *Biofuels Engineering Process Technology.* New York: McGraw-Hill.

Environmental Protection Agency (EPA). 2006. *Municipal Solid Waste Generation, Recycling, and Disposal in the United States: Facts and Figures for 2006.* Washington, DC: U.S. Environmental Protection Agency.

Goswami, Y., Kreith, F., and Kreider, J. F. 2000. *Principles of Solar Engineering*, 2nd ed. New York: Taylor & Francis.

Haynie, D. T. 2008, *Biological Thermodynamics*, 2nd ed. Cambridge, UK: Cambridge University Press.

Himmel, M., ed. 2008. *Biomass Recalcitrance: Deconstructing the Plant Cell Wall for Bioenergy.* Hoboken, NJ: Wiley-Blackwell.

Klass, D. L. 1998. *Biomass for Renewable Energy, Fuels, and Chemicals.* San Diego, CA: Academic Press.

Kreith, F. 2007, "Cellulosic Ethanol: Answer to the Biofuels Challenges?" *Solar Today,* http://www.solartoday.org/2007/may_june07/Oil_Addiction_MJ07_Kreith.pdf.

Kreith, F., and West, R. 1997. *CRC Handbook of Energy Efficiency.* Boca Raton, FL: CRC Press.

Kreith, F., and West, R. 2007. "The Road Not Yet Taken," *Mechanical Engineering*, April.

Kruger, P. 2006. *Alternative Energy Resources: The Quest for Sustainable Energy*. Hoboken, NJ: John Wiley.

Perlack, R. D., Wright, L. L., Turhollow, A. F., Graham, R. L., Stokes, B. J., and Erbach, D. C. 2005. *Biomass as Feedstock for a Bioenergy and Bioproducts Industry: The Technical Feasibility of a Billion-Ton Annual Supply*. DOE/GO-102005-2136, ORNL/TM-2005/66. Also available at www.eere.energy.gov/afdc/pdfs/fueltable.pdf.

Pimentel, D. 2003. "Ethanol Fuels: Energy Balance, Economics, and Environmental Impacts Are Negative," *Natural Resources Research* 12(2), June, 127–134.

Reyes, S. C., Sinfelt, J. H., and Feeley, J. S. 2003. "Evolution of Processes for Synthesis Gas Production: Recent Developments in an Old Technology," *Industrial and Engineering Chemistry Research* 42, 1588–1597.

Schobert, H. H. 2002. *Energy and Society*. New York: Taylor & Francis.

Spath, P. L., and Dayton, D. C. 2003. *Preliminary Screening—Technical and Economic Assessment of Synthesis Gas to Fuels and Chemicals with Emphasis on the Potential for Biomass-Derived Syngas*. NREL/TP-510-34929. Golden, CO: NREL.

Speight, J. 2008. *Synthetic Fuels Handbook: Properties, Process, and Performance*. New York: McGraw-Hill.

Tester, J. W., Drake, E. M., Driscoll, M. J., Golay, M. W., and Peters, W. A. 2005. *Sustainable Energy*. Cambridge, MA: MIT Press.

Tidwell, J. W., and Weir, A. D. 1986. *Renewable Energy Resources*. New York: Taylor & Francis.

Tyson, K. S., Bozell, J., Wallace, R., Petersen, E., and Moens, L. 2004. *Biomass Oil Analysis: Research Needs and Recommendations*. NREL/TP-510-34796. Golden CO: NREL.

Uhrig, R. E., Schultz, K. R., and Bogart, S. L. 2007. "Implementing the 'Hydrogen Economy' with Synfuels," *The Bent of Tau Beta Pi* 97(3), Spring, 18–22.

Van Gerpen, J., Shanks, B., Pruszko, R., Clements, D., and Knothe, G. 2004. *Biodiesel Production Technology*. NREL/SR-510-36244. Golden, CO: NREL.

Van Loo, S., and Koppejan, J. 2008. *The Handbook of Biomass, Combustion, & Co-firing*. London: Earthscan.

Wright, L., Boundy, B., Perlack, B., Davis, S., and Saulsbury, B. 2006. *Biomass Energy Data Book: Edition 1*. ORNL/TM-2006-571. Oak Ridge, TN: ONRL. Available only at www.cta.ornl.gov/bedb.

WEBSITES

www.berr.gov.uk/files/file39295.pps
www.eere.energy.gov
www.eere.energy.gov/RE/bio_fuels.html
www.epa.gov/lmop/index
www.fischer-tropsch.org
www.genomics.energy.gov
www.pinellascounty.org/utilities/wte
www.potb.org/methane-energy.htm
www.wheelabratortechnologies.com
www.wte.org

Geothermal Energy

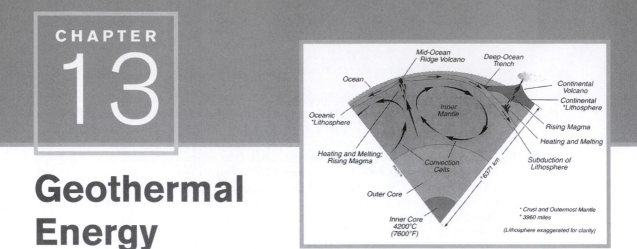

13.1   INTRODUCTION

Geothermal energy is energy from the Earth. Traditionally, geothermal energy has been viewed as harvesting energy whose sources lie deep within the Earth. In recent years, the use of ground-source heat pumps, which use soil as a reservoir, has become common; see www.geokiss.com and Section 13.5 for more information. In Figure 13.1 is a telling statement about the pervasiveness of geothermal energy. But harvesting this energy poses problems of both access and technology. One of the goals of this chapter is to impart an understanding of the challenges associated with geothermal energy systems.

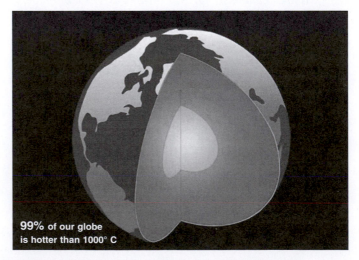

Figure 13.1 A geothermal energy factoid (Haring GeoProject).

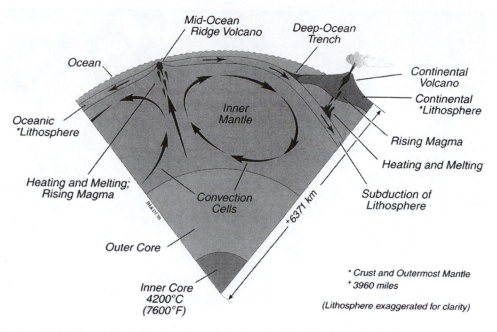

Figure 13.2 Schematic of the structure of the Earth (NREL/TP-840-40665).

The geological structure of the Earth is a good starting point for examining geothermal energy. Consider the cut away view of the Earth's composition in Figure 13.2. At the center is a molten inner core, surrounded by an outer core. An inner mantle surrounds the outer core. The lithosphere is a thin, rigid crust that serves as the outermost mantle. The lithosphere is fractured into tectonic plates—12 large ones and a number of smaller ones. These plates move relative to each other driven by gravitational and convection forces in the mantle. This geological process, known as plate tectonics, creates fault lines where the plates meet and produces volcanic and seismic activity. Figure 13.3 illustrates the major plates and their boundaries. Where the plates move apart, pools of molten rock—magma—rise, leading to volcanic activity. Where the plates collide, one plate typically moves underneath the other, a process called subduction. In the subduction zone, temperatures become high enough to melt rock, and seismic and volcanic activities are common. Much of the geothermal energy is inaccessible because of its great depths, but along the plate boundaries, geothermal activity is close enough to the surface to be accessible. The active geothermal zones are indicated in Figure 13.3. The active geothermal zones are, not surprisingly, the regions with the most geothermal activity.

The next section examines the estimated extent of geothermal energy resources and provides an introduction to the harvesting procedures appropriate for geothermal energy.

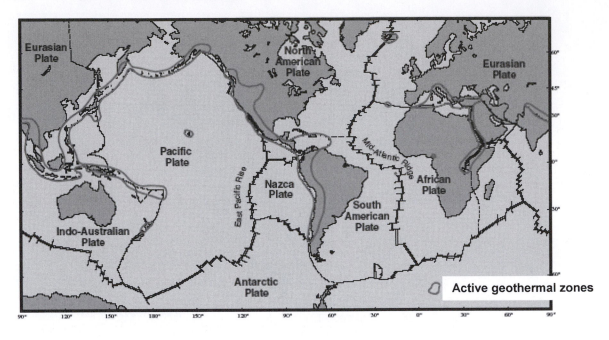

Figure 13.3 Major tectonic plates and geothermal zones (NREL/TP-840-40665).

13.2 GEOTHERMAL RESOURCES

As can be inferred from Figure 13.1, geothermal resources are extensive, but the real problems are accessibility and technology. Gupta and Roy (2007) provide a contemporary assessment of geothermal resources and the geothermal energy outlook for the entire Earth. Energy harvesting procedures for geothermal energy vary widely and depend on the local geology. Geothermal resources are characterized by their thermal and compositional characteristics as follows:

1. Hydrothermal or geohydrothermal
2. Geopressurized
3. Magma
4. Enhanced geothermal systems (or hot, dry rock)

Different harvesting strategies are required for the different classes of geothermal resources. But before exploring the harvesting strategies, we will examine general characteristics of each class.

"Hydrothermal" refers to conditions in which water is heated and/or evaporated by direct contact with hot porous rock. The porous or permeable rock in a hydrothermal site is contained (bounded) by rock strata of low permeability. Water, provided by aquifers, trickles through the porous rock and is heated and perhaps vaporized (to form steam), and is discharged at the surface. Hydrothermal systems that produce steam are labeled as vapor dominated, while hydrothermal systems that produce hot water or a mixture of hot water and steam are called liquid dominated. Energy from hydrothermal sources can be harvested relatively easily. Many existing geothermal power plants are associated with hydrothermal systems. But rock formations with hydrothermal characteristics are rare, and hydrothermal resources are the most limited of the four categories.

Geopressurized resources are associated with sediment-filled reservoirs that contain hot water confined under pressures much greater than hydrostatic. The fluids contained in geopressurized systems are in the temperature range 150–180° C and may have pressures up to 600 bar (~9000 psi). In many geopressurized systems, the hot water may also contain methane in solution, as well as very high levels of dissolved solids—up to 100,000 ppm. These brine solutions are both highly corrosive and difficult to handle.

Magma, or molten rock, at accessible depths is contained in pools under active volcanoes. Magma temperatures are typically in excess of 650° C, making magma attractive from a resource standpoint. Although the energy density of molten rock is quite high, proposed methods for harvesting magma energy are at best speculative.

Hot, dry rock (HDR) suitable as a geothermal resource is characterized by temperatures in excess of 200° C, but, as the name implies, has little in the way of naturally contained liquids. Recently HDR systems have been referred to as enhanced geothermal systems (EGS), and the U.S. Department of Energy has adopted the EGS classification. The basic idea behind harvesting energy from EGS resources is to inject water under pressure to fracture the rock, and to use the steam formed from the injected water to drive a turbine to generate electricity.

Table 13.1 presents estimates of the accessible geothermal resources in terms of the aforementioned classifications. The values in the table are order-of-magnitude estimates and are listed in order of increasing availability of geothermal resources: hydrothermal, geopressurized, magma, and EGS (hot, dry rock). The energy economy of the United States is about 100 quad (see Chapter 1) and that of the world about 400 quad. Thus, at the current usage rates, hydrothermal resources in the United States could last about 100 years, but the other geothermal resource classifications offer an essentially unlimited quantity of energy. The problem is, of course, how to harvest and use all of these resources.

TABLE 13.1 Geothermal resource estimates (adapted from Tester et al. 2005)

Resource	United States (1000 quad)	World (1000 quad)
Hydrothermal	10	130
Geopressurized	170	540
Magma	1,000	5,000
Hot, dry rock	30,000	105,000

Figure 13.4 illustrates regions in the United States with accessible geothermal resources. In this figure, geothermal resources are expressed in terms of end-use capability. Resources above 100° C have the potential to be utilized for electric power generation, while resources below 100° C can be used for direct thermal energy. Most of these geothermal resources are located in the mountain and coastal areas of the West that conform generally to the active geothermal zones of Figure 13.3. As briefly mentioned in the introduction, ground-source heat pumps are also considered users of geothermal energy—albeit it at low temperatures with shallow (meters rather than kilometers) access. The entire continental United States offers the possibility of geothermal energy for ground-source heat pump applications.

Different geothermal resource classifications require different harvesting technologies. Geothermal system details are examined in the next section.

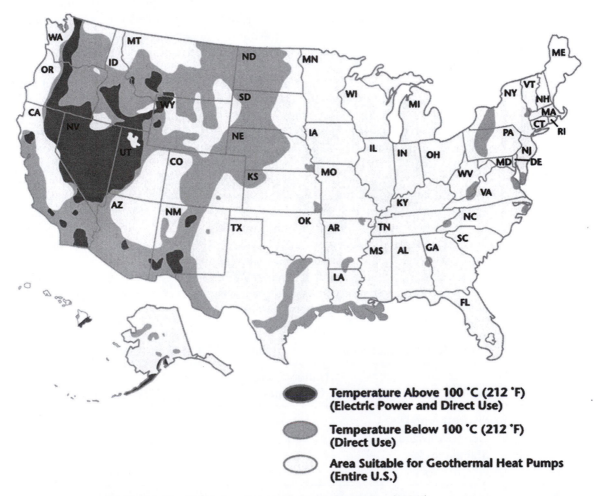

Figure 13.4 Geothermal resources in the United States (DOE).

GEOTHERMAL ENERGY SYSTEMS

Different system configurations have been proposed and investigated for harvesting geothermal energy resources according to the different classifications. DiPippo (2007) provides a seminal presentation and case studies for geothermal energy and is recommended reading for understanding the engineering aspects of geothermal energy. Several geothermal systems will be examined in detail next.

Hydrothermal

Hydrothermal sources have either hot water or steam available and represent the easiest geothermal sources to harvest. Dissolved solids, entrained solid particles, and non-condensable gases are problems with most geothermal resources. The entrained solids are generally removed by centrifugal separators at the well head, and filters are placed in the system to ensure that solid particles are removed. Non-condensable gases abound in some locations and pose many problems for geothermal systems. Many of the non-condensable gases form acids under wet conditions, necessitating the use of stainless steel or other expensive materials. Additionally, these gases pose environmental hazards if released to the atmosphere. Hydrothermal sources are usually considered as either vapor dominated or liquid dominated.

Vapor-dominated hydrothermal sources are the most suitable for generation of electricity, but are the rarest source of geothermal energy. As the name implies, in vapor-dominated systems steam is the available resource. Vapor-dominated systems require steam at temperatures >175° C (Kutscher 2000). Conditions at the surface seldom exceed 205° C and 8 bar (El-Wakil 1984). A system schematic for vapor-dominated systems is presented in Figure 13.5. The steam from the well passes

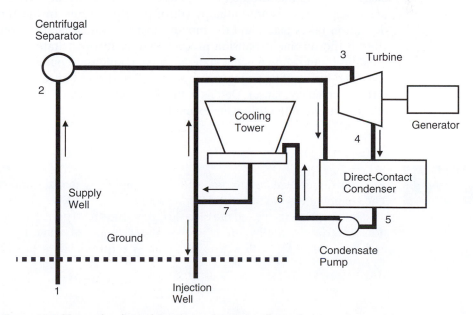

Figure 13.5 Vapor-dominated geothermal system schematic.

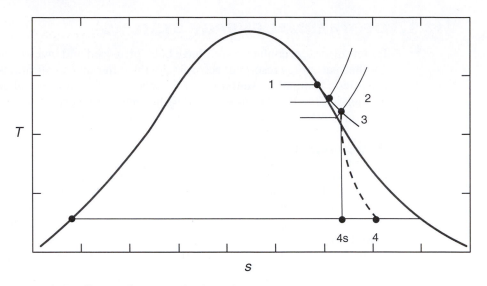

Figure 13.6 *T-s* diagram for vapor-dominated systems.

through a centrifugal separator and enters the turbine. After exiting the turbine, the steam is condensed, cooled in a cooling tower, and re-injected. In the figure, the thermodynamic states are indicated as 1–7. The process path is indicated on the *T-s* diagram presented as Figure 13.6. State 1 is saturated vapor (steam) at the well bottom. The vapor goes through a centrifugal separator (state 2) for removal of solid particles and enters the turbine (state 3) as a superheated vapor. Processes 1–2 and 2–3 are essentially constant-enthalpy (throttling) processes. In the turbine, some condensation takes place, and the mixture exits at state 4. State 4s represents the exit state if the turbine expansion process were isentropic. State 5 is the exit from the condenser, and state 6 is the state after the condensate pump at the cooling tower. The fluid leaves the cooling tower at state 7 and is either re-injected or used in the direct-contact condenser. Details of the thermodynamics of vapor-dominated systems are presented in the following example.

Example 13.1

A vapor-dominated geothermal system is supplied with saturated steam at 3 MPa. The steam enters the turbine at 0.5 MPa and exits at 15 kPa. The turbine isentropic efficiency is 82 percent, and the electrical generator is 90 percent efficient. If re-injection occurs at the cooling tower, analyze the system performance (thermal efficiency and heat rate). What flow rate of steam is required for a power generation of 10 MW?

Solution Saturated vapor at 3 MPa possesses the following properties:

$$h_1 = 2804.2 \frac{kJ}{kg} \quad \text{and} \quad T_1 = 233.9 \text{ K}$$

The processes from states 1 to 3 are constant enthalpy, so

$$h_1 = h_2 = h_3$$

As illustrated in Figure 13.6, a constant-enthalpy process from states 1 to 3 results in superheated vapor. At 0.5 MPa ($T_{sat} = 151.9°$ C) and $h_3 = 2804.2$ kJ/kg, the remaining properties of interest are

$$T_3 = 176.7°\text{C} \quad \text{and} \quad s_3 = 6.945 \frac{kJ}{kg \text{ K}}$$

with 24.8° C of superheat. The process path 3–4s is isentropic, so

$$s_{4s} = s_3 = 6.945 \frac{kJ}{kg \text{ K}} \quad \text{and} \quad P_4 = 15 \text{ kPa}$$

with state 3 in the superheated region and 4s "under the dome." At 15 kPa,

$$s_f = 0.7549 \frac{kJ}{kg \text{ K}} \quad s_{fg} = 7.2536 \frac{kJ}{kg \text{ K}} \quad h_f = 225.94 \frac{kJ}{kg} \quad h_{fg} = 2373.1 \frac{kJ}{kg}$$

The quality at state 4s can be found as

$$s_{4s} = 6.945 \frac{kJ}{kg \text{ K}} = 0.7549 \frac{kJ}{kg \text{ K}} + x \cdot 7.2536 \frac{kJ}{kg \text{ K}}$$

$$x = 0.853$$

With the quality known, the enthalpy can be obtained by

$$h_{4s} = 225.94 \frac{kJ}{kg} + 0.853 \cdot 2373.1 \frac{kJ}{kg} = 2251 \frac{kJ}{kg}$$

Then the ideal (isentropic) work extracted becomes

$$W_{ideal} = h_3 - h_{4s} = 2804.2 \frac{kJ}{kg} - 2251 \frac{kJ}{kg} = 554 \frac{kJ}{kg}$$

and the actual work is

$$W_{act} = \eta_e W_{ideal} = 0.82 \cdot 554 \frac{kJ}{kg} = 454 \frac{kJ}{kg}$$

At 20° C, the enthalpy of the liquid is

$$h_7 = 83.96 \frac{kJ}{kg}$$

The thermal efficiency of electricity production is the electrical energy produced per kg divided by the available energy of the liquid (well to ambient), or

$$\eta_{\text{thermal}} = \frac{\eta_{\text{elec}} W_{\text{act}}}{h_1 - h_7} = \frac{0.9 \cdot 454 \, \dfrac{\text{kJ}}{\text{kg}}}{2804.2 \, \dfrac{\text{kJ}}{\text{kg}} - 83.96 \, \dfrac{\text{kJ}}{\text{kg}}} = 0.15$$

from which the heat rate can be computed as

$$\text{Heat rate} = \frac{3412}{\eta_{\text{thermal}}} \, \frac{\text{Btu}}{\text{kWh}} = 22{,}730 \, \frac{\text{Btu}}{\text{kWh}}$$

Because the thermal efficiency is so low, the heat rate is significantly higher than those associated with the prime movers examined in Chapter 12.

The mass flow rate of steam required for 10 MW of electricity production is

$$\dot{m}_{\text{steam}} = \frac{10 \, \text{MW}}{\eta_{\text{elec}} \cdot W_{\text{act}}} = \frac{10 \, \text{MW}}{0.9 \cdot 454 \, \dfrac{\text{kJ}}{\text{kg}}} = 24.5 \, \frac{\text{kg}}{\text{sec}}$$

The thermal efficiency of only 15 percent does not include losses and pumping requirements. Indeed, the accepted thermal efficiency of vapor-dominated geothermal systems is about 10 percent. One ramification of low thermal efficiencies is that system mass flow rates and sizes must be large to generate significant power.

Liquid-dominated geothermal resources are more abundant than vapor-dominated sources. In liquid-dominated systems, water is available at 150–315° C. If the pressure is reduced, the water will flash into a two-phase mixture of relatively low quality—hence the name "liquid dominated." Three system configurations are possible for liquid-dominated systems: flash, binary, and total. Each will be examined in turn.

In flash systems, hot brine from a geothermal well is brought to the surface, where the pressure is reduced and some of the liquid is flashed into vapor. The vapor is expanded through a turbine to produce electricity, and the liquid is re-injected through another bore. A schematic of a flash liquid-dominated system is presented in Figure 13.7, and a *T-s* diagram of the process is provided in Figure 13.8.

Water from the well bore is supplied at state 1 and enters the flash separator at state 2 after a constant-enthalpy process. In the flash separator, vapor at state 3g is directed into the turbine, and liquid at state 3f is re-injected. State 3g is expanded to state 4 in the turbine. The isentropic expansion process to state 4s is also indicated in Figure 13.8. The mixture is condensed in the direct-contact condenser to state 5,6. Part of the condensate is re-injected, and part is directed to the cooling tower and

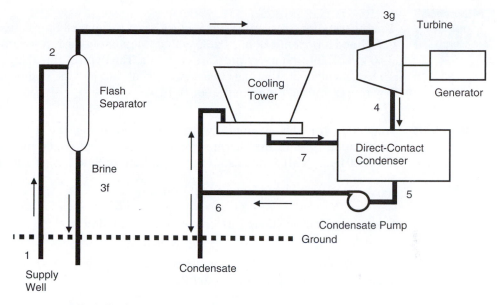

Figure 13.7 Flash liquid-dominated geothermal system schematic.

subsequently used in the direct-contact condenser as the cooling fluid. One of the disadvantages of the flash system is that the brine (state 3f), which contains significant energy, is re-injected. Energy is extracted only from the vapor phase. The next example illustrates flash liquid-dominated geothermal systems.

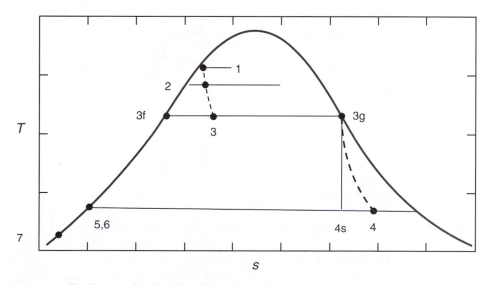

Figure 13.8 *T-s* diagram for flash liquid-dominated systems.

Example 13.2

A geothermal power plant is supplied with water at a well bottom temperature of 225° C and a pressure of 8 MPa. The fluid flows into a flash separator maintained at 40 kPa. The turbine exit pressure is 10 kPa. The overall efficiency of the turbine is 0.83. Calculate the thermal efficiency, the heat rate, and the steam and water mass flow rates required for an output of 10 MW.

Solution

Properties for this example were computed using the IAPWS Industrial Formulation 1997 for the Thermodynamic Properties of Water and Steam (Palmer et al. 2004). At 8 MPA and 225° C, the well bottom fluid has an enthalpy of

$$h_1 = 968.18 \frac{kJ}{kg}$$

The processes from states 1 to 3 are constant enthalpy, so

$$h_1 = h_2 = h_3$$

As illustrated in Figure 13.8, a constant-enthalpy process from 1 to 3 results in a mixture of liquid and vapor. At 40 kPa,

$$h_f = 317.62 \frac{kJ}{kg} \quad h_{fg} = 2318.5 \frac{kJ}{kg} \quad h_g = 2636.1 \frac{kJ}{kg} \quad s_g = 7.667 \frac{kJ}{kg \ K}$$

The quality at state 3 can be found as

$$h_3 = 968.18 \frac{kJ}{kg} = 317.62 \frac{kJ}{kg} + x \cdot 2318.5 \frac{kJ}{kg}$$

$$x = 0.281$$

Since only vapor ($x = 1$) enters the turbine, $h_3 = 2636.1$ kJ/kg and $s_3 = 7.667$ kJ/kg K. At the turbine exit, $P_4 = 10$ kPa, and the remaining properties of interest are

$$s_f = 0.6492 \frac{kJ}{kg \ K} \quad s_{fg} = 7.4998 \frac{kJ}{kg \ K} \quad h_f = 191.81 \frac{kJ}{kg} \quad h_{fg} = 2392.1 \frac{kJ}{kg}$$

The isentropic process, 3–4s, has $s_3 = s_{4s} = 7.667$ kJ/kg K, so the quality at state 4s becomes

$$s_{4s} = 7.667 \frac{kJ}{kg \ K} = 0.6492 \frac{kJ}{kg \ K} + x \cdot 7.4998 \frac{kJ}{kg \ K}$$

$$x = 0.936$$

And the enthalpy at state 4s is

$$h_{4s} = 191.81 \frac{kJ}{kg} + 0.936 \cdot 2392.1 \frac{kJ}{kg} = 2431 \frac{kJ}{kg}$$

The ideal (isentropic) work extracted becomes

$$W_{ideal} = h_3 - h_{4s} = 2636 \frac{kJ}{kg} - 2431 \frac{kJ}{kg} = 205 \frac{kJ}{kg}$$

and the actual work is

$$W_{act} = \eta_e W_{ideal} = 0.83 \cdot 205 \, \frac{kJ}{kg} = 170 \, \frac{kJ}{kg}$$

At 20° C, the enthalpy of the liquid is $h_7 = 83.91$ kJ/kg.

The mass flow rate of steam required for 10 MW of electricity production is

$$\dot{m}_{steam} = \frac{10 \, MW}{W_{act}} = \frac{10 \, MW}{170 \, \frac{kJ}{kg}} = 58.69 \, \frac{kg}{sec}$$

and the water required from the geothermal well is

$$\dot{m}_{water} = \frac{\dot{m}_{steam}}{x} = \frac{58.670}{0.281} \, \frac{kg}{sec} = 209.2 \, \frac{kg}{sec}$$

Because of the low quality of the working fluid at the steam separator, the required water mass flow rate from the geothermal well is high for the amount of power produced.

The thermal efficiency of the electricity production is the electrical energy produced per kg of steam divided by the available energy (enthalpy) of the liquid (well to ambient), or

$$\eta_{thermal} = \frac{W_{act}}{h_1 - h_7} = \frac{170 \, \frac{kJ}{kg}}{968.18 \, \frac{kJ}{kg} - 83.91 \, \frac{kJ}{kg}} = 0.193$$

from which the heat rate can be computed as

$$Heat \ rate = \frac{3412}{\eta_{thermal}} \, \frac{Btu}{kWh} = 17{,}710 \, \frac{Btu}{kWh}$$

Compared with most of the systems examined in this textbook, the heat rate is quite high since the thermal efficiency is low. However, the thermal efficiency calculated using only the steam does not consider the energy of the re-injected water. The "well bottom" thermal efficiency of the electricity production is the electrical energy produced by the steam divided by the available energy (enthalpy) of the liquid extracted from the well bottom, or

$$\eta_{well\text{-}bottom} = \frac{\dot{m}_{steam} W_{act}}{\dot{m}_{water}(h_1 - h_7)} = \frac{x \, W_{act}}{h_1 - h_7} = x \cdot \eta_{thermal} = 0.281 \cdot 0.193 = 0.054$$

As with the thermal efficiency, the efficiency based on the well-bottom water flow rate does not include losses and the auxiliary equipment energy required to operate the facility. Thus, in the flash separator system the electrical energy generation per kg of extracted water from the well bottom is quite low.

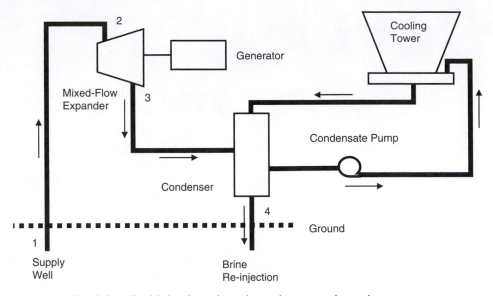

Figure 13.9 Total-flow liquid-dominated geothermal system schematic.

As demonstrated in Example 13.2, the flash liquid-dominated geothermal system has a low well bottom efficiency since so much energy is re-injected back into the site. A proposed solution is the total-flow concept for liquid-dominated systems, in which the turbine is replaced by a mixed-flow expander that extracts energy from the vapor-liquid mixture—thus permitting the total well head flow to be expanded to the condenser pressure. A suggested schematic and *T-s* diagram are presented in Figures 13.9 and 13.10, respectively.

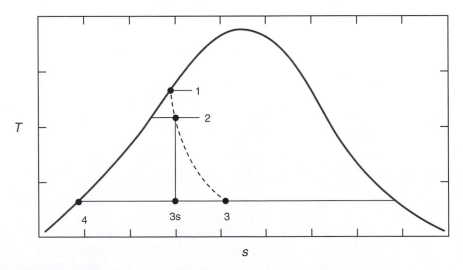

Figure 13.10 *T-s* diagram for total-flow liquid-dominated system.

Instead of undergoing the flash process, the fluid from the supply well enters a mixed-flow expander at state 2 and exits the expander at state 3. All of the fluid from the well is directed to the expander. In principle the process is simple; however, most geothermal fluids contain significant concentrations of corrosive solid and dissolved gases, and the design of turbines expanding from a near-saturated liquid to two-phase flow is difficult. Indeed, an examination of the literature shows much interest in mixed-flow expanders for geothermal operation during the 1970s and 1980s, with waning interest after that, followed by increased interest currently. The following example is a variation of Example 13.2, but uses a mixed-flow expander.

Example 13.3

Consider a total-flow geothermal system using the same well bottom conditions as Example 13.2. The fluid enters the expander as a saturated liquid and is discharged at 10 kPa. Determine the thermal efficiency, the heat rate, and the flow rate for a 10-MW system if the mixed-flow expander possesses an isentropic efficiency of 0.50.

Solution

Thermodynamic properties for this example are taken from Cengel and Boles (1998). For the prescribed conditions, state 2 is the saturated state; hence in Figure 13.10, states 1 and 2 occur on the saturation line and are the same. The saturation pressure for a saturation temperature of 225° C is 2.548 MPa. Thus at state 2

$$h_f = 966.78 \frac{kJ}{kg} \quad h_{fg} = 1836.5 \frac{kJ}{kg} \quad sf_2 = 2.5639 \frac{kJ}{kg\ K} \quad sfg_2 = 3.6863 \frac{kJ}{kg\ K}$$

The process 2–3s is isentropic, so that

$$s_{3s} = sf_2 = 2.5369 \frac{kJ}{kg\ K}$$

The mixture exits the expander at 10 kPa. The corresponding properties for state 3 are

$$s_f = 0.6493 \frac{kJ}{kg\ K} \quad s_{fg} = 7.5009 \frac{kJ}{kg\ K} \quad h_f = 191.83 \frac{kJ}{kg} \quad h_{fg} = 2392.8 \frac{kJ}{kg}$$

The quality at the exit of the expander becomes

$$s_{3s} = 2.5369 \frac{kJ}{kg\ K} = 0.6493 \frac{kJ}{kg\ K} + x \cdot 7.5009 \frac{kJ}{kg\ K}$$

$$x = 0.255$$

And the enthalpy at state 3s is

$$h_{3s} = 191.83 \frac{kJ}{kg} + 0.255 \cdot 2392.8 \frac{kJ}{kg} = 802.6 \frac{kJ}{kg}$$

The ideal (isentropic) work extracted becomes

$$W_{ideal} = h_2 - h_{3s} = 966.78 \frac{kJ}{kg} - 802.6 \frac{kJ}{kg} = 164.2 \frac{kJ}{kg}$$

and the actual work is

$$W_{act} = \eta_e W_{ideal} = 0.50 \cdot 164.2 \, \frac{kJ}{kg} = 82.1 \, \frac{kJ}{kg}$$

At 20° C, the enthalpy of the liquid is $h_{ambient} = 83.96$ kJ/kg.

The mass flow rate required for 10 MW of electricity production is

$$\dot{m}_{steam} = \frac{10 \text{ MW}}{W_{act}} = \frac{10 \text{ MW}}{82.1 \, \dfrac{kJ}{kg}} = 121.8 \, \frac{kg}{sec}$$

The thermal efficiency of the electricity production is the electrical energy produced divided by the available energy (enthalpy) from the well, or

$$\eta_{thermal} = \frac{W_{act}}{h_3 - h_2} = \frac{82.1 \, \dfrac{kJ}{kg}}{967 \, \dfrac{kJ}{kg} - 83.96 \, \dfrac{kJ}{kg}} = 0.093$$

from which the heat rate can be computed as

$$\text{Heat rate} = \frac{3412}{\eta_{thermal}} \, \frac{\text{Btu}}{\text{kWh}} = 36,690 \, \frac{\text{Btu}}{\text{kWh}}$$

The heat rate is higher than for the flashed system, but the mass flow rate is significantly reduced since all of the extracted fluid traverses the mixed-flow expander. This example glosses over the many problems associated with a mixed-flow expander for the total-flow geothermal system, but it does illustrate why such an arrangement is advantageous.

The other common system for liquid-dominated geothermal applications is the binary system. A number of existing geothermal power plants utilize such systems. A schematic illustrating the important components of a binary liquid-dominated system is provided in Figure 13.11.

In a binary geothermal system, two fluids are involved: the hot brine from the geothermal well and a working fluid (generally a hydrocarbon) that circulates in the closed portion of the system containing the turbine, condenser, and heat exchangers. In the binary system, the hot brine for the well supplies the hot fluid to the heat exchanger in which the organic fluid undergoes evaporation. The organic vapor is expanded through the turbine and condensed in the condenser and, hence, is in a closed system. The cooling tower is used as the cold medium in the condenser. Strictly speaking, the only geothermal components of the binary system are the supply/re-injection process with the well and the heat exchanger. The working fluids used in binary geothermal systems include propane (C_3H_8), isobutene (2-methyl propane, C_4H_{10}), isopentane (2-methyl butane, C_5H_{12}), and water-ammonia. The working fluids all have

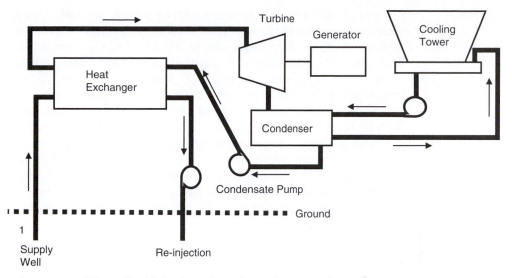

Figure 13.11 Binary liquid-dominated geothermal system schematic.

boiling points lower than that of water. Binary systems have advantages over flash liquid-dominated systems—not the least of which is that only the heat exchanger and hot brine transport components are exposed to the generally harsh and corrosive brine conditions.

Geopressurized

Geopressurized resources may have pressures up to 1000 bar (~ 15,000 psi), pressures that are much greater (by as much as a factor of two) than hydrostatic, and are in the temperature range 150–180° C. In the United States the primary location of such resources is in the gulf coast region in Texas and Louisiana. Figure 13.12 shows, in the cross-hatched regions, the geopressurized resource locations.

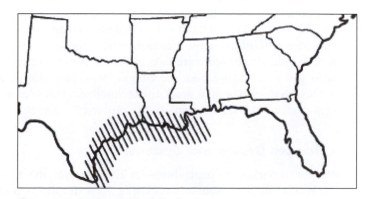

Figure 13.12 Geopressurized resource locations.

Geopressurized resource sites are 2000 to 9000 m in depth. Common features of many geopressurized resources are the high content of dissolved methane and the high levels of dissolved solids—up to 100,000 ppm. Geopressurized brine typically contains 30–80 ft^3/barrel of methane. Indeed, Smil (2003) estimates that global resources of geopressurized brine gas could be more than 100 times the current known natural gas reserves and that the Gulf of Mexico region may contain more brine-gas methane than the current natural gas reserves. These brine solutions are both highly corrosive and difficult to handle. Most proposed schemes for harvesting geopressurized resources involve the recovery of both energy and methane. The combination of the depth and the content of geopressurized brine makes harvesting energy and methane for this resource difficult and capital intensive. A number of feasibility studies of geopressurized systems, some with economic components, have been performed over the years, but the technical problems and high costs have precluded anything but pilot studies.

Magma

Magma is molten rock (with temperatures typically in excess of 650° C) contained in pools 3–10 km below active volcanoes. Although the energy density of molten rock is high and the estimated resources are quite extensive (see Table 13.1), methods for harvesting magma energy are at best speculative. The website www.magma-power.com contains a contemporary discussion of extracting energy from magma formations. The good news is that, in addition to producing energy, the high temperature and pressure associated with magma pools also produce desirable gaseous fuels, such as hydrogen, carbon monoxide, and methane. The bad news is that the high-temperature environment and the high concentration of corrosive gases and solids exacerbate material compatibility and longevity problems.

At 600° C iron oxide reacts with water to produce hydrogen according to

$$2FeO + H_2O \rightarrow 2FeO_{1.5} + H_2 \tag{13-1}$$

The quantity of hydrogen produced can be enhanced by introducing biomass into the injected water. In a chemical process similar to gasification, appreciable quantities of carbon monoxide, carbon dioxide, and methane along with steam and hydrogen result.

Figure 13.13 reproduces a sketch of how a magma power/fuel plant might appear. In addition to utilizing the steam produced for generating electrical power, gaseous fuels would also be recovered. As indicated in the inset, magma pools are quite large, with a single magma pool capable of providing energy and fuel for many years. Although the potential is great, the technological problems are immense, and the general consensus is that magma-based power/fuel plants are years away.

Enhanced Geothermal Systems

Proposed harvesting procedures for EGS (or hot, dry rock) systems involve injecting water into the resource, circulating the water through the rock as the energy capture mechanism, and bringing the heated water to the surface for use. Because of

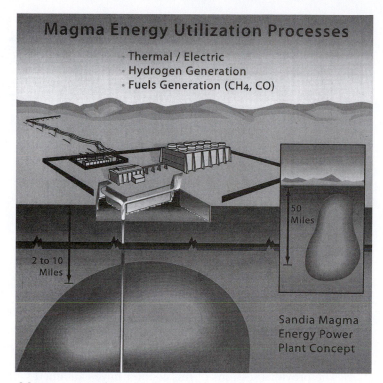

Figure 13.13 Magma power plant visualization (Sandia National Laboratory).

the low thermal conductivity of the rock, large surface areas are needed for extraction of meaningful amounts of energy. Thus, a major problem in EGS systems is how to obtain the large surface areas in the hot rock required for effective harvesting of thermal energy. Most proposed procedures for the EGS resource advocate hydrofracturing the rock by injecting water at very high pressure (200 atm) into the strata. In EGS systems with sufficient surface area, the use of techniques developed for enhanced oil recovery may be employed. For an interesting tidbit of EGS history, search for the topic "Project Plowshare" on the Internet.

EGS facilities require both injection bores and production bores. A proposed schematic for EGS systems is provided in Figure 13.14. In the figure, the fractured hot rock is illustrated, as are the injection and production bores. El-Wakil (1984) reported a Los Alamos National Laboratory study as stating that water would be pumped in at 1600 psia and 65°C and retrieved at 2000 psia and 280°C. In 2008, the U.S. Department of Energy completed an 8-month assessment of the technology developments needed for EGS to become commercially viable. The results of the study are available at http://www1.eere.energy.gov/geothermal/egs_technology.html. Pacing technology requirements include site selection and characterization, reservoir creation, wellfield development, system operation, and drilling and power conversion technologies. As with some other geothermal resources, economical and safe implementation of EGS will require significant development and capital.

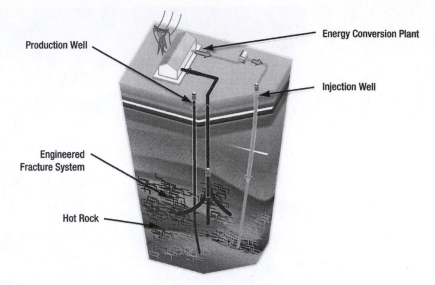

Figure 13.14 Schematic of an EGS power plant (DOE).

13.4   GEOTHERMAL EXAMPLES

The previous sections of this chapter have examined various aspects of geothermal energy systems, so some real-life examples are in order. The United States geothermal electrical generating capacity was 2800 MW (U.S. Department of Energy) in 2006, with California, Nevada, Utah, and Hawaii obtaining significant amounts of energy from geothermal resources. The website www.geoheat.oit.edu contains a listing of and some details on all geothermal power plants in the United States. A detailed compilation of geothermal energy use in the United States is provided by Lund et al. (2005). The geothermal plants in the United States are dry steam (hydrothermal), binary, or flash (some are double flash).

The Geysers Geothermal Area, north of San Francisco, contains 17 dry steam geothermal power plants. Figure 13.15 is a photo of the Big Geysers 75-MW hydrothermal power plant. The facility has been in operation since 1980.

The Mammoth Pacific power plants, in the Sierra Nevada Mountains, utilize brine from the Casa Diablo Hot Springs. Mammoth Pacific I, shown in Figure 13.16, was built in 1984. It generates 10 MW and is a binary geothermal plant.

Figure 13.17 shows an 85-MW double-flash geothermal power plant located at Coso Junction, CA. In a double-flash geothermal power plant, the liquid exiting the flash separator is flashed to produce low-pressure steam that can be utilized by a second low-pressure turbine or injected at an appropriate stage of the main turbine. Navy I has been online since 1987.

The primary focus of this chapter to this point has been geothermal power systems using steam and/or hot brine to produce electric power. As alluded to in section 13.1, geothermal heat pumps or ground-source heat pumps can also be considered users of geothermal energy. Ground-source heat pumps are covered in the next section.

Figure 13.15 One of the Geysers Geothermal Area's hydrothermal power plants (EERE).

Figure 13.16 Mammoth Pacific I, a binary geothermal power plant (EERE).

Figure 13.17 Navy I double-flash geothermal power plant (EERE).

13.5 GROUND-SOURCE HEAT PUMPS

Ground-source heat pumps (GSHP) or geothermal heat pumps utilize the ground (soil) for heat rejection in summer and as a heat source in winter. Conventional heat pumps reject heat to the atmosphere in summer and extract energy from the atmosphere in winter. A few feet below the surface, the ground temperature remains nearly constant throughout the year. Using the ground as a heat source/sink for a heat pump allows improved performance over a heat pump using the atmosphere as a heat source/sink. The Carnot efficiency is the maximum possible thermal efficiency for a heat engine operating between two temperatures and producing work and is stated as

$$\eta_{\text{Carnot}} = \frac{T_H - T_C}{T_H} \tag{13-2}$$

where T_H and T_C are the absolute temperatures of the high-temperature and low-temperature reservoirs, respectively. Refrigerators, air conditioners, and heat pumps are heat engines operating in reverse—transferring energy from a low-temperature reservoir to a high-temperature reservoir by the addition of work. For heat engines operating in reverse, the coefficient of performance (COP) is the usual performance metric. For a heat pump the desired energy result is the addition of heat to the hot reservoir, and the coefficient of performance is defined as the heat added to the high-temperature reservoir divided by the work input required. The COP for a Carnot heat pump is

$$\text{COP}_{\text{heat pump}} = \frac{\text{heat addition to high-temperature reservoir}}{\text{work input}} = \frac{T_H}{T_H - T_C} \tag{13-3}$$

In a similar fashion, the COP for a Carnot refrigerator or air conditioner is the heat removed from the low-temperature reservoir divided by the work input or

$$\text{COP}_{\text{refrig}} = \frac{\text{heat removed from low-temperature reservoir}}{\text{work input}} = \frac{T_L}{T_H - T_C} \quad (13\text{-}4)$$

The reason for the difference is the desired outcome—heat addition for the heat pump and heat removal for the refrigerator. The two COPs are related as

$$\text{COP}_{\text{heat pump}} = \text{COP}_{\text{refrig}} + 1 \quad (13\text{-}5)$$

Since $\text{COP}_{\text{refrig}}$ is always positive, $\text{COP}_{\text{heat pump}}$ is always greater than 1. Although an actual heat pump is significantly different from a reversed Carnot engine, the thermodynamic implications of Eqs. (13-3) and (13-5) for a heat pump are valid— namely, the smaller the temperature difference between the high- and low-temperature reservoirs, the higher the COP.

In the winter, heat is extracted from the ground instead of the air so that T_C is the temperature of the ground instead of the outside air temperature, thus significantly increasing the $\text{COP}_{\text{heat pump}}$. Conventional (air-source) heat pumps possess COP values of nearly 3, while GSP heat pumps have COP values approaching 4. In the summer, heat is rejected to the ground, not the outside air, such that the $\text{COP}_{\text{refrig}}$ is increased. The use of the ground as a source/sink is thus advantageous for the GSHP. The general configuration of a GSHP is featured in Figure 13.18. The heat

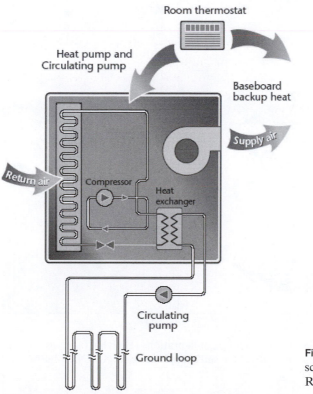

Figure 13.18 GSHP system schematic (Canadian Renewable Energy Network).

exchanger transfers energy between the compressor refrigerant and the ground via a circulating pump in the ground loop.

The ground can be utilized as a heat sink/source for GSHPs in several different configurations: (a) closed-loop, horizontal; (b) closed-loop, vertical; (c) closed-loop, lake/pond; and (d) open-loop. Figure 13.19 presents an example of each configuration.

The most common configuration for residences is the closed-loop, horizontal in Figure 13.19(a). The pipes are arranged in a helical coil shape and buried 4 to 6 feet deep in a trench 2 feet wide. A disadvantage of this method is the amount of soil that

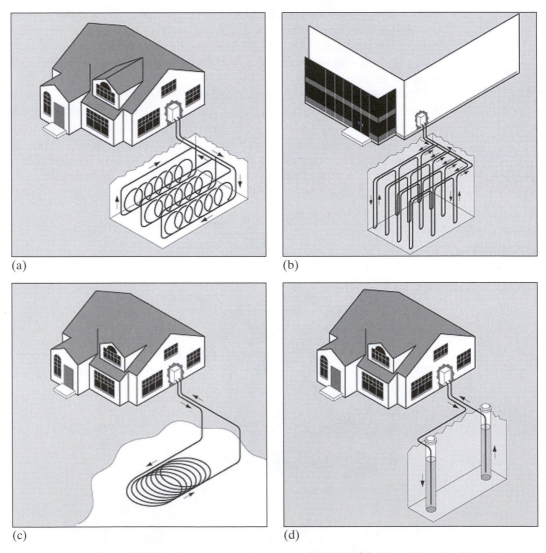

(a) (b) (c) (d)

Figure 13.19 GSHP ground loop configurations (EERE). (a) Closed-loop, horizontal. (b) Closed-loop, vertical. (c) Closed-loop, lake/pond. (d) Open-loop.

must be moved/disturbed. Closed-loop, vertical arrangements, Figure 13.19(b), are quite common in government/commercial/industrial sites and are becoming more common for residences. In vertical systems, holes 4 inches in diameter are drilled about 20 feet apart and 100–400 feet deep. In each vertical hole, two pipes are connected at the bottom to form a "U" shape. The pipes are manifolded together and connected to the heat pump in the conditioned space of the building. If a body of water, such as a lake or large pond, is available, the water can be used as the heat source/sink, as in Figure 13.19(c). A helical or slinky arrangement is again used, but the pipe must be coiled at least 8 feet under the surface for freeze protection. Finally, an open-loop system, as in Figure 13.19(d), can be utilized. Water from a well or a pond is circulated through the GSHP and then returned to the ground via a recharge well. The discharge must conform to local codes and regulations for groundwater discharge.

Since the ground temperature is an important consideration for GSHPs, some discussion of geography is warranted. Depending on the location, at 6 feet underground the near-constant temperature ranges from 45° to 75°F. Figure 13.20 illustrates the "yearly average" ground temperature distribution in the United States. Not surprisingly, the ground temperature correlates reasonably well with the "heating degree-days" representations in Figure 8.6—that is, the lower the heating-degree days, the higher the ground temperature.

ORNL 2001-03084/abh

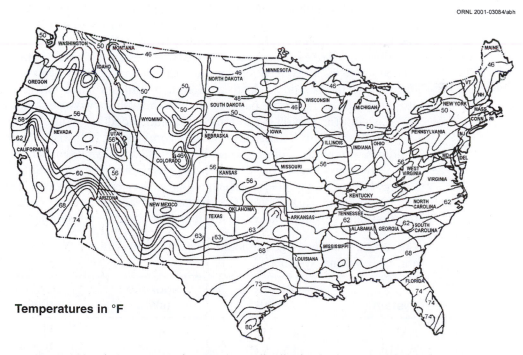

Temperatures in °F

Figure 13.20 Average ground temperature distribution in the United States (Oak Ridge National Laboratory).

Corrosion-inhibitive fluids with antifreeze are used as heat transfer fluids. The general recommendation is that antifreeze sufficient for protection to 10°F below the minimum expected temperature be used. The fluids should be biodegradable and nontoxic and possess relatively low viscosities. The two most commonly used fluids are ethylene or propylene glycol and an alcohol-water mixture. Methyl , ethyl, or iso-propyl alcohols have all been used in GSHP applications. Bio-inhibitive water may be used in climates where freezing is not a problem.

Residential systems can be equipped with a "desuperheater" for heating domestic hot water. In the summer, heat that is taken from the house and rejected into the ground loop is used to heat the water. In the winter, the desuperheater can reduce water heating costs by about half, with a conventional water heater meeting the rest of the hot water needs.

Obviously, GSHP systems are more expensive than conventional systems. The question then becomes, how much energy can be saved by using a GSHP? U.S. Department of Energy studies suggest that approximately 70 percent of the energy used in a GSHP system comes from the ground. The Environmental Protection Agency now designates GSHPs as EnergyStar®-rated products. Rafferty (2001) reports the results of a study by Kavanaugh of energy usage by the same type of residence in three different climates. The results are summarized in Table 13.2. Four systems were studied: air-source heat pump (ASHP), air-source heat pump with variable-speed drive (ASHPvs), standard GSHP, and high-efficiency GSHP. The total kWh usages for cooling, heating, and domestic hot water are presented. The savings for the GSHP options are significant compared to either conventional system. The GSHP systems demonstrate significant electricity savings.

The general rule-of-thumb cost for GSHP systems is $2500/ton installed. This price is about twice what a conventional residential system would cost. Thus, the energy cost savings and the possible incentives offered by utilities make GSHP economically attractive with a reasonable payback period. The website www.geokiss.com, contains extensive information on GSHPs and is recommended for additional information.

GSHPs are also extensively used for commercial, industrial, and government buildings, where they offer the same savings and advantages as for residences. Details, information, and case studies are available from the U.S. Department of Energy Office of Geothermal Technologies, www.eren.doe.gov/geothermal; from the Geothermal Heat Pump Consortium, www.geoexchange.org; and from the International Ground Source Heat Pump Association, www.igshpa.okstate.edu.

TABLE 13.2 Energy usage for cooling, heating, and domestic hot water for GSHP and conventional systems

System	Atlanta (kWh)	Spokane (kWh)	Portland (kWh)
ASHP	14,925	16,458	11,299
ASHPvs	12,159	13,850	9,111
GSHP std.	9,455	9,163	7,354
GSHP high eff.	8,098		

13.6 **CLOSURE**

This chapter has reviewed geothermal energy, primarily as a source of heat for power generation, but also for use in GSHP systems. As with many of the energy sources reviewed in this textbook, geothermal has much potential, but research and development issues need to be resolved and significant capital will be required to fully realize this potential.

REVIEW QUESTIONS

1. What is geothermal energy?

2. What are four types of geothermal energy sources?

3. Why is hot, dry rock of interest as a geothermal energy source?

4. What is the typical thermal efficiency expected from a geothermal system?

5. How does the heat rate for a geothermal system compare with that of a gas turbine?

6. Why is the thermal efficiency of a geothermal system such a relatively small value?

7. What are the characteristics of a hydrothermal energy source?

8. What are the characteristics of a geopressured energy source?

9. What are the characteristics of an EGS energy source?

10. Which geothermal resource is the easiest to utilize?

11. Which geothermal resource is the most abundant? What problems does it raise?

12. What is a binary-cycle geothermal system?

13. What is an advantage of a binary-cycle geothermal system?

14. What is the status of commercialization of geothermal energy?

15. What is a geothermal heat pump?

16. How does a ground-source heat pump differ from a conventional heat pump? What advantage does a ground-source heat pump have over an air-source heat pump?

17. What are the three principal components of a geothermal heat pump?

18. Is a geothermal heat pump currently competitive in the economic sense? Explain.

19. In Atlanta, how many kWh per year is a high-efficiency ground-source heat pump estimated to save over a conventional air-source heat pump in a typical residential application?

EXERCISES

1. A vapor-dominated geothermal system is supplied with saturated steam at 2 MPa. The steam enters the turbine at 0.3 MPa and exits at 15 kPa. The turbine isentropic efficiency is 85 percent, and the electrical generator is 95 percent efficient. If re-injection occurs at the cooling tower, analyze the system performance (thermal efficiency and heat rate).

2. A geothermal power plant is supplied with water at a well bottom temperature of 260° C and a pressure of 10 MPa. The fluid flows into a flash separator maintained at 0.5 bar. The turbine exit pressure is 0.02 bar. The overall efficiency of the turbine is 0.83. Calculate the thermal efficiency, the heat rate, and the steam and water mass flow rates required for an output of 5 MW.

3. Consider a total-flow geothermal system using the same well bottom conditions as in Exercise 2. The fluid enters the expander as a saturated liquid and is discharged at 0.02 bar. The efficiency, η_e, of the total flow expander is 0.50. Determine the thermal efficiency, the heat rate, and the flow rate for a 5-MW system.

REFERENCES

Cengel, Y. A., and Boles, M. A. 1998. *Thermodynamics*, 3rd ed. New York: McGraw-Hill.

Decher, R. 1994. *Energy Conversion*. New York: Oxford University Press.

DiPippo, R. 2007. *Geothermal Power Plants: Principles, Applications and Case Studies*, 2nd ed. Oxford, UK: Elsevier Ltd.

El-Wakil, M. M. 1984. *Powerplant Technology*. New York: McGraw-Hill.

Green, B. D., and Nix, R. G. 2006. *Geothermal—The Energy Under Our Feet: Geothermal Resource Estimates for the United States*. NREL/TP-840-40665. Golden, CO: NREL.

Gupta, H., and Roy, H. 2007. *Geothermal Energy: An Alternative Resource for the 21st Century*. Amsterdam: Elsevier.

Kasuda, T., and Archenbach, P. R. "Earth Temperature and Thermal Diffusivity at Selected Stations in the United States," *ASHRAE Transactions* 71(1), 1965.

Kruger, P. 1976. "Geothermal Energy." In *Annual Review of Energy*, Vol. 1. Palo Alto, CA: Annual Reviews.

Kutscher, C. F. 2000. *The Status and Future of Geothermal Electric Power*. NREL/CP-550-28204. Golden, CO: NREL.

Lund, J. W., Bloomquist, R. G., Boyd, T. L., and Renner, J. 2005. "The United States of America Country Update." In *Proceedings of the World Geothermal Congress, 2005, Anatalya, Turkey*.

Mock, J. E., Tester, J. W., and Wright, P. M. 1997. *Annual Review of Energy and the Environment*, Vol. 22. Palo Alto, CA: Annual Reviews.

Palmer, D. A., Fernandez-Prini, R., and Harvey, A. H., eds. 2004. *Aqueous Systems at Elevated Temperatures and Pressures*. Amsterdam: Elsevier.

Rafferty, K. 2001. An *Information Survival Kit for the Prospective Geothermal Heat Pump Owner*. Klamath Falls, OR: Geo-Heat Institute.

Smil, V. 2003. *Energy at the Crossroads: Global Perspectives and Uncertainties*. Boston: MIT Press.

Sorensen, H. A. 1983. *Energy Conversion Systems*. New York: John Wiley.

Tester, J. W., Drake, E. M., Driscoll, M. J., Golay, M. W., and Peters, W. A. 2005. *Sustainable Energy*. Cambridge, MA: MIT Press.

U.S. Department of Energy, 2008. *An Evaluation of Enhanced Geothermal Systems Technology*, available at http://www1.eere.energy.gov/geothermal/egs_technology.html.

WEBSITES

www.geothermal.id.doe.gov
www.eere.energy.gov/geothermal
www.energyquest.ca.gov
www.geoheat.oit.edu
www.geothermal.org
www.geo-energy.org
www.geokiss.com
www.geoexchange.org
www.smu.edu/geothermal

CHAPTER

14

Ocean Energy

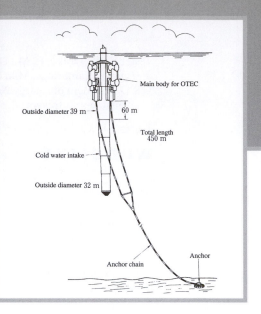

INTRODUCTION

The oceans house an enormous amount of thermal and mechanical energy. The thermal energy comes from the mass and temperature of the oceans, and the mechanical energy from the waves and tidal actions of the oceans. The oceans can be viewed as a very large solar thermal collector. The solar irradiation is incident on the surface, where the solar energy is absorbed and the surface layer of the water heated. Over the years a number of proposals have been made for harvesting energy resulting from the temperature difference between the solar-heated surface and the deep ocean water—a general concept called ocean thermal energy conversion (OTEC). The actions of waves and tides have long been of interest, and various systems have been proposed, built, and/or operated to harvest wave and tidal energy. This chapter will explore OTEC and mechanical energy from the oceans.

14.2 OCEAN THERMAL ENERGY CONVERSION (OTEC)

The solar irradiation incident on the surface of the oceans heats the layer of water near the surface, but deep ocean convective currents transport cooler water to the lower depths of the equatorial regions. On a typical day, the energy equivalent of 250 billion barrels of oil (www.nrel.gov/otec/what.html) is absorbed! The net result is temperature stratification between the hot surface water and cooler water at depths greater than about 1 km. Surface temperatures can be in excess of 27° C, and the temperature at depths greater than 1 km can be as low as 4° C. Thus, between the surface and the deep water, a temperature difference of 23° C or more can exist. If the surface is viewed as a hot reservoir and the deep water as a cold reservoir, then a heat engine can extract useful work. OTEC systems exploit this temperature difference to

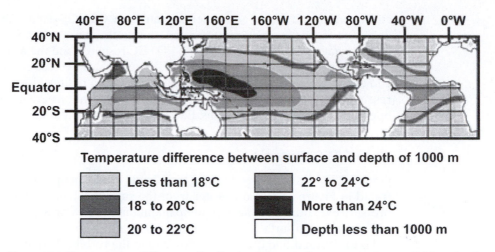

Temperature difference between surface and depth of 1000 m

Less than 18°C	22° to 24°C
18° to 20°C	More than 24°C
20° to 22°C	Depth less than 1000 m

Figure 14.1 Temperature difference distribution between the ocean surface and 1-km depth (U.S. Department of Energy).

produce useful work. Depending on the OTEC system configuration, other useful products and processes may also result.

As shown, the temperature difference is an important metric in determining the feasibility of an OTEC system. Generally, OTEC systems are deemed economically feasible if the surface-to-depth temperature difference is greater than 20° C. Figure 14.1 illustrates the distribution of surface-to-depth (1-km) temperature differences in the oceans. The western equatorial Pacific Ocean region contains a large area with temperature differences in excess of 24° C. Much of the ocean between ±20° latitude and the equator possesses surface-to-depth temperature differences of 20° C to 24° C. However, much of the OTEC-applicable areas are in mid-ocean and/or far removed from both population centers and high-energy-consumption locations. Thermodynamic considerations, examined next, also place limits on the feasibility of OTEC systems.

Since an OTEC system is essentially a heat engine, the Carnot efficiency represents an upper bound on the thermal efficiency. Consider the thermal efficiency of an OTEC system operating with a surface temperature of 27° C and a temperature difference of 20° C. The Carnot efficiency can be computed as

$$\eta_{carnot} = \frac{T_H - T_C}{T_H} = \frac{\Delta T}{T_H} = \frac{20 \text{ K}}{(27 + 273) \text{ K}} = 0.067 \tag{14-1}$$

Hence, the maximum possible efficiency is only 6.7 percent. In reality, the efficiency projected or obtained for most OTEC facilities is around 3 percent. The implications of such a low efficiency are at least twofold: (1) high flow rates and large components are required for significant power extraction, and (2) OTEC economics are very sensitive to irreversibilities and losses in components. OTEC systems with high flow rates and large component sizes lead to very large capital expenditures for construction and operation. If losses are underestimated, an OTEC system may be subject to deficit economics or may require more energy to operate than is extracted. These realities have, to a large extent, limited OTEC research and the availability of funds for construction and operation of OTEC systems. However, as energy costs escalate and energy availability becomes more and more of a political issue, OTEC will represent a more attractive opportunity. OTEC configurations are generally classed as open-flow or closed-flow systems. Each will be examined in turn.

Open OTEC Systems

Open OTEC systems operate using the Claude cycle. An open OTEC system is illustrated in Figure 14.2. Warm water from the surface (state 1) is admitted to a low-pressure evaporator. A vacuum pump maintains the low pressure in the evaporator, where the warm surface water is flashed into a low-quality mixture. The warm liquid is discharged, and the vapor is directed to the turbine. Energy is extracted in the turbine, and the resulting liquid-vapor mixture is discharged at state 3. In the direct-contact condenser, cold water at state 5 is used to condense the mixture from the turbine. State 4 is the state at the condenser exit. Because of the low quality of the mixture in the evaporator, most of the warm surface water entering the evaporator is discharged.

The *T-s* diagram corresponding to the schematic of Figure 14.2 is presented in Figure 14.3. The warm water enters the evaporator at state 1 and is flashed to state 2, with 2f the liquid state and 2g the vapor state. The warm liquid is discharged, and the vapor (quality = 1) enters the turbine at 2g and exits at state 3 as a mixture. State 3s

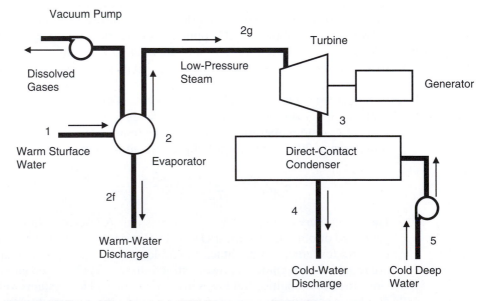

Figure 14.2 Open (Claude cycle) OTEC system configuration.

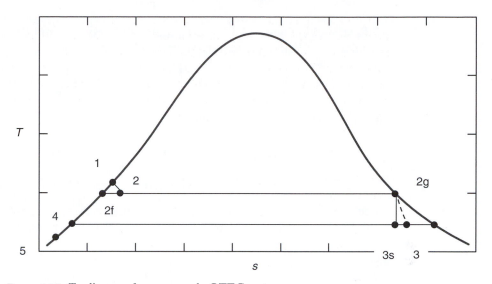

Figure 14.3 *T-s* diagram for open-cycle OTEC system.

represents the exit state if the turbine process were isentropic. Cold water at state 5 is used in the direct-contact condenser. The condenser water exits at state 4. Because the quality at state 2 is low and the quality at state 3 is near unity, the mass flow rate through the turbine differs significantly from the warm-water discharge and the cold-water intake. Example 14.1 examines an open OTEC system performance metrics.

Example 14.1	An open-cycle OTEC system operates with warm surface water at 27° C and surface condenser water at 13° C from the deep cold water at 11° C. The evaporator pressure is 0.0317 bar, which corresponds to a saturation temperature of 25° C; the condenser pressure and temperature are 15° C and 0.017 bar. The turbine efficiency is 0.80. If the turbine is to extract 100 kW, determine the system efficiency and the warm-water, cold-water, and turbine mass flow rates.

Solution	Properties for this example were computed using the IAPWS Industrial Formulation 1997 for the Thermodynamic Properties of Water and Steam (Palmer et al. 2004). At 27° C, the warm surface water possesses an enthalpy of

$$h_1 = 113.19 \, \frac{\text{kJ}}{\text{kg}}$$

The process from states 1 to 2 is constant enthalpy, so that

$$h_1 = h_2 = 113.19 \, \frac{\text{kJ}}{\text{kg}}$$

As illustrated in Figure 14.3, a constant enthalpy process from states 1 to 2 results in a mixture of liquid and vapor. At 25° C,

$$h_f = 104.83 \frac{\text{kJ}}{\text{kg}} \quad h_{fg} = 2441.9 \frac{\text{kJ}}{\text{kg}} \quad h_g = 2546.72 \frac{\text{kJ}}{\text{kg}} \quad s_g = 8.5573 \frac{\text{kJ}}{\text{kg K}}$$

The quality at state 2 can be found as

$$h_2 = 113.19 \frac{\text{kJ}}{\text{kg}} = 104.83 \frac{\text{kJ}}{\text{kg}} + x_2 \cdot 2441.9 \frac{\text{kJ}}{\text{kg}}$$

$$x_2 = 0.003424$$

Since only vapor $(x = 1)$ enters the turbine, $h_{3g} = 2546.72 \text{ kJ/kg}$ and $s_{3g} = 8.5573 \text{ kJ/kg K}$. At the turbine exit, $T_4 = 15°$ C. The remaining properties of interest are

$$s_f = 0.2245 \frac{\text{kJ}}{\text{kg K}} \quad s_{fg} = 8.5565 \frac{\text{kJ}}{\text{kg K}} \quad h_f = 62.98 \frac{\text{kJ}}{\text{kg}} \quad h_{fg} = 2465.6 \frac{\text{kJ}}{\text{kg}}$$

The isentropic process, 2g–3s, has $s_2 = s_{3s} = 8.5573 \text{ kJ/kg K}$, so the quality at state 3s becomes

$$s_{3s} = 8.5573 \frac{\text{kJ}}{\text{kg K}} = 0.22454 \frac{\text{kJ}}{\text{kg K}} + x \, 8.5565 \frac{\text{kJ}}{\text{kg K}}$$

$$x = 0.974$$

and the enthalpy at state 3s is

$$h_{3s} = 62.98 \frac{\text{kJ}}{\text{kg}} + 0.974 \cdot 2465.6 \frac{\text{kJ}}{\text{kg}} = 2464 \frac{\text{kJ}}{\text{kg}}$$

The ideal (isentropic) work extracted can be calculated as

$$W_{\text{ideal}} = h_{2g} - h_{3s} = 2546.72 \frac{\text{kJ}}{\text{kg}} - 2464 \frac{\text{kJ}}{\text{kg}} = 82.6 \frac{\text{kJ}}{\text{kg}}$$

and the actual work is

$$W_{\text{act}} = \eta_e W_{\text{ideal}} = 0.80 \cdot 82.6 \frac{\text{kJ}}{\text{kg}} = 66.08 \frac{\text{kJ}}{\text{kg}}$$

The enthalpy at state 3 is thus

$$h_3 = h_{2g} - W_{\text{act}} = 2546.72 \frac{\text{kJ}}{\text{kg}} - 66.08 \frac{\text{kJ}}{\text{kg}} = 2481 \frac{\text{kJ}}{\text{kg}}$$

The turbine mass flow rate required to extract 100 kW from the turbine is

$$\dot{m}_{\text{turbine}} = \frac{100 \text{ kW}}{W_{\text{act}}} = \frac{100 \text{ kW}}{66.08 \dfrac{\text{kJ}}{\text{kg}}} = 1.513 \frac{\text{kg}}{\text{sec}}$$

The thermal efficiency of the system is the energy produced per kg of turbine flow divided by the enthalpy difference between the steam entering the system and exiting the condenser, or

$$\eta_{\text{thermal}} = \frac{W_{\text{act}}}{h_3 - h_4} = \frac{W_{\text{act}}}{h_3 - h_{3f}} = \frac{66.08 \, \dfrac{\text{kJ}}{\text{kg}}}{2481 \dfrac{\text{kJ}}{\text{kg}} - 62.98 \dfrac{\text{kJ}}{\text{kg}}} = 0.027$$

The warm-water flow rate required is the turbine mass flow rate divided by the quality at state 2

$$\dot{m}_{\text{warm-water}} = \frac{\dot{m}_{\text{turbine}}}{x_2} = \frac{1.513 \, \dfrac{\text{kg}}{\text{sec}}}{0.003424} = 442.03 \frac{\text{kg}}{\text{sec}}$$

In the direct-contact condenser, cold water at state 5 (13° C) is used to condense the vapor at state 3. At state 5, the enthalpy is $h_5 = 54.60$ kJ/kg. An energy balance yields

$$\dot{m}_{\text{turbine}} \left(h_3 - h_4 \right) = \dot{m}_{\text{cold-water}} \left(h_4 - h_5 \right)$$

The cold-water mass flow rate then becomes

$$\dot{m}_{\text{cold-water}} = \dot{m}_{\text{turbine}} \frac{\left(h_3 - h_4 \right)}{\left(h_4 - h_5 \right)} = 1.513 \, \frac{\text{kg}}{\text{sec}} \, \frac{2481 \dfrac{\text{kJ}}{\text{kg}} - 62.98 \dfrac{\text{kJ}}{\text{kg}}}{62.98 \dfrac{\text{kJ}}{\text{kg}} - 54.60 \dfrac{\text{kJ}}{\text{kg}}} = 436.596 \frac{\text{kg}}{\text{sec}}$$

The low quality in the evaporator results in a large warm-water flow rate compared with the turbine mass flow rate, and the quality near unity at the turbine exit results in a high cold-water flow rate.

The efficiency and flow rates represented in the preceding example are realistic for OTEC systems and explain, in part, the thermodynamic difficulties of open-system OTEC systems. The high specific volume of the turbine working fluid, 43.40 m³/kg for Example 14.1, is startlingly different from that of modern steam power plants, at 0.021 m³/kg. Thus, the mature technology of steam turbines is not generally applicable to open-cycle OTEC turbines. El-Wakil (1984) discusses turbine requirements and difficulties for application of open-cycle OTEC systems. In a 1979 Westinghouse study (www.nrel.gov/otec/electric_turbines), a turbine 43.6 m in diameter operating at 200 rpm was proposed for a 100-MW OTEC system.

Closed OTEC Systems

One way to avoid turbine issues in OTEC systems is to use a binary system with a working fluid other than water. Closed-cycle OTEC (Anderson cycle) systems using working fluids such as ammonia and propane have been proposed. Figure 14.4

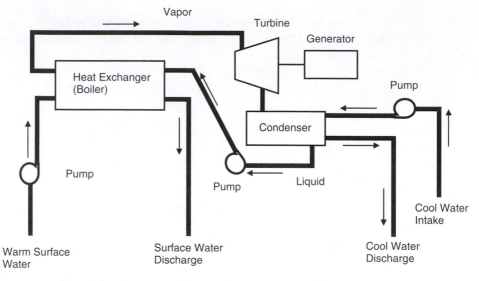

Figure 14.4 Closed (Anderson cycle) OTEC system configuration.

provides a schematic of a closed-cycle binary OTEC system. The warm surface water provides the energy to vaporize the working fluid that is directed to the turbine. After exiting the turbine, the cold water is used to condense the working fluid.

| Example 14.2 | A closed-cycle OTEC system uses propane as the working fluid. The warm surface water enters and leaves the evaporator at 28° C and 25° C, respectively. The cold water enters and leaves the condenser at 5° C and 8° C, respectively. The propane evaporating temperature is 22° C, and the condensing temperature is 11° C. The plant output is 125 MW, and the turbine efficiency is 0.90. The overall heat transfer coefficient for both the evaporator and the condenser is 1400 W/m² K. Determine the propane flow rate, the evaporator and condenser surface areas, the warm- and cold-water mass flow rates, and the overall thermal efficiency. |

Solution Propane properties were obtained from Perry and Chilton (1973). The evaporator operates at 22° C, and the condenser operates at 11° C. Table 14.1 shows the saturation properties at these temperatures.

TABLE 14.1 Propane properties at 22° C and 11° C

T (°C)	P (kPa)	h_f (kJ/kg)	h_{fg} (kJ/kg)	h_g (kJ/kg)	s_f (kJ/kg K)	s_{fg} (kJ/kg K)	s_g kJ/kg K
11	649.9	546.2	361.7	907.9	4.359	1.273	5.633
22	877.4	573.8	344.4	918.2	4.456	1.165	5.621

In the evaporator the warm fluid vaporizes the propane to a saturated vapor at 22° C. The vapor enters the turbine at 22° C with $h_g = 918.2$ kJ/kg and $s_g = 5.621$ kJ/kg K. At the turbine exit, the propane temperature is 11° C, and for isentropic expansion through the turbine, the entropy liquid-vapor mixture at the exit would be 5.621 kJ/kg K. The quality at the turbine exit for isentropic expansion becomes

$$s_{\text{turbine–exit}} = 5.621 \frac{\text{kJ}}{\text{kg K}} = 4.359 \frac{\text{kJ}}{\text{kg K}} + x \cdot 1.273 \frac{\text{kJ}}{\text{kg K}}$$

$$x = 0.991$$

and the enthalpy for the isentropic expansion is

$$h_{\text{isentropic}} = 546.2 \frac{\text{kJ}}{\text{kg}} + 0.991 \cdot 361.7 \frac{\text{kJ}}{\text{kg}} = 904.5 \frac{\text{kJ}}{\text{kg}}$$

The ideal (isentropic) work extracted is the change in enthalpy across the turbine,

$$W_{\text{ideal}} = h_{\text{entrance}} - h_{\text{isentropic}} = 918.2 \frac{\text{kJ}}{\text{kg}} - 904.5 \frac{\text{kJ}}{\text{kg}} = 13.7 \frac{\text{kJ}}{\text{kg}}$$

and the actual work is

$$W_{\text{act}} = \eta_{\text{turbine}} \, W_{\text{ideal}} = 0.90 \cdot 13.7 \frac{\text{kJ}}{\text{kg}} = 12.31 \frac{\text{kJ}}{\text{kg}}$$

The enthalpy at the exit of the turbine is thus

$$h_{\text{exit}} = h_{\text{entrance}} - W_{\text{act}} = 918.2 \frac{\text{kJ}}{\text{kg}} - 12.31 \frac{\text{kJ}}{\text{kg}} = 905.9 \frac{\text{kJ}}{\text{kg}}$$

The propane mass flow rate can be calculated as

$$\dot{m}_{\text{propane}} = \frac{\text{power}_{\text{extracted}}}{W_{\text{act}}} = \frac{125 \text{ MW}}{12.31 \frac{\text{kJ}}{\text{kg}}} = 10{,}150 \frac{\text{kg}}{\text{sec}}$$

Propane enters the evaporator as a saturated liquid at 11° C and exits the evaporator as a saturated vapor at 22° C. The energy required is thus

$$\dot{Q}_{\text{evap}} = \dot{m}_{\text{propane}} \cdot \left[h\big((22°\text{C})_g - h(11°\text{C})_f\big) \right]$$

$$= 10{,}150 \frac{\text{kg}}{\text{sec}} \cdot \left(918.2 \frac{\text{kJ}}{\text{kg}} - 546.2 \frac{\text{kJ}}{\text{kg}} \right) = 3.78 \times 10^6 \text{ kW}$$

Since most of the energy transfer is required for the phase change, which occurs at 22° C, the log mean temperature difference (LMTD) for the evaporator becomes

$$\text{LMTD}_{\text{evap}} = \frac{\Delta T_2 - \Delta T_1}{\ln\left(\dfrac{\Delta T_2}{\Delta T_1}\right)} = \frac{(28°\text{C}-22°\text{C})-(25°\text{C}-22°\text{C})}{\ln\left(\dfrac{28°\text{C}-22°\text{C}}{25°\text{C}-22°\text{C}}\right)} = \frac{6°\text{C}-3°\text{C}}{\ln\left(\dfrac{6°\text{C}}{3°\text{C}}\right)} = 4.33°\text{C}$$

For a heat exchanger,

$$\dot{Q}_{evap} = U \cdot A_{evap} \cdot LMTD_{evap}$$

from which the evaporator surface area can be computed as

$$A_{evap} = \frac{\dot{Q}_{evap}}{U \cdot LMTD_{evap}} = \frac{3.78 \cdot 10^6 \text{ kW}}{1400 \frac{W}{m^2 \, ^\circ C} \cdot 4.33 ^\circ C} = 6.234 \times 10^5 \text{ m}^2$$

The saturated liquid water enthalpies for 28° C and 25° C are 117.35 kJ/kg and 104.81 kJ/kg, respectively. An energy balance then yields the warm-water mass flow rate:

$$\dot{m}_{warm\text{-}water} = \frac{\dot{Q}_{evap}}{h(22 ^\circ C)_f - h(11 ^\circ C)_f} = \frac{3.78 \cdot 10^5 \text{ kW}}{117.35 \frac{kJ}{kg} - 104.81 \frac{kJ}{kg}} = 3.01 \times 10^5 \frac{kg}{sec}$$

Propane enters the condenser as a mixture at 11° C and exits the condenser as a saturated liquid at 11° C. The energy required to condense the vapor is thus

$$\dot{Q}_{evap} = \dot{m}_{propane} \cdot \left[h_{exit} - h(11 ^\circ C) \right]$$

$$= 10{,}150 \frac{kg}{sec} \cdot \left(905.9 \frac{kJ}{kg} - 546.2 \frac{kJ}{kg} \right) = 3.65 \times 10^6 \text{ kW}$$

The log mean temperature difference (LMTD) for the condenser becomes

$$LMTD_{cond} = \frac{\Delta T_2 - \Delta T_1}{\ln\left(\frac{\Delta T_2}{\Delta T_1}\right)} = \frac{(11 ^\circ C - 5 ^\circ C) - (11 ^\circ C - 8 ^\circ C)}{\ln\left(\frac{11 ^\circ C - 5 ^\circ C}{11 ^\circ C - 8 ^\circ C}\right)} = \frac{6 ^\circ C - 3 ^\circ C}{\ln\left(\frac{6 ^\circ C}{3 ^\circ C}\right)} = 4.33 ^\circ C$$

For the condenser,

$$\dot{Q}_{cond} = U \cdot A_{cond} \cdot LMTD_{cond}$$

from which the condenser surface area can be computed as

$$A_{cond} = \frac{\dot{Q}_{cond}}{U \cdot LMTD_{cond}} = \frac{3.65 \cdot 10^6 \text{ kW}}{1400 \frac{W}{m^2 \, ^\circ C} \cdot 4.33 ^\circ C} = 6.02 \times 10^5 \text{ m}^2$$

The saturated liquid water enthalpies for 8° C and 5° C are 21.05 kJ/kg and 33.55 kJ/kg, respectively. An energy balance then yields the cold-water mass flow rate:

$$\dot{m}_{cold\text{-}water} = \frac{\dot{Q}_{cond}}{h(8 ^\circ C)_f - h(5 ^\circ C)_f} = \frac{3.65 \cdot 10^5 \text{ kW}}{33.55 \frac{kJ}{kg} - 21.05 \frac{kJ}{kg}} = 2.92 \times 10^5 \frac{kg}{sec}$$

The thermal efficiency is the turbine power output divided by the energy transferred in the evaporator, or

$$\eta_{\text{thermal}} = \frac{\text{Power}_{\text{out}}}{\dot{Q}_{\text{evap}}} = \frac{125\,\text{MW}}{3.78 \cdot 10^6\,\text{kW}} = 0.033$$

As would be expected because of the small temperature differences involved in this closed-cycle OTEC system, the thermal efficiency is low. The thermodynamic disadvantages of small temperature differences in heat engines apply to both OTEC configurations. Although the closed cycle does not have the turbine availability (or suitability) problem of the open-cycle OTEC system, the closed-cycle system requires very large heat exchangers. The surface area required for both the evaporator and condenser is in excess of 600,000 m^2!

Hybrid OTEC Systems

A hybrid OTEC system possesses features of both the open-cycle and closed-cycle OTEC arrangements. In a hybrid OTEC system, the warm surface water is flashed into steam in a process similar to that for the open-cycle OTEC configuration. The steam is then used to vaporize the working fluid, typically ammonia, in a closed loop containing a turbine. Desalinated water is provided by condensing the steam.

OTEC System Outputs

In addition to generating electricity, OTEC systems can provide useful outputs such as

1. *Desalinated water.* The condensation of steam as in the proposed hybrid OTEC systems could provide potable water.

2. *Nutrients for mariculture applications.* The cold-water stream for an OTEC system contains abundant nutrients (and is relatively free of pathogens) that can be used in commercial mariculture applications.

3. *Air conditioning for buildings and moderate-temperature refrigeration.* The cold-water stream (5° C or 41° F) can be used in conjunction with heat exchangers to provide conditioned air for cooling or chilled water for moderate-temperature refrigeration applications.

4. *Mineral extraction.* Since OTEC systems have a high flow rate, the extraction of minerals dissolved in the ocean water has been proposed as an additional OTEC application.

OTEC Assessment

Examples 14.1 and 14.2 demonstrate the size and efficiency problems associated with OTEC systems. Indeed, except for a few pilot plants, no commercial OTEC facility has been built and operated. To date the biggest success has been the pilot plant in Hawaii that operated from 1993 to 1998 (Vega 2002). The facility produced a net power of 103 kW with 255 kW of gross power extracted. The warm surface water was 26° C, and the cold deep water was 6° C. Vega (2002) provides a contemporary assessment of OTEC and makes several suggestions for future development.

Since the 1970s, a number of large OTEC systems have been proposed and a few have reached the detailed engineering stage. Most of these systems are quite large and are to be moored in deep ocean waters. Figures 14.5 and 14.6 reproduce concept sketches of how a large OTEC facility might appear. The two reproductions are similar in general layout and configuration. Figure 14.5, a TRW design, was reported in 1976 and appears in Kayton (1981), and Figure 14.6, a Lockheed concept, appeared at about the same time and has been discussed in Rudiger and Smith (1977), Baker (2006), and Takahashi (2000). Additional information on both of these concepts is available in the four aforementioned references. An interesting feature of the TRW design is that the expelled sea water is used to provide a thrust in a dynamic

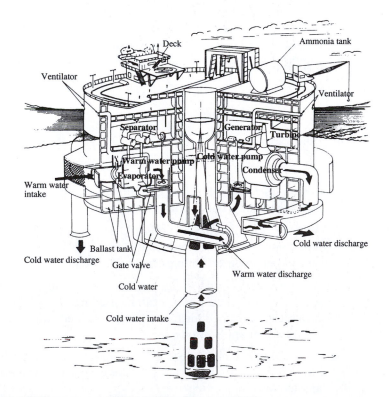

Figure 14.5 TRW proposed OTEC facility (Kayton, 1981).

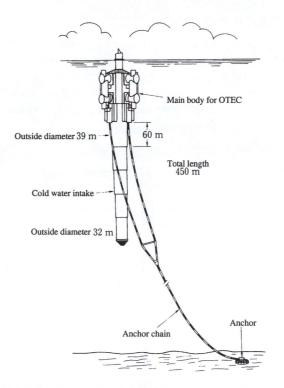

Main body for OTEC

Outside diameter 39 m

60 m

Total length
450 m

Cold water intake

Outside diameter 32 m

Anchor chain

Anchor

Figure 14.6 Lockheed proposed OTEC facility (Lockheed, 1977).

positioning strategy that helps the facility to hold station (position and alignment). The Lockheed OTEC design possessed a generating capacity of 160 MW and would have used 260,000 tons of concrete!

Although the U.S. Department of Energy no longer supports significant OTEC research, it does provide a website (www.nrel.gov.otec/research) that summarizes the current status of OTEC research. The list of needed components is quite long and includes pacing items in cold-water pipes, direct-contact condensers, evaporators, system integration, surface condensers, and turbines.

Areas especially suited for initial OTEC system applications include the small island nations in the south Pacific Ocean, American territories (such as Guam) in the Pacific, and the state of Hawaii. All of these locations use diesel-powered generators for electricity and have water potability issues.

Recently, OTEC plant ships have been proposed (Vega 2002 and others). These ocean-going ships would contain OTEC systems and would move slowly in OTEC-favorable regions to harvest energy and minerals. The electric power from the harvested energy would be used in on-board manufacturing and fuel (such as hydrogen) production. Figure 14.7, from Vega (2002), presents the layout of a proposed pilot facility with a 5-MW electrical output.

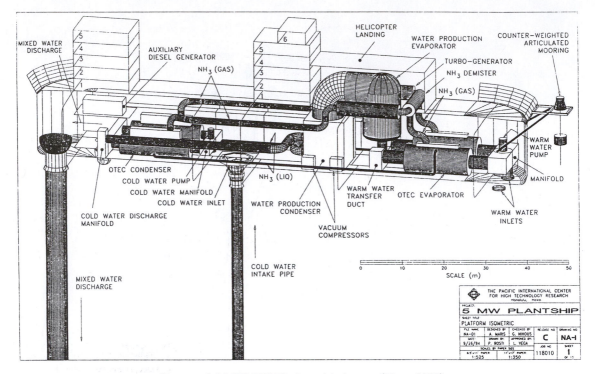

Figure 14.7 Proposed 5-MW OTEC plant ship layout (Vega, 2002).

Thus far in this chapter, OTEC has been examined, but ocean energy harvesting from tides and waves is also possible. The next sections of this chapter investigate these concepts.

14.3 TIDAL ENERGY

Tidal energy uses the rise and fall of the tides to indirectly generate electricity. In principle the harvesting of energy from the tides is similar to hydroelectric power. At high tide a reservoir is charged, and at low tide the reservoir is discharged. In both the charging and discharging processes, a turbine can be used to generate electricity as in other hydroelectric facilities. However, the temporal aspects of the tides add other considerations. The simplest tidal power system is called the ebb generating system and uses a dam, usually called a barrage, across an estuary. Sluice gates open and close to regulate the flow into and out of the reservoir. Two-way tidal energy systems generate electricity with flow into and out of the reservoir, but are more complex and costly than "single-direction" generation since the turbine and supporting structure must be more robust and complex.

The tides are periodic but not constant, since they are governed by the moon, and to some extent the sun. Tides are described based on their schedule and range. The range is the elevation difference between high and low tide, and the schedule includes the times of occurrence of the high and low tides. A "lunar month" is 29.5 days, so a lunar day is 24 hours, 50 minutes. During a lunar day, the tides rise and fall twice—a tidal cycle is thus 12 hours, 25 minutes. This means that high and low tides do not occur at the same time every day, but progress 25 minutes per cycle, or 50 minutes per lunar day. Moreover, the range is not constant but varies during the lunar month, with the maximum range occurring at the time of the new and full moons and the minimum range occurring at the first and third quarter moons. At a given location, the average ranges are complicated but predictable. The maximum range is called the spring tide, and the minimum range the neap tide. A graphical representation, such as Figure 14.8, is a good way to understand the details of tidal motions. The figure shows the daily tidal cycle, the monthly tidal cycles, the tidal range variations, and the relationship of the phases of the moon with the spring and neap tides.

The observation that the high and low tides occur at different times during different days and that the range varies from day to day (spring to neap to spring tides) means that control strategies to maximize tidal energy benefits (typically power produced) are more involved than strategies for hydroelectric facilities. The consideration of how to distribute the generating time based on the tidal range for a given day impacts the size (output) of the turbine and the system total output. A typical

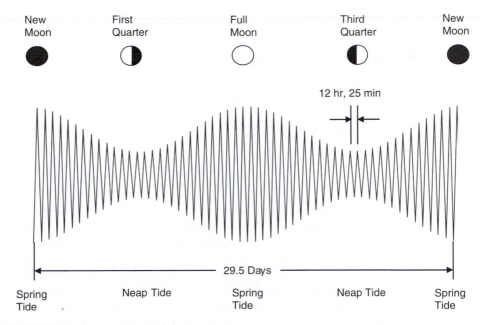

Figure 14.8 Typical monthly tidal ranges.

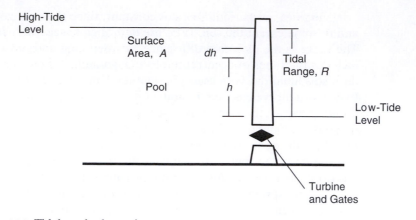

Figure 14.9 Tidal pool schematic.

arrangement for a tidal pool used to generate power is illustrated in Figure 14.9. The difference in elevation due to the tidal range determines the energy stored in the pool. The energy stored in a differential element of height dh located above the low-level tide is

$$dW = g\,dm\,h \qquad (14\text{-}2)$$

where dm is the mass in the differential element and $dm = \rho\,A\,dh$, where A is the horizontal area, so that

$$dW = g\,\rho\,A\,h\,dh \qquad (14\text{-}3)$$

Then for the tidal pool,

$$W = \int_0^R dW = \int_0^R g\,\rho\,A\,h\,dh \qquad (14\text{-}4)$$

If the surface area variation with elevation in the tidal pool is small, then A is sensibly constant and

$$W = \frac{1}{2}\,g\,\rho\,A\,R^2 \qquad (14\text{-}5)$$

Equation (14-5) demonstrates that the available energy in a tidal pool, referenced to the low-tide datum, is proportional to the surface area and the square of the tidal range. Hence, the tidal range is an important metric in assessing tidal energy potential.

The energy indicated by Eq. (14-5) is based on the low-tide datum; hence, the energy is attainable only when discharged to that datum. This observation raises the question of how a tidal energy system might be operated to maximize the usefulness of the energy harvested. Consider the conditions illustrated in Figure 14.9, namely, the tidal pool at its maximum height and the low-tide datum at its minimum. If all the tidal pool content were discharged rapidly, then the energy attainable would be the maximum; however, if the pool were gradually discharged, then the low-tide datum would increase and the high-tide elevation would decrease, and the available energy would be decreased. Two protocols are thus associated with tidal discharge strategies: (1) a simple tidal pool, in which the discharge is rapid and the conditions of Eq. (14.5) are fulfilled; and (2) the modulated simple tidal pool, in which the flow is varied over time and the energy recovered is reduced from the maximum. Although protocol 2 results in less available energy than protocol 1, protocol 2 has major advantages. Protocol 2 requires smaller turbines and results in a more uniform power level applied to the grid or for local use. Protocol 1 results in large "spikes" of electrical power that occur at high and low tides.

El-Wakil (1984) investigates protocol 2, as illustrated in Figure 14.10, which is adapted from his book. A single tidal cycle with a period of 6.2083 hours is indicated. Generation is initiated at time t_1, when the elevation difference between the pool and the tide is sufficient for efficient generation, and is stopped at time t_2, when the elevation difference becomes too small for efficient generation. The determination of the times is facility dependent, but to reduce the "spikedness" of the generation, the generation time should be an appreciable portion of the tidal cycle. During generation, the gates to the pool are controlled in such a fashion that the pool fills linearly with time as indicated by the dashed line in Figure 14.10. The tide is represented by a sine function,

$$H = \frac{R}{2} \sin\left(\frac{\pi t}{6.2083 \text{ h}}\right) \tag{14-6}$$

and the pool depth is a linear function of time:

$$y = aR(t-t_1) \tag{14-7}$$

where a controls the slope of the tidal pool filling and R is the tidal range. Over a tidal cycle El-Wakil determined the energy available for harvesting to be

$$W = g\rho \, A \, R^2\left[0.988a\left(\cos\left(\frac{\pi t_1}{6.2083}\right) - \cos\left(\frac{\pi t_2}{6.2083}\right)\right) - \frac{a^2}{2}\left(t_2^2 - t_1^2\right)\right] \tag{14-8}$$

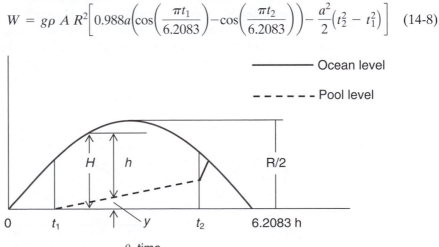

Figure 14.10 Nomenclature for modulated single-pool tidal system.

Example 14.3 answers the question of how much protocol 2 reduces the energy.

Example 14.3	Investigate the energy available from protocol 1 and protocol 2 per km² of surface area if the tidal range is 12 m, a is 0.0625 h^{-1}, and energy is harvested between $t_1 = 1$ h and $t_2 = 4$ h. How does a influence the energy harvested?

Solution   The nominal density of sea water is 1025 kg/m³. For protocol 1, Eq. (14-5) represents the energy available and has the value

$$W_1 = \frac{1}{2} g \rho A R^2 = \frac{1}{2} \cdot 9.807 \frac{m}{sec^2} \cdot 1025 \frac{kg}{m^3} \cdot 1km^2 \cdot (12\ m)^2 = 201{,}040\ kWh$$

Equation (14-8) represents the energy availability with modulated flow, and appears as

$$W_2 = g\rho A R^2 \left(0.988a \left(\cos\left(\frac{\pi t_1}{6.2083} \right) - \cos\left(\frac{\pi t_2}{6.2083} \right) \right) - \frac{a^2}{2}(t_2^2 - t_1^2) \right)$$

$$= 9.807 \frac{m}{sec^2} \cdot 1025 \frac{kg}{m^3} \cdot 1km^2 \cdot (12m)^2$$

$$\cdot \left\{ \begin{array}{l} 0.988 \cdot h \cdot 0.0625\ h^{-1} \cdot \left(\cos\left(\frac{\pi \cdot 1\ h}{6.2083\ h} \right) - \cos\left(\frac{\pi \cdot 4\ h}{6.2083\ h} \right) \right) \\ -\dfrac{(0.0625\ h^{-1})^2}{2} \cdot \left[(4\ h)^2 - (1\ h)^2 \right] \end{array} \right\}$$

$$= 20{,}811\ kWh$$

Hence, the modulated flow, protocol 2, has the potential to produce only about one-tenth of the energy that protocol 1 can. This is a dramatic decrease in the energy available for harvesting, but energy harvested is spread over a four-hour time period instead of a very brief generating interval, and the turbine size (power output) required is dramatically smaller. The primary control variable in the modulated scenario is the value of a; hence, a legitimate question is, how does the energy harvest availability depend on a? If the calculation for W_2 is repeated for a range of values of a and the results are plotted, Figure 14.11 results. As a is increased for a small value, the energy availa-

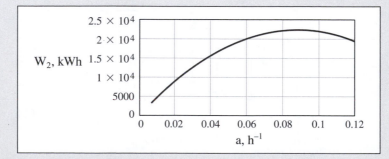

Figure 14.11 Energy availability as a function of a.

bility increases, reaches a maximum, and then gradually decreases. The value of a for maximum energy availability is near $0.09 \, \text{h}^{-1}$, and the value of $a = 0.0625 \, \text{h}^{-1}$ prescribed in the problem statement results in nearly as much possible energy recovery.

In Example 14.3 and in Eqs. (14-5) and (14-8), the energy harvested is just the energy extracted from the flow and not the gross energy output of the tidal system. Turbine efficiency as well as losses in the system need to be considered in order to obtain the net system output. El-Wakil (1984) suggests that the overall efficiency may be as low as 30 percent.

The expressions in Eqs. (14-5) and (14-8) have R^2 in the leading term, demonstrating the importance of the tidal range in determining energy harvesting potential for tidal systems. Tidal systems are attractive in regions where the tidal range is high and the coastal geometry permits the development of large-surface-area tidal pools. Gorlov (2001) provides a list of regions with high tidal ranges, from which Table 14.2 is adapted. Locations with large tidal ranges are limited in number, and the total available tidal resources are very limited compared with most of the alternative energy sources considered in this textbook.

As of early 2008 only four tidal energy systems were operational, with an additional one under consideration. Table 14.3, also adapted from Gorlov, contains details of the four operating tidal systems. Only one, La Rance, generates large quantities of electricity and uses an appreciable surface area.

TABLE 14.2 Locations with high tidal ranges

Site	Country	Tidal Range, R (m)
Bay of Fundy	Canada	16.2
Severn Estuary	United Kingdom	14.5
Port of Ganville	France	14.7
La Rance	France	13.5
Puerto Rio Gallegos	Argentina	13.3
Bay of Mezen	Russia	10.0
Sea of Okhotsk	Russia	13.4

TABLE 14.3 Existing tidal energy facilities

Site	Country	Power (MW)	Basin Area (km^2)	Mean Tide (m)
La Rance	France	240.0	22.0	8.6
Kislaya Guba	Russia	0.4	1.1	2.3
Annapolis	Canada	18.0	15.0	6.4
Jiangxia	China	3.9	1.4	5.1

In 2005 South Korea announced plans to construct a tidal energy facility at Sihwa Lake. The Sihwa Tidal Power Plant (Schmid 2005), with an output of 260 MW, will be larger than La Rance, the current largest in the world. The facility will utilize 10 bulb-type turbines, each with an output of 26 MW. The plant will use the elevation difference between high tide and a reservoir to generate the 260 MW of electricity. The facility will generate power in one flow direction, from the ocean to Sihwa Lake, with up to 60 billion tons of sea water circulated annually.

The La Rance Tidal Power Plant, near Saint-Michel in Brittany (www.edf.fr/html/en/decouvertes/voyage/usine/usine.html), is clearly the most outstanding example of such facilities. La Rance became operational in 1967, and it has operated without any major breakdowns and annually generates more than 600 million kWh. The facility has 24 turbines, each producing 10 MW. The dam, the barrage, is 750 m long and 13 m high, and, as indicated in Table 14.3, the pool has an area of 22 km^2. La Rance has operated more than 160,000 hours and has produced electricity that is economically competitive. Electricite de France has recently been carrying out a general overhaul of all the equipment, including the installation of three new turbines per year until the originals have all been replaced.

In addition to their large capital costs, tidal energy systems raise significant environmental and ecological questions. Effects on the environment are site specific and difficult to quantify. These effects include aquatic and shoreline ecosystem impacts, silting, and water quality issues.

14.4 WAVE ENERGY

The final topic examined in this chapter is wave energy. Ocean waves are caused by wind, which is an indirect effect of uneven solar heating, and by the motion of the Earth. As with other alternative energy resources, the quantity of energy available is enormous, but the techniques and technologies for harvesting the energy are still under development. Even moderate wave motion possesses an energy density in excess of the incident solar energy density. However, before exploring the energy content of waves, we will consider the usual description of a traveling wave:

$$y = a \sin\left(\frac{2\pi}{\lambda}x - \frac{2\pi}{\tau}t\right) \tag{14-9}$$

where a is the amplitude, λ is the wavelength, and τ is the period; y is the height above mean sea level, t is the time, and x is the horizontal coordinate. Wave nomenclature for a traveling wave is presented in Figure 14.12. The wavelength, λ, the distance of a complete cycle in the x-direction, and the crest, a, are illustrated in the figure. Since the wave is periodic, when $t = \tau$ the wave shape is identical to the shape at $t = 0$.

Although the wave motion is in the x-direction, with a speed of $c = \lambda/\tau$, the motion of the water is not the same as the motion of the wave. The water rotates in place with an elliptical path in the plane of the wave propagation (El-Wakil 1984).

The total energy of a wave is the sum of the potential and kinetic energies (details in El-Wakil 1984). The potential energy density per unit area is

$$\frac{PE}{A} = \frac{1}{4}\rho \cdot a^2 \cdot g \tag{14-10}$$

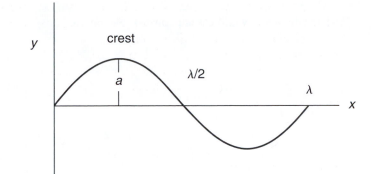

Figure 14.12 Wave nomenclature.

and the kinetic energy density per unit area is also

$$\frac{KE}{A} = \frac{1}{4}\rho \cdot a^2 \cdot g \qquad\qquad (14\text{-}11)$$

so that the total energy is the sum of the potential and kinetic energies or

$$\frac{TE}{A} = \frac{1}{2}\rho \cdot a^2 \cdot g \qquad\qquad (14\text{-}12)$$

The power density is the energy density times the frequency, f, or

$$\frac{PD}{A} = \frac{1}{2}\rho \cdot f \cdot a^2 \cdot g \qquad\qquad (14\text{-}13)$$

El-Wakil (1984) approximates the relationship between the wavelength and period of an ocean wave to be

$$\lambda = 1.56 \cdot \tau^2 \text{ m} = 5.12 \cdot \tau^2 \text{ ft} \qquad\qquad (14\text{-}14)$$

with τ in seconds. Example 14.4 illustrates how to use the preceding calculations.

Example 14.4	A 2-m ocean wave has a period of 5 seconds. Find the wavelength, the wave velocity, and the energy and power densities for this wave.

Solution   With a wave height of 2 m, the amplitude is one-half of the wave height, or $a = 1$ m. The relationship between the period and wavelength is provided by Eq. (14-14), and the wavelength becomes

$$\lambda = 1.56 \cdot \tau^2 \text{ m} = 1.56 \cdot \left(\frac{5 \cdot \text{sec}}{\text{sec}}\right)^2 \cdot \text{m} = 39 \text{ m}$$

The wave speed is calculated as

$$c = \frac{\lambda}{\tau} = \frac{39 \text{ m}}{5 \text{ sec}} = 7.8 \frac{\text{m}}{\text{sec}}$$

and the frequency is the reciprocal of the period, or

$$f = \frac{1}{\tau} = \frac{1}{5 \text{ sec}} = 0.2 \frac{1}{\text{sec}}$$

The total energy (TE) and power (PD) densities are calculated using Eqs. (14-12) and (14-13), respectively:

$$\frac{TE}{A} = \frac{1}{2} \rho \cdot a^2 \cdot g = \frac{1}{2} \cdot 1025 \frac{\text{kg}}{\text{m}^3} \cdot (1 \text{ m})^2 \cdot 9.807 \frac{\text{m}}{\text{sec}^2} = 503 \frac{\text{J}}{\text{m}^2}$$

$$\frac{PD}{A} = \frac{1}{2} \rho \cdot a^2 \cdot f \cdot g$$

$$= \frac{1}{2} \cdot 1025 \frac{\text{kg}}{\text{m}^3} \cdot (1 \text{ m})^2 \cdot 0.2 \frac{1}{\text{sec}} \cdot 9.807 \frac{\text{m}}{\text{sec}^2} = 1005 \frac{\text{W}}{\text{m}^2}$$

For these wave conditions, the power density of 1005 W/m² is high—greater than the solar irradiation absorbed by the water or incident on the water surface.

Example 14.4 illustrates that a reasonable wave possesses a higher power density than the solar energy that might be absorbed by the ocean. Considering the extent of the oceans and the wave conditions present at any given time, the energy available from waves is enormous. The questions then become: (1) How can wave energy be harvested? (2) Why are wave energy devices not in widespread use?

Most of the proposed and demonstrated wave energy machines harvest mechanical rather than thermal energy. Two common tactics for harvesting wave energy are to use the wave as an "air compressor" to power a turbine or to use the wave motion to charge a hydraulic reservoir that powers a hydraulic motor. Examples of each will be reviewed.

A schematic of an air-powered turbine is shown in Figure 14.13. The periodic motion of the wave causes the air trapped in the lower chamber to be forced through a venturi in the upper chamber to the turbine, which extracts energy to drive an elec-

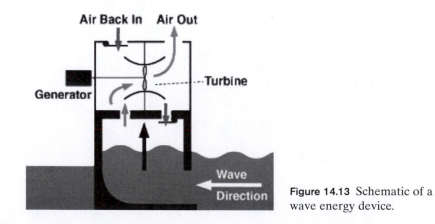

Figure 14.13 Schematic of a wave energy device.

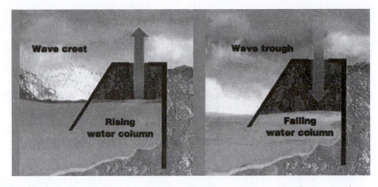

Figure 14.14 Wavegen oscillating water column system illustration (Wavegen).

trical generator. When the wave recedes, valves in the upper chamber and lower chamber open to admit air into the lower chamber, and the process is repeated with the next wave. To harvest significant energy, the wave machines must be large and there must be many of them. For example, Wavegen, a Voith and Siemens company, has installed the first grid-connected, commercial-scale wave energy plant off the coast of western Scotland. The plant uses an inclined oscillating water column system and serves as a full-scale test bed for component validation and development. An illustration from Wavegen's website (www.wavegen.co,uk) is presented in Figure 14.14. The inclined side of the chamber is a dominant feature.

The only large commercially available wave energy device is marketed by Pelamis Wave Power, formerly Ocean Power Delivery. The company's web address is www.pelamiswave.com. The Pelamis Wave Energy Converter is a semi-submerged, articulated system composed of cylindrical sections connected by hinged joints. The wave-induced motion drives hydraulic rams that pump high-pressure fluid through hydraulic motors that drive electrical generators. The power is fed via a transmission line to shore. Several devices can be linked to form a wave energy farm. The Pelamis P-750 is 140 m long and 3.5 m in diameter, and is rated at 750 kW. Specifications for the P-750 are given in Table 14.4. A Pelamis P-750 will produce a yearly average of 25–40 percent of the rated power. Figure 14.15 shows a photograph of the P-750. The articulated nature of the design is perhaps the most salient feature.

Figure 14.15 Pelamis P-750 articulated wave energy converter (Pelamis).

TABLE 14.4 P-750 specifications (Pelamis brochure)

Overall length	150 m
Diameter	3.5 m
Displacement	700 tonnes (metric, 1000 kg)
Nose	5 m long, drooped conical
Power takeoff	Three power conversion units
	Hydraulic rams (2 in heave, 2 in sway)
Ram speed	0–0.2 m/sec
Working pressure	100–350 bar
Power conversion	Two variable-displacement motors
Generator	Two 157 kVa/125 kW
Speed	1500 rpm
Power	750 kW
Energy	2.7 GWh/year
Wave power	Nominal 55 kW/m

14.5 CLOSURE

OTEC, tidal, and wave energy have been explored in this chapter. The energy potential is enormous for OTEC, and for wave energy in particular, but the technical, operational, and financial problems are daunting. Ocean energy devices must cope with large size, high capital costs, corrosive environments, and uncertain and severe weather conditions. The potential is present, however, and as energy costs continue to escalate, more resources are being directed toward ocean energy as at least a promising part of the energy solution.

REVIEW QUESTIONS

1. What does OTEC mean?
2. Describe the principle involved in energy harvesting for an OTEC system.
3. What is the appeal of OTEC?
4. What is the expected thermal efficiency of an OTEC system?
5. On a per kW basis, why are OTEC facilities expected to be large in comparison with most other energy sources?
6. What are two location problems associated with OTEC facilities?
7. Why will commercial OTEC power generation facilities have very large salt water flow rates?
8. In addition to energy, what is expected to be a major source of income for OTEC facilities?

9. What is the temperature difference (surface to deep ocean) in OTEC systems?
10. What is the general strategy for harvesting wave energy?
11. What is the general strategy for harvesting tidal energy?
12. What is the current state of power generation from OTEC energy sources?
13. How do the thermal efficiencies of fuel cells, geothermal systems, and OTEC system differ?

EXERCISES

1. Plot the maximum thermal efficiency for an OTEC facility operating with a temperature difference range of $10°C < \Delta T < 24°C$. How important is ΔT to the feasibility of an OTEC system?

2. An open-cycle OTEC system operates with warm surface water at 24° C and surface condenser water at 14° C with the deep cold water at 12° C. The evaporator pressure is 0.0264 bar, which corresponds to a saturation temperature of 22° C, and the condenser pressure and temperature are 16° C and 0.0182 bar. The turbine efficiency is 0.83. If the turbine is to extract 100 kW, determine the system efficiency and the warm-water, cold-water, and turbine mass flow rates. Contrast these results with the results of Example 14.1.

3. A closed-cycle OTEC system uses propane as the working fluid. The warm surface water enters and leaves the evaporator at 25° C and 22° C, respectively. The cold water enters and leaves the condenser at 7° C and 10° C, respectively. The propane evaporating temperature is 19° C, and the condensing temperature is 13° C. The plant output is 100 MW, and the turbine efficiency is 0.83. The overall heat transfer coefficient for both the evaporator and the condenser is 1400 W/m² K. Determine the propane flow rate, the evaporator and condenser surface areas, the warm- and cold-water mass flow rates, and the overall thermal efficiency.

4. The Annapolis, Canada, tidal energy system is specified as producing 18 MW with an area of 15 km² and a tidal range of 6.4 m. Investigate the energy available from protocol 1 and protocol 2 of the Annapolis system if energy is harvested between $t_1 = 1$ h and $t_2 = 4$ h.

5. A 1-m ocean wave has a period of 5 sec in the ocean. Find the wavelength, the wave velocity, and the energy and power densities for this wave. Contrast the results of this exercise with those of Example 14.4. How important is wave height in determining wave energy?

REFERENCES

Avery, W. H., and Berl, W. G. 1997. "Solar Energy from the Tropical Oceans," *Issues in Science and Technology* 14 (Winter).
Avery, W. H., and Wu, Chih. 1994. *Renewable Energy from the Ocean: A Guide to OTEC.* New York: Oxford University Press.

Baker, J. B. 2006. "Energy from the Sea," *Technology Review* 96, July/August.

Cruz, J., ed. 2008. *Ocean Wave Energy*. Berlin: Springer-Verlag.

El-Wakil, M. M. 1984. *Powerplant Technology*. New York: McGraw-Hill.

Gorlov, A. M. 2001. "Tidal Energy." *In Encyclopedia of Science and Technology*, 2955–2960.

Kayton, M. 1981. "Steady-State and Dynamic Performance of an Otec Plant," *IEEE Transactions on Power Apparatus and Systems* PAS-100 (3), March, 1148–1153.

Palmer, D. A., Fernandez-Prini, R., and Harvey, A. H. 2004. *Aqueous Systems at Elevated Temperatures and Pressures: Physical Chemistry in Water, Steam and Hydrothermal Solutions*. New York: Academic Press.

Perry, R. H., and Chilton, C. H. 1973. *Chemical Engineer's Handbook*, 5th ed. New York: McGraw-Hill.

Rudiger, C. E., and Smith, L. O. 1977. "OTEC—An Emerging Program of Signification to the Marine Community," Proceedings of the 3rd Oceans '77 Annual Combined Conferences, Los Angeles, Oct. 1977, Volume 2, p. 41C-1 to 41C-7.

Schmid, H. 2005. "Sihwa in the Mix," *Power Engineering International, 13*(9), September.

Vega, L. A. 2002. "Ocean Thermal Energy Conversion Primer," *Marine Technology Society Journal* 6(4) 25–35.

WEBSITES

www.nrel.gov/otec/electric_turbines

www.nrel.gov.otec/research

www.nrel.gov/otec/what.html

www.edf.fr/html/en/decouvertes/voyage/usine/usine.html

www.wavegen.co,uk

www.pelamiswave.com

CHAPTER

15

Nuclear Energy

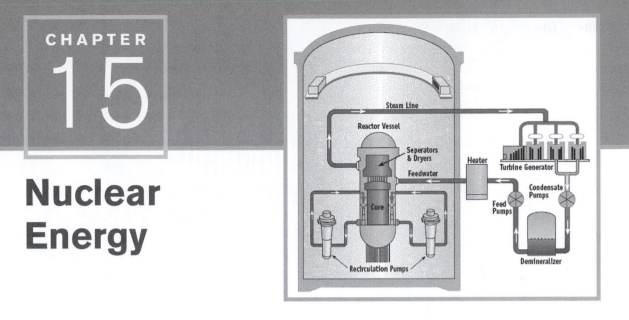

15.1   INTRODUCTION

In the United States, nuclear power has been a controversial issue for more than a generation. It was controversial even before the Three Mile Island incident in 1979, with legal barriers and maneuvers effectively halting additional penetration into the electric utility industry. After Three Mile Island, construction of new nuclear power plants in the United States essentially ceased. Internationally, nuclear power has achieved greater acceptance, with western Europe and Japan now generating much of their electricity from nuclear energy. Poor design choices and operational procedures in the former USSR led to the 1986 Chernobyl disaster, which further clouded nuclear power generation safety issues. However, with the increasing concern over greenhouse gas–driven climate change as well as energy independence, nuclear power is again under active consideration in the United States, and for the first time in almost 30 years plans are ongoing for new nuclear power plant construction.

An interesting graphic on nuclear power is provided by the World Nuclear Association on its website (www.world-nuclear.org), and is reproduced as Figure 15.1. Currently, about 16 percent of electricity is generated with nuclear power at a near-constant level of 2700 TWh the last few years. Even after the moratorium on construction of nuclear power plants in the United States, the worldwide share of nuclear power and its output in kWh generated per year, as evidenced by the figure, continued to increase. Nuclear power produces about 20 percent of the electricity in the United States. By any measure, nuclear power is vital to the production of electricity and is hardly an "alternative energy" source in the generally accepted connotation of the phrase. However, nuclear power is included in this textbook because of the important role it currently plays and because the growing interest in nuclear power, especially in the United States, portends that it will become more important, not less so. Indeed, many individuals in the electric power industry believe that nuclear power may become dominant in this century.

383

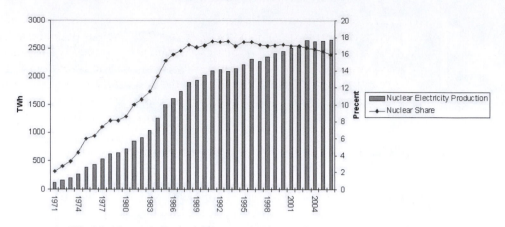

Figure 15.1 Worldwide nuclear electricity production and percentage share (World Nuclear Association).

This chapter discusses the fundamentals and basics of nuclear energy, examines the various configurations of nuclear power plants, and reviews the potential and technical problems of fusion power production.

15.2 FUNDAMENTALS OF NUCLEAR ENERGY

An atom consists of a heavy, positively charged nucleus with much lighter electrons in orbit about the nucleus (see Chapter 9). The nucleus is composed of electrically neutral neutrons and positively charged protons. The charge on the proton is equal, but opposite, to the charge on the electron. Since an atom is electrically neutral, the number of protons is equal to the number of electrons. The three primary subatomic particles are the electron, the neutron, and the proton. The masses of these particles, expressed in atomic mass units, or amu (1 amu = 1.660539×10^{-27} kg), are as follows (Shultis and Faw 2002):

Neutron mass, m_n = 1.008665 amu

Proton mass, m_p = 1.007276 amu

Electron mass, m_e = 0.0005486 amu

The proton and neutron, the particles that make up the nucleus, are called nucleons. By convention, the number of protons in the nucleus defines the atomic number, Z, of an element, and the sum of the neutrons and protons in the nucleus defines the mass number, A. The usual notation for an atom of element X is $_Z X^A$ so that the atomic number and mass number are easily tracked. Most of the mass of an atom is in the nucleus. Isotopes of an element have the same number of protons but a different number of neutrons and, thus, different mass numbers. For example, uranium has three common isotopes: $_{92}U^{238}$, $_{92}U^{235}$, and $_{92}U^{234}$. In this protocol, "ordinary" hydrogen appears as $_1H^1$, and because the hydrogen nucleus contains a single proton and no neutrons, $_1H^1$ is also a proton. The nucleus of deuterium contains a proton and a neutron and appears as $_1H^2$ and is sometimes written as $_1D^2$. The helium nucleus contains a pair of neutrons and protons and is cast as $_2He^4$. Since the electron mass is quite small compared to the mass of a neutron or a proton, an electron is often represented as $_{-1}e^0$. A neutron, which carries no charge, then becomes $_0n^1$.

In a chemical reaction whole atoms are involved, as in the hydrogenation reaction

$$C + 2\,H_2 \rightarrow CH_4 \tag{15-1}$$

In this reaction atoms of carbon and hydrogen combine to form methane, but the atoms retain their identities; indeed, the number of carbon and hydrogen atoms on the left (reactant) and right (product) sides of the equation must be equal. In a nuclear reaction the reactant nuclei do not show up in the products, but other nuclei or different isotopes of the reactants are present. Consider a nuclear reaction involving elements A, B, C, and D:

$$_{Z1}A^{A1} + {}_{Z2}B^{A2} \rightarrow {}_{Z3}C^{A3} + {}_{Z4}D^{A4} \tag{15-2}$$

Conservation of mass number and conservation of atomic number (protons) demand

$$A1 + A2 = A3 + A4 \quad \text{and} \quad Z1 + Z2 = Z3 + Z4 \tag{15-3}$$

An example is ordinary aluminum capturing in its nucleus a helium nucleus, $_2He^4$. The nucleus with the addition of the helium nucleus is unstable and decays:

$$_{13}Al^{27} + {}_2He^4 \rightarrow {}_{14}Si^{30} + {}_1H^1 \tag{15-4}$$

The result of the nuclear reaction is silicon and a proton (hydrogen nuclei). The sums of the mass numbers (31) and the atomic numbers (15) are the same on the two sides of the equation.

Each isotope of each element has a specific mass that is dependent on the mass number and the atomic number. The atomic mass unit is defined to be one-twelfth the mass of a neutral ground-state atom of $_6C^{12}$. All other isotopes are referenced to this standard. The atomic mass units of all 3200 isotopes are provided in atomic mass tables. Atomic mass tables are available in textbooks on nuclear engineering or nuclear physics and on a number of websites; www.ie.lbl.gov/toi2003/MassSearch.asp is an especially useful site. In this chapter, the atomic mass table presented in Shultis and Faw (2002) is used. From this table, the following values were abstracted:

$$\text{mass of } {}_{13}Al^{27} = 29.981538 \text{ amu}$$
$$\text{mass of } {}_2He^4 = 4.002603 \text{ amu}$$
$$\text{mass of } {}_{14}Si^{30} = 29.973770 \text{ amu}$$
$$\text{mass of } {}_1H^1 = 1.007825 \text{ amu}$$

For the nuclear reaction of Eq. (15-4) the mass of the reactants is 29.981538 amu + 4.002603 amu = 30.984141 amu, and the mass of the products is 29.73770 amu + 1.007825 amu = 30.981595 amu. Mass was not conserved! The product mass is less than the reactant mass. The difference, $\Delta m = -0.002546$ amu, represents the mass converted to energy via Einstein's famous expression

$$\Delta E = \Delta mc^2 \tag{15-5}$$

where c is the speed of light, 3×10^6 m/sec. Thus, the energy is

$$\Delta E = \Delta mc^2 = -0.002546 \text{ amu} \cdot \left(3 \cdot 10^8 \frac{\text{m}}{\text{sec}} \right)^2 = -2.2914 \cdot 10^{14} \text{ amu} \cdot \frac{\text{m}^2}{\text{sec}^2}$$

$$= -2.2914 \cdot 10^{14} \cdot \text{amu} \cdot \frac{\text{m}^2}{\text{sec}^2} \cdot \frac{1.660539 \cdot 10^{-27} \text{ kg}}{\text{amu}} \cdot \frac{\text{N} \cdot \text{sec}^2}{\text{kg m}} \cdot \frac{\text{J}}{\text{N m}}$$

$$= -3.8053 \cdot 10^{-13} \text{ J} \cdot \frac{\text{eV}}{1.602176 \cdot 10^{-19} \text{ J}} \cdot \frac{\text{MeV}}{10^6 \text{ eV}}$$

$$= -2.375 \text{ MeV} \tag{15-6}$$

Hence, the nuclear reaction of Eq. (15-4) produces 2.375 MeV of energy by the conversion of mass to energy. Were the energy produced by a chemical reaction to be examined, it would prove to be orders of magnitude less—hence the interest in nuclear reactions as an energy source.

When Z protons and $(A-Z)$ neutrons come together to form a nucleus, energy is emitted. This energy, called the binding energy, comes from the conversion of some of the mass of the neutrons and protons. The mass of the resulting nucleus is reduced from the mass of the individual neutrons and protons by the mass equivalent [Eq. (15-5)] of the binding energy.

Three types of nuclear reactions are important in the production of energy: fission, fusion, and radioactivity. Each will be examined in turn. In fission, a "heavy" (one with a high atomic number) element is split into two or more lighter nuclei; in fusion, two or more light nuclei are fused to form a heavier nucleus. Radioactivity occurs when an isotope spontaneously undergoes decay or disintegration into a different nucleus, usually by emission of one or more smaller particles. Currently, all large, commercial nuclear reactors utilize fission, so fission is of primary interest in this chapter.

As El-Wakil (1984) points out, neutrons are the only particles that can be practically used in fission reactors, and only a few isotopes are fissionable by neutrons: $_{92}U^{235}, _{94}Pu^{239}, _{92}U^{233}, _{92}U^{238}, _{90}Th^{232}$, and $_{94}Pu^{240}$. Nuclides that undergo fission when a neutron is absorbed by the nucleus are called fissile. Consider two fission reactions involving $_{92}U^{235}$.

$$_{92}U^{235} + _{0}n^1 \rightarrow _{54}Xe^{140} + _{38}Sr^{94} + 2 \, _{0}n^1$$
$$_{92}U^{235} + _{0}n^1 \rightarrow _{56}Ba^{137} + _{36}Kr^{97} + 2 \, _{0}n^1 \tag{15-7}$$

These reactions result in two nuclei different from uranium and the emission of two neutrons. If a reaction is to be sustained, more neutrons must be emitted than absorbed. Example 15.1 examines the energy release from the first of the nuclear reactions in Eq. (15-7).

Example 15.1	Compute the energy resulting from the $_{92}U^{235} + _{0}n^{1} \rightarrow _{54}Xe^{140} + _{38}Sr^{94} + 2\,_{0}n^{1}$ nuclear reaction.

Solution   The atomic mass tables (Shultis and Faw 2002) provide the amu values for the reactants and products of the reaction:

$$\text{mass of } _{92}U^{235} = 235.043923 \text{ amu}$$
$$\text{mass of } _{0}n^{1} = 1.008665 \text{ amu}$$
$$\text{mass of } _{54}Xe^{140} = 139.921640 \text{ amu}$$
$$\text{mass of } _{36}Sr^{94} = 93.915360 \text{ amu}$$

The masses of the reactants and products are

$$\text{Reactants} = 235.043923 \text{ amu} + 1.008665 \text{ amu} = 236.053 \text{ amu}$$
$$\text{Products} = 139.921640 \text{ amu} + 93.915360 \text{ amu} + 2 \cdot 1.008665 \text{ amu} = 235.854 \text{ amu}$$

The change in mass is

$$\Delta\text{mass} = 235.854 \text{ amu} - 236.053 \text{ amu} = -0.198 \text{ amu}$$

And the energy released is

$$\Delta E = \Delta mc^2 = -0.198 \text{ amu} \left(3 \cdot 10^8 \frac{m}{sec}\right)^2 = -2.963 \cdot 10^{-11} J = -184.93 \text{ MeV}$$

Nuclides that are not themselves fissile but can be converted to fissile nuclides upon adsorption of a neutron are called fertile. For example, when $_{92}U^{238}$, a non-fissile isotope, absorbs a neutron, it eventually becomes $_{94}Pu^{239}$, a fertile isotope.

In fusion two or more light nuclei are combined to form a heavier nucleus. Such reactions are possible only if the nuclei possess sufficient kinetic energy to overcome the repulsive forces and are able to reach each other. This topic will be examined in detail later in the chapter. A typical fusion reaction and the energy released are as follows:

$$_{1}H^{2} + _{1}H^{2} \rightarrow _{1}H^{3} + _{1}H^{1} + 4.03 \text{ MeV} \tag{15-8}$$

As with fission reactions, fusion reaction energy release results from the conversion of mass to energy.

Radioactivity, the spontaneous decay of a nucleus by emission of one or more smaller particles, is the last of the three nuclear reactions to be examined. A key word in this definition is "spontaneous" since, unlike fission and fusion reactions, the instability of the nucleus of a radioactive isotope is the causative effect. Isotopes that occur naturally are stable in most elements, but elements with atomic numbers greater than 84 have no stable isotopes, and a few isotopes of elements with atomic numbers less than 84 are also unstable. Isotopes artificially produced in particle accelerators and nuclear reactors are all unstable and number in the thousands. The time required for an unstable nucleus to decay is indeterminate, but among a large cohort of the same unstable nuclei, there is a statistical probability that a certain fraction will decay within a specified time. If N is the number of radioactive isotopes present at time t, then the rate of decay is proportional to N. Mathematically, this is expressed as

$$\frac{dN}{dt} = -\lambda N \qquad (15\text{-}9)$$

where λ is the decay constant. Equation (15-9) is variable separable; if N_o isotopes are present at $t = 0$, then

$$\int_{N_o}^{N} \frac{dN}{N} = -\int_{0}^{t} \lambda \, dt \quad \text{or} \quad \ln\frac{N}{N_o} = -\lambda t \quad \text{or} \quad \frac{N}{N_o} = \exp(-\lambda t) \quad (15\text{-}10)$$

Decay is thus exponential in time. An important metric in radioactive decay is the half-life. By definition, the half-life is the time required for one-half of the radioactive nuclei to decay. Using Eq. (15-10), the half-life can be expressed as

$$\frac{N_o/2}{N_o} = \exp(-\lambda t_{\text{half-life}}) \quad \text{or} \quad t_{\text{half-life}} = \frac{\ln(1/2)}{-\lambda} = \frac{0.69315}{\lambda} \quad \text{or} \quad \lambda = \frac{0.69315}{t_{\text{half-life}}}$$

$$(15\text{-}11)$$

Equation (15-10) can then be cast as

$$\frac{N}{N_o} = \exp\left[-0.69315\frac{t}{t_{\text{half-life}}}\right] \qquad (15\text{-}12)$$

Equation (15-12) is plotted in Figure 15.2; the abscissa is $t/t_{\text{half-life}}$, the number of half-lives. The exponential decay is rapid, such that by slightly over 3 half-lives the nuclei have diminished to only 10 percent; by 10 half-lives, the nuclei have been reduced to less than 0.1 percent.

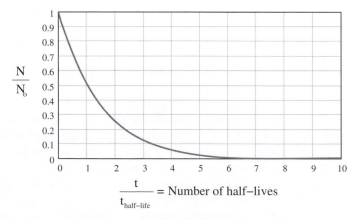

Figure 15.2 Exponential decay as a function of half-life.

The half-lives of different isotopes differ by many orders of magnitude. For example, the half-life of $_{84}Po^{212}$ is only 0.298 sec, while the half-life of $_{92}U^{238}$ is 4.51×10^9 years. Most textbooks on nuclear engineering contain abbreviated tables on radionuclides that include half-life and other useful data. However, a number of websites provide extensive listings; www.nucleardata.nuclear.lu.se/nucleardata/toi/nucSearch.asp is especially user friendly and is a partial implementation of the contents of the eighth edition of *Table of Isotopes* (1999).

Shultis and Faw (2002) identify four mechanisms that result in the decay of a natural (*not* man-made) isotope. We next consider each of the four.

Alpha Decay

An alpha (α) particle is the nucleus of a helium atom and is written as $_2He^4$. Alpha decay occurs when an α particle is ejected. An example is

$$_{88}Ra^{226} \rightarrow {}_{86}Rn^{222} + {}_2He^4 \qquad (15\text{-}13)$$

In alpha decay, Z is reduced by 2 and A by 4; in the reaction of Eq. (15-13), Ra is the parent and Rn is the daughter. Alpha particles give up their kinetic energy by ionizing the matter through which they pass. Most of the energy is dissipated close to the emitting nucleus. Alpha particles possess limited penetrating ability—a few centimeters in air and a much shorter distance (about the thickness of a sheet of paper) in solids. Unless alpha emitters are ingested, they pose little health danger.

Beta Decay

Beta (β) decay is the transformation of a neutron in the nucleus into a proton and an electron, which results in the emission of the electron in order to maintain the charge balance of the nucleus. Beta particles are thus electrons, $_{-1}e^0$. An example of beta decay is

$$_{82}Pb^{214} \rightarrow {}_{83}Bi^{214} + {}_{-1}e^0 + \nu \qquad (15\text{-}14)$$

where ν represents the neutrino, a small, neutrally charged particle that carries about 5 percent of the energy of the reaction. In many cases, the neutrino in Eq. (15-14) is ignored. The β particle has more penetrating ability than the α particle. In beta decay, the atomic number is increased by 1, while the mass number is unchanged.

Positron Decay

In positron decay, the nucleus is unstable because of an excess of protons, and a proton is converted into a neutron by the emission of a positron, a subatomic particle with the mass of an electron but with a positive charge; the symbol is $_{+1}e^0$. An example is

$$_{11}Na^{22} \rightarrow {}_{10}Ne^{22} + {}_{+1}e^0 + \nu \qquad (15\text{-}15)$$

As with beta decay, a neutrino is also emitted. Shortly after emission, the positron will capture an electron, and the mass of the positron electron pair will be completely converted to energy, with the results being two photons traveling in opposite directions, each with an energy of 0.511 MeV.

Gamma Decay

Gamma radiation is electromagnetic radiation of extremely short wavelength and very high frequency—implying significant energy content. As Table 6.1 indicates, gamma rays have wavelengths, λ, less than $10^{-4}\ \mu$m. Gamma decay does not change either the mass number or the atomic number, but represents a reduction in the energy level of the nucleus. In gamma decay, a high-energy photon is emitted from the nucleus. These rays have much more penetrating ability than either alpha or beta particles, and external exposure to γ rays can be very hazardous.

| **Example 15.2** | A radioactive isotope, $_{53}I^{135}$, decays by beta emission into $_{54}Xe^{135}$. Determine the time required for the number of $_{53}I^{135}$ atoms to be reduced by 90 percent. |

Solution From the *Table of Isotopes* (1999), the half-life of $_{53}I^{135}$ is 6.57 h. If the number of nuclides is to be reduced by 90 percent, then only 10 percent of the original number will remain, or

$$\frac{N}{N_o} = 0.10$$

from which

$$\frac{N}{N_o} = 0.1 = \exp\left(-0.69315\frac{t}{t_{\text{half-life}}}\right) = \exp\left(-0.69315\frac{t}{6.57\ \text{h}}\right)$$

Solving for t yields

$$t = -\frac{t_{\text{half-life}}}{0.69315}\ln\left(\frac{N}{N_o}\right) = -\frac{6.57\ \text{h}}{0.69315}\ln(0.1) = 21.82\ \text{h}$$

15.3 NUCLEAR POWER

In this textbook, the discussion of nuclear power is limited to the generation of electricity. However, the United States Navy has a long and rich history of using nuclear power for shipboard operations, including propulsion as well as electricity distribution. The Navy has designed, built, operated, and decommissioned nuclear reactors, maintaining an astonishing safety and performance record and thus representing a significant repository of U.S. nuclear engineering expertise. This section will examine the use of nuclear reactors for the generation of electricity.

Worldwide as of 2007, 439 nuclear power plants with a generating capacity of 384,000 MW (384 GW) were in commercial operation (Nuclear Engineering International 2007). The United States Nuclear Regulatory Commission provides a list of domestic operating nuclear power units at www.nrc.gov/reactors/operating/list-power-reactor-units. In addition, several hundred nuclear reactors are used in the naval vessels of various countries. A general rule of thumb is that the fission of 1 gram of uranium provides the energy equivalent of 10 tons of oil. An overview of nuclear reactors is thus appropriate.

In any discussion of nuclear reactors, the concept of the neutron economy is central. Section 15.2 reviewed several aspects of fission reactions, including the energy released by such reactions. The fundamental idea behind a nuclear reactor is to provide radioactive fuel in an environment such that nuclear reactions can proceed in a safe and dependable manner to provide power to generate electricity. For that to occur, a sufficient quantity of neutrons must be produced to sustain a controlled nuclear reaction sufficient to provide the power required. Uranium, for example, is a common fuel for many reactors and will fission in many different ways, resulting in various fission products. A general nuclear reaction for a large cohort of $_{92}U^{235}$ atoms will result in the "average" reaction (Murray 1961),

$$_{92}U^{235} + _{0}n^{1} \rightarrow _{92}U^{236} \rightarrow _{Z1}FF^{A1} + _{Z2}FF^{A2} + 2.43 \, _{0}n^{1} \qquad (15\text{-}16)$$

where $_{Z1}FF^{A1}$ and $_{Z2}FF^{A2}$ are the fission fragments of the reaction. The mass numbers of the fission fragments and the neutrons emitted must sum to 236. On average, 2.43 neutrons are produced. If the reaction is to be sustained, then at least one of the 2.43 neutrons must be absorbed and result in the fission of the nucleus of another $_{92}U^{235}$ atom.

The nuclear reaction presented in Eq. (15-16) yields on average about 200 MeV per fission. An important consideration for any power-producing machine, such as a nuclear reactor, is how much fuel is used to produce a specified amount of energy. A convenient metric is 1 MWd (megawatt-day) = 24,000 kWh. So a reasonable question is, how much $_{92}U^{235}$ is required to produce 1 MWd? Example 15.3 addresses this question.

Example 15.3

How much $_{92}U^{235}$ is required to produce 1 MWd if the average energy yield is 200 MeV per fission and the electrical conversion efficiency is 35 percent?

Solution

Since the energy yield per fission is 200 MeV, the number of fissions per second for 1 MW can be calculated as

$$1MW = 1,000,000 \text{ W} = 1,000,000 \, \frac{J}{sec} \cdot \frac{MeV}{1.602 \cdot 10^{-13} J} \cdot \frac{fission}{200 \text{ MeV}}$$

$$= 3.121 \cdot 10^{16} \frac{fissions}{sec}$$

The mass of one atom of $_{92}U^{235}$ can be computed: since 235 g of $_{92}U^{235}$ contains 6.023×10^{23} atoms/g mol,

$$m_{atom} = \frac{235 \text{ g mol}}{6.023 \cdot 10^{23} \text{ atom mol}} = 3.902 \cdot 10^{-22} \frac{g}{atom}$$

The amount of uranium required to produce 1MWd of energy then becomes

$$1MWd = 3.212 \cdot 10^{16} \frac{\text{fissions}}{\text{sec}} \cdot \frac{86,400 \text{ sec}}{\text{day}} \cdot \frac{3.902 \cdot 10^{-22} \text{ g}}{\text{atom}} \cdot \frac{1 \text{ atom}}{\text{fission}} = 1.052 \frac{g}{day}$$

Shultis and Faw (2002) state that about 15 percent of the neutrons absorbed by $_{92}U^{235}$ result in something other than fission. The mass per day for 1 MWd at an electrical conversion efficiency of 35 percent is thus

$$m_U = 1.052 \cdot \frac{g}{day} \cdot \frac{1}{0.85 \cdot 0.35} = 3.536 \frac{g}{day}$$

Hence, in one day 3.54 grams of $_{92}U^{235}$ would be consumed to produce 1 MWd (24,000 kWh).

Most neutrons resulting from the reactions such as the one indicated by Eq. (15-16) are fast neutrons—indicating high kinetic energies (~2 MeV) and, hence, high speeds. A fast neutron is difficult for a $_{92}U^{235}$ nucleus to absorb, and unless it is absorbed, it will not initiate another nuclear reaction. A slow (or thermal) neutron has a much higher probability of being absorbed than a fast neutron. A thermal neutron is one that is in thermal equilibrium with its surroundings. Thus, fast neutrons need to be slowed down to become thermal neutrons to enhance their chances of capture so that nuclear reactions can be sustained. Fast neutrons that are slowed down are said to be scattered. The material in a nuclear reactor that is used to slow down or thermalize the fast neutrons is called the moderator. A moderator should have small nuclei with high neutron scattering and low potential for neutron absorption (or capture). Materials that fit these specifications include light water (containing $_1H^1$), heavy water (containing deuterium, $_1H^2$), graphite (C), and beryllium (Be or BeO). The moderator material also plays a role in the fuel requirements of a nuclear reactor.

As mined and processed, natural uranium is composed of 99.2 percent $_{92}U^{238}$ and 0.7 percent $_{92}U^{235}$. If the moderator is graphite or heavy water, then natural uranium can be used as a fuel. However, if light water is the moderator, then the amount of $_{92}U^{235}$ in the fuel must be increased to from 0.7 percent to 3.5–5 percent. Enriching uranium is an expensive and energy-consuming process; Shultis and Faw (2002) discuss enrichment procedures.

The neutron economy is also a factor in reactor control. For a steady-state reactor operation an increase in neutron production results in an increase in the fission rate and in the power output. If neutron production were to continue to increase, the reactor core would eventually melt down. To change the power level of a reactor, neutron production must be altered (increased or decreased) until the required new power level is reached. The neutron production in a reactor can be altered by means of control rods or chemical shims. Control rods are made of materials that absorb neutrons and thus reduce the number of neutrons that can be absorbed by fissile nuclei. Control rods can be moved in or out to affect the neutron economy. Materials that absorb neutrons include boron, cadmium, and indium. A chemical shim involves the use of a soluble neutron absorber, usually boron, in the coolant of a reactor. Chemical shims allow intermediate time control and help to reduce the movement of control rods.

Control rods or chemical shims alter the production of neutrons from one fission generation to the next. The effective neutron multiplication factor, k_{eff}, is defined as the ratio of the neutrons produced from one fission generation to the next fission generation, or

$$k_{eff} = \frac{\text{neutrons produced in generation } n + 1}{\text{neutrons produced in generation } n} = \frac{n_{n+1}}{n_n} \tag{15-17}$$

where n is the number of neutrons. Since $k_{eff} = 1$ represents steady-state operation, the excess of k_{eff} is used to define the reactivity, ρ:

$$\rho = \frac{k_{eff} - 1}{k_{eff}} \tag{15-18}$$

If n_n is the number of neutrons produced in generation n and n_{n+1} the number produced in generation $n + 1$, then $n_{n+1} = k_{eff} n_n$ and $\Delta n = n_{n+1} - n_n = n_n \cdot (k_{eff} - 1)$. If the time between successive generations is t_{avg}, then (Murray 1961; Sorensen 1983)

$$\frac{\Delta n}{\Delta t} = \frac{dn}{dt} = \frac{n \cdot (k_{eff} - 1)}{t_{avg}} = \frac{n \cdot \rho \cdot k_{eff}}{t_{avg}} \tag{15-19}$$

Integrating Eq. (15-19) with $n = n_o$ at $t = 0$ yields

$$\ln \frac{n}{n_o} = \frac{\rho \cdot k_{eff}}{t_{avg}} t \quad \text{or} \quad n = n_o \cdot \exp\left(\frac{\rho \cdot k_{eff}}{t_{avg}} t\right) \tag{15-20}$$

Hence, the number of neutrons undergoes exponential growth with time. Example 15.4 illustrates how rapidly power can increase in a nuclear reactor with $k_{eff} > 1$.

Example 15.4	At $t = 0$ in a reactor, $k_{eff} = 1.001$ with $t_{avg} = 0.0001$ sec. How much has the power increased at $t = 1$ sec?

Solution   With $k_{eff} = 1.001$, the reactivity is

$$\rho = \frac{k_{eff} - 1}{k_{eff}} = \frac{1.001 - 1}{1.001} = 0.000999$$

The ratio of neutrons produced is n/n_o:

$$\frac{n}{n_o} = \exp\left(\frac{\rho \cdot k_{\text{eff}}}{t_{\text{avg}}} t\right) = \exp\left(\frac{0.000999 \cdot 1.001}{0.0001 \text{ sec}} \cdot 1 \text{ sec}\right) = 22{,}024$$

In one second, the number of neutrons increases by a factor of 22,024, and since the power is proportional to the number of neutrons, the power increases by the same factor. With such a small neutron magnification factor, the power increase in a single second is astonishingly large—the implication being that under these circumstances, control would be virtually impossible. However, control is possible because of delayed neutrons.

Delayed neutrons are produced with the radioactive decay of some products from the fission of $_{92}U^{235}$. Out of the ~2.43 neutrons produced in the fission reaction of Eq. (15-16), about 0.65 percent come from the decay of fission products such as Br and I. The times for the emission of the delayed neutrons are very long compared to the conventional fission cycle time. The net result is that t_{avg} is much longer for these than for the 99.35 percent of the neutrons released at the time of the reaction, the prompt neutrons. Murray (1961) states that the average life of the delayed neutrons is 12.7 seconds. Instead of the $t_{\text{avg}} = 0.0001$ sec for the prompt neutrons, when the delayed neutrons are considered in conjunction with prompt neutrons, $t_{\text{avg}} \sim 0.1$ sec. For Example 15.4, if $t_{\text{avg}} = 0.1$ sec, after one second the ratio of neutrons produced is 1.1—quite a contrast to the result of 22,024 for $t_{\text{avg}} = 0.0001$ sec, and a more rational number for nuclear reactor control.

As with many energy systems, different configurations of nuclear reactors have different advantages and disadvantages, and for various reasons, one configuration is preferred over another. Nuclear reactor configurations are based on the following considerations: (1) fuel, (2) moderator, (3) coolant, (4) control rods, (5) cost, and (6) containment. Table 15.1, adapted from www.uic.com.au, provides a convenient summary of the most prevalent reactor configurations. The salient features of each type will be reviewed. The Nuclear Regulatory Commission website www.nrc.gov/reading-rm/basic-ref/teachers/0.4pdf and Nero (1979) contain a variety of schematics and explanations for the various reactor types and are good sources of qualitative information about nuclear reactors and subsystems.

Pressurized Water Reactor (PWR)

PWRs represent the largest number of reactors, with an estimated 264 for commercial power and about 200 in service as naval propulsion systems. A simplified schematic of a PWR is given in Figure 15.3, where the containment vessel and the reactor vessel are shown. The concept behind PWRs is that the reactor vessel is pressurized to a sufficient pressure so that the coolant (and moderator) is maintained in a liquid state. Typically PWRs operate at about 300–325° C and are pressurized to

TABLE 15.1 Nuclear reactor types and characteristics

Type	Number	GW	Fuel	Coolant	Moderator
Pressurized water (PWR)	264	251	Enriched UO_2	Water	Water
Boiling water (BWR)	91	86	Enriched UO_2	Water	Water
Pressurized heavy water (PHWR) "CANDU"	43	24	Natural UO_2	Heavy water	Heavy water
Gas-cooled	18	11	Natural U, Enriched UO_2	CO_2	Graphite
Light water graphite	12	12	Enriched UO_2	Water	Graphite
Fast neutron (FBR)	4	1	PuO_2 UO_2	Liquid sodium	None

15 or 16 MPa. Details of the reactor vessel are presented in Figure 15.4. The reactor core is located in the lower portion of the vessel, and the control rods and their drive mechanism are depicted in the figure. In addition to the high operating pressure of the reactor vessel, a primary feature of a PWR is the pair of fluid loops. The high-pressure coolant loop circulates fluid within the reactor core and to a steam generator. In the steam generator the hot, high-pressure fluid is used to make steam at a

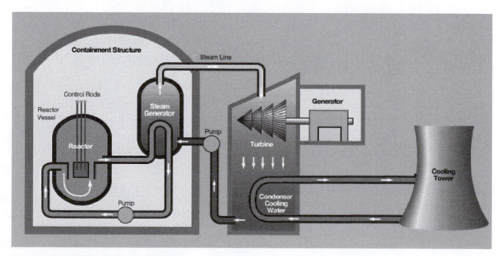

Figure 15.3 Schematic of a PWR.

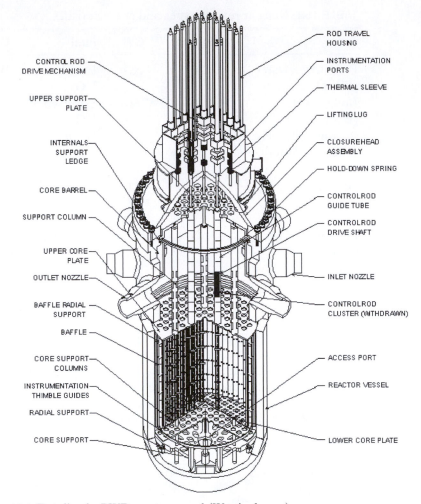

Control Rod Drive Mechanism
Upper Support Plate
Internals Support Ledge
Core Barrel
Support Column
Upper Core Plate
Outlet Nozzle
Baffle Radial Support
Baffle
Core Support Columns
Instrumentation Thimble Guides
Radial Support
Core Support

Rod Travel Housing
Instrumentation Ports
Thermal Sleeve
Lifting Lug
Closurehead Assembly
Hold-Down Spring
Controlrod Guide Tube
Controlrod Drive Shaft
Inlet Nozzle
Controlrod Cluster (Withdrawn)
Access Port
Reactor Vessel
Lower Core Plate

Figure 15.4 Details of a PWR reactor vessel (Westinghouse).

much lower pressure (6 MPa) compared with the reactor coolant loop that is circu-lated outside the containment structure to the turbine and generator. The steam exit-ing the turbine is condensed in the cooling tower and returned to the steam generator. Because the steam to the turbine is not exposed to the reactor core, the turbine does not need to be shielded.

Westinghouse, Babcock and Wilcox, and Combustion Engineering are the designers of many of the existing PWR systems. After these reactors were built, Westinghouse and Combustion Engineering combined their nuclear assets with those of British Nuclear Fuels Limited to form Westinghouse BNFL. The French-German-owned firm Framatome ANP acquired much of the nuclear technology of Babcock and Wilcox.

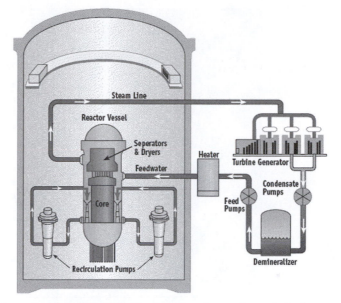

Figure 15.5 Schematic of a BWR (U.S. Nuclear Regulatory Commission).

Boiling Water Reactor (BWR)

The second most popular reactor type is the boiling water reactor (BWR). A simplified schematic of a BWR is provided in Figure 15.5. In the schematic, the containment vessel and the reactor vessel are shown. A BWR differs from a PWR in several important features: (1) the pressure in the reactor vessel is much lower (7 MPa) since steam is the desired fluid, (2) a BWR essentially contains only a single fluid loop, and (3) steam is generated in the reactor vessel and directed outside the containment structure to the turbine. BWRs operate at about 285° C. Details of a General Electric BRW/6 reactor vessel are presented in Figure 15.6. A steam-water mixture leaves the top of the core and flows into the separator, where the water droplets are removed and the steam is directed to the turbine. The steam separator assembly is in the upper portion of the vessel. Any steam remaining after exit from the turbine is condensed, and the liquid is returned to the reactor vessel. The recirculation pumps permit the coolant flow rate in the core to be varied to change the power output of the reactor.

A comparison of Figures 15.4 and 15.6 illustrates the significant differences between the PWR and BWR reactor vessels. The steam (vaporized coolant) in a BWR is radioactive since it is in direct contact with the reactor core; thus, the turbine and piping must be shielded.

Pressurized Heavy Water Reactor (PHWR)

A number of pressurized heavy water reactors (PHWR) are presently in use. As the name implies, heavy water is used as the moderator. The reactor temperature of a PHWR is about 290° C. The most prevalent design of this type is called the

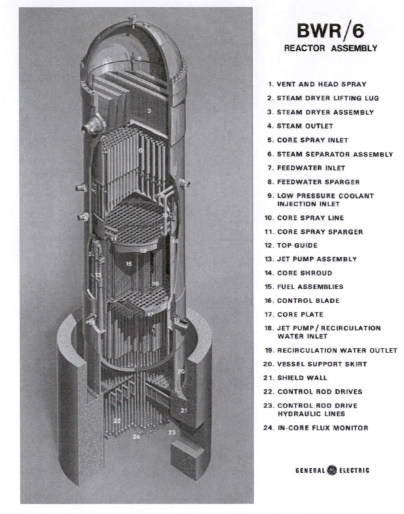

BWR/6
REACTOR ASSEMBLY

1. VENT AND HEAD SPRAY
2. STEAM DRYER LIFTING LUG
3. STEAM DRYER ASSEMBLY
4. STEAM OUTLET
5. CORE SPRAY INLET
6. STEAM SEPARATOR ASSEMBLY
7. FEEDWATER INLET
8. FEEDWATER SPARGER
9. LOW PRESSURE COOLANT INJECTION INLET
10. CORE SPRAY LINE
11. CORE SPRAY SPARGER
12. TOP GUIDE
13. JET PUMP ASSEMBLY
14. CORE SHROUD
15. FUEL ASSEMBLIES
16. CONTROL BLADE
17. CORE PLATE
18. JET PUMP / RECIRCULATION WATER INLET
19. RECIRCULATION WATER OUTLET
20. VESSEL SUPPORT SKIRT
21. SHIELD WALL
22. CONTROL ROD DRIVES
23. CONTROL ROD DRIVE HYDRAULIC LINES
24. IN-CORE FLUX MONITOR

GENERAL ⊕ ELECTRIC

Figure 15.6 Details of a BWR reactor vessel (General Electric).

CANDU after its country of origin, Canada. PHWRs offer the following advantages over the PWR and BWR: (1) since heavy water is used as the moderator, the uranium fuel does not have to be enriched; (2) they can be refueled without shutting down; and (3) they are not as expensive to build and operate as some reactor types. A schematic of a PHWR is presented in Figure 15.7. The reactor vessel is often called a calandria. The PHWR shares many characteristics with a PWR. However, a large amount of heavy water is required, and heavy water is expensive. But since the uranium fuel does not have to be enriched, fuel costs are lower. An issue of concern is that the spent fuel contains $_{94}\text{Pu}^{239}$, an isotope that can be used to produce nuclear weapons.

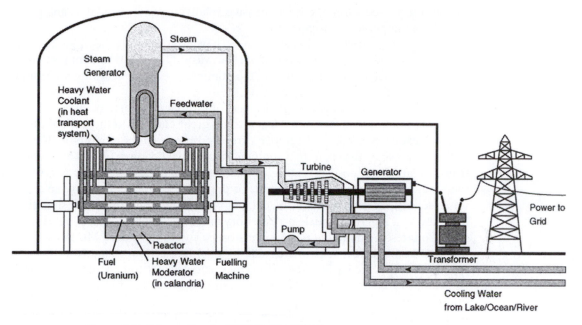

Figure 15.7 Schematic of a PHWR (CANDU).

Gas-Cooled Reactor (GCR)

A gas-cooled reactor (GCR) uses carbon dioxide as the coolant and graphite as a moderator. GCRs allow the use of natural uranium as fuel. A schematic of a GCR is presented in Figure 15.8. Carbon dioxide is an effective coolant and does not absorb neutrons. The CO_2 reaches 650° C in the reactor core. Gases are not good moderators,

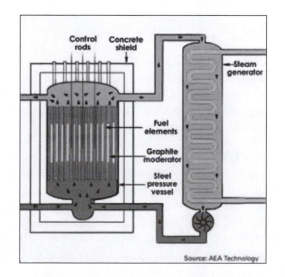

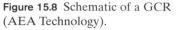

Figure 15.8 Schematic of a GCR (AEA Technology).

so graphite is used. Graphite is inexpensive, relatively widely available, and possesses good high-temperature properties.

Issues of concern for nuclear power have traditionally been safety, nuclear weapon proliferation, and disposal of nuclear waste. Dorf (2001) examines issues concerned with nuclear power. Arguably, the dominant issues concerning nuclear power have been radioactive waste and safety.

The issue of handling/storage/disposal of radioactive waste from nuclear power reactors (and military programs) has been of great concern in the United States for many decades. Nuclear fuel removed from power reactors exhibits high levels of radioactivity since the spent fuel contains radioactive fission products and other neutron-activated isotopes. The extracted fuel, generally in the form of rods, is stored on the reactor site in secure water ponds or dry casts. Permanent storage in a nuclear waste repository is the long-term goal, but in the United States, the designation and licensing of a long-term nuclear repository has been an emotional political issue. In 1985 Congress mandated a permanent storage site in the Yucca Mountains, 160 km north of Las Vegas, NV. However, due to legal battles and maneuvers, 2016 (as of January 2008) is the earliest date that any waste is likely to be stored. One problem associated with spent fuel is that reprocessing is prohibited in the United States; in other countries, spent nuclear fuel is reprocessed and the recovered uranium is sent back for enrichment and reuse, significantly reducing the volume and half-life of spent fuel nuclear waste. One technique for long-term storage of radioactive waste is to mix the waste with pulverized glass and then solidify the glass mixture. The resulting canisters leach only about 0.1 percent (per volume basis) per 10,000 years and are considered environmentally benign.

Although Three Mile Island generated much adverse publicity and resulted in an unbelievably expensive cleanup effort, very little radiation was released to the atmosphere and no public health issues have ever been raised. The disaster at Chernobyl, however, caused significant radiation release to the atmosphere and is worth examining.

Chernobyl

The Chernobyl nuclear reactor disaster of 1986 is infamous as a catastrophe to both people and the environment. The causes of Chernobyl serve as a litany of how *not* to do things. Details are provided in Deutch and Lester (2004) and at the websites www.world-nuclear.org/info/chernobyl/inf07.html and www.uic.com.au/nip22app.htm. The location of Chernobyl is indicated in Figure 15.9.

The reactor was a light-water graphite reactor, a type extensively used in the countries of the former USSR. The general consensus is that the accident was the result of a flawed reactor design coupled with serious operator mistakes in a system characterized by minimal training, nonexistent safety culture, and Cold War isolation. The Chernobyl reactor was a 1000-MW reactor of the RBMK type, a boiling-water-cooled, graphite-moderated design that was the workhorse of the USSR nuclear reactor program. A schematic of the RBMK nuclear reactor is shown in Figure 15.10. The RBMK 1000 uses uranium enriched to about 2 percent $_{92}U^{235}$. The enriched ura-

Figure 15.9 The location of Chernobyl (www.world-nuclear.org).

nium is contained in 1700 individual pressure tubes, each about 7 m long, through which cooling water flows and steam is generated.

Since graphite is the moderator, excessive boiling reduces the cooling (and neutron absorption by the coolant) without inhibiting neutron absorption since the moderator is unaltered by the extra boiling. This positive-feedback behavior, termed a positive void coefficient, is one of the problems with the RBMK nuclear reactors, with the effect that at low power levels an increase in steam results in an increase in the reactivity. Conventional BWRs have a negative void coefficient—a decline in reactivity with an increase in boiling, since water is the also the moderator in BWRs.

In addition to the inherent stability problems associated with a positive void coefficient, the Chernobyl accident occurred during a test of a safety system. As part of

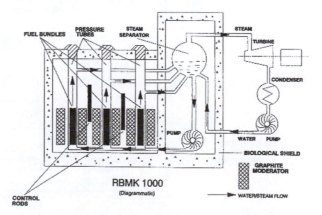

Figure 15.10 RBMK reactor schematic (www.world-nuclear.org).

this test, the flow of cooling water through the core was interrupted, resulting in the reactivity, ρ, becoming greater than unity. The power started to increase and was estimated to have reached 100 times the nominal full-power output within four seconds! The rapid increase in power meant that not enough time was available to insert the control rods. In any case, after four seconds the enormous power surge had destroyed the fuel channels, effectively blocking any further insertion of the control rods. The fuel then overheated, melted, and fragmented. When steam came into contact with the fragmented fuel, a steam explosion occurred that literally lifted the 1000-ton reactor upper plate. Unlike most reactors (especially in the Western nations), the RBMK reactor was not surrounded by a containment structure, so that when the explosion occurred and lifted the upper cover plate, the reactor core was open to the atmosphere. Hydrogen, produced by reactions between the steam, the fuel, and the graphite, then caused a second explosion, which resulted in significant amounts of radioactive material being discharged into the atmosphere. To exacerbate matters, the hot graphite ignited and fission continued, producing additional heat and releasing even more radioactive material into the atmosphere. A number of emergency response workers attempting to stabilize the situation and put out the fire eventually died of radiation poisoning. Figure 15.11 depicts the current state of the reactor at Chernobyl.

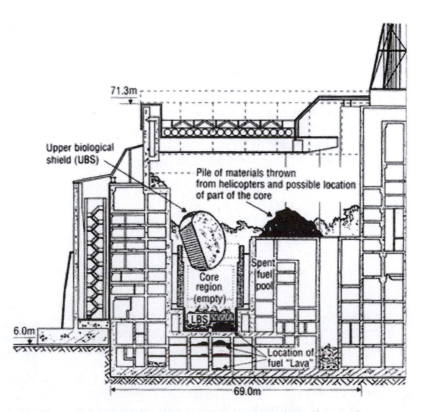

Figure 15.11 Chernobyl reactor after the accident (www.world-nuclear.org).

About 5000 metric tonnes of boron, dolomite, sand, clay, and lead were dropped onto the burning core by helicopters to control the burning and to limit the release of radioactive materials to the atmosphere.

The adverse effects of Chernobyl have been many. So far, more than 50 deaths have been directly attributed to the disaster. At least 5 percent of the radioactive core ultimately found its way into the atmosphere. The USSR's nuclear program was not canceled, but no new RBMK reactors were constructed, and existing RBMK reactors have been retrofitted to make them safer. Safety and training for nuclear reactor personnel have become more important in the former USSR states. Some adjacent towns were permanently evacuated, and long-term health effects on a large number of citizens continue to be of great concern. Indeed, some early estimates place the ultimate death toll in the tens of thousands (Deutch and Lester 2004). However, more recent studies have concluded that the long-term radiation-related health issues are not nearly as significant as early projections indicated (www.world nuclear.org).

Nuclear Power in the United States in the Twenty-first Century

After a hiatus of almost 30 years, new nuclear reactor designs have either been certified or are in the certification process by the United States Nuclear Regulatory Commission (NRC). The Energy Information Administration (EIA) maintains a list of new reactor designs on its website www.eia.doe.gov/cneaf/nuclear/page/analysis/nucenviss2.html. EIA lists 12 new reactor designs that have been certified or are in some phase of the certification process. Table 15.2 contains summary details of new designs. Additional information on these new designs is available on the websites of the various companies. The website for the Nuclear

TABLE 15.2 Status of selected new nuclear reactor designs (EIA)

Name	Manufacturer	Power (MW)	Type	Certification Date or Status
AP600	Westinghouse	600	PWR	1999
AP1000	Westinghouse	1000	PWR	2005
ABWR	General Electric	1400	BWR	1997
System 80+	Westinghouse	1300+	PWR	1997
ESBWR	General Electric	1550	BWR	In progress
EPR	UniStar Nuclear	1600	PWR	Preliminary
PMBR	Westinghouse	165	Pebble bed	Preliminary
IRIS	Westinghouse	165	Gas cooled	Preliminary
US-APWR	Mitsubishi	1700	PWR	In progress
ACR Series	Atomic Energy	700–1200	CANDU	On hold (?)
GT-MHR	General Atomics	285	Gas cooled	Preliminary
4S	Toshiba	10+	Gas cooled	Preliminary

Energy Institute, www.nei.org, maintains an updated list of construction/operating license applications for new nuclear power plants in the United States. As of January 2008, this website noted 22 potential sites for new fission reactors, with 19 of these sites actively seeking licenses. The International Atomic Energy Agency (www.iaea.org) maintains a similar worldwide list.

The increasing activity and interest surrounding new fission nuclear power plants are reflected in the increased number of recent books on the subject. White (2005) and Sweet (2006) advocate nuclear power as an alternative to fossil fuel electrical generation. Shultis and Faw (2002) and LaMarsh and Baratta (2001) are recent editions of classic nuclear engineering textbooks.

Adams (2005) provides a lucid, documented assessment of the use of fission power in a sustainable scenario and concludes that fission power has the potential to provide a large proportion of the world's energy for many years. He recommends the use of high-conversion reactors, reactors in which the production of fissile isotopes exceeds the consumption of fissile isotopes (breeder nuclear reactors) and the reprocessing and "reburning" of nuclear fuel. If a majority of the heavy isotopes can be made to fission, then most of the remaining radioactive waste would decay to a radioactive level less than that of the original ore in 300 years. Adams points out that significant technical and political challenges must be met for this minimum-waste, sustainable scenario to be realized.

15.4   FUSION POWER

Fusion power has long been viewed as a panacea for the world's energy problems, and much effort and research over an extended period of time have been devoted to fusion power. A number of relatively recent books (Freidberg 2007; McCracken and Stott 2005; Harms 2002) and research projects demonstrate the optimism about the eventual feasibility of fusion power. A brief overview of fusion is thus appropriate here.

Fusion was briefly mentioned in Section 15.2 as one of the three categories of nuclear reactions, and Eq. (15-8) was presented as an example of a fusion reaction. A number of reactions are candidates for power production from fusion reactors. The most likely ones and their energy releases are summarized as

$$_1H^2 + {}_1H^2 \rightarrow {}_1H^3 + {}_1H^1 + 4.03 \text{ MeV}$$
$$_1H^2 + {}_1H^2 \rightarrow {}_2He^3 + {}_0n^1 + 3.27 \text{ MeV}$$
$$_1H^2 + {}_1H^3 \rightarrow {}_2He^4 + {}_0n^1 + 17.59 \text{ MeV} \qquad (15\text{-}21)$$
$$_1H^2 + {}_2He^3 \rightarrow {}_2He^4 + {}_1H^1 + 18.35 \text{ MeV}$$
$$_1H^3 + {}_1H^3 \rightarrow {}_2He^4 + 2\,{}_0n^1 + 11.33 \text{ MeV}$$

As indicated in the fusion reactions in Eq. (15-21), energy is released. The energy release results from the difference in the masses of the reactants and products, as with fission nuclear reactions. Consider the following example, which is based on the second reaction listed in Eq. (15-21).

Example 15.5	Compute the energy resulting from $_1H^2 + {_1}H^2 \rightarrow {_2}He^3 + {_0}n^1$, a fusion reaction.

Solution   The atomic mass tables (Shultis and Faw 2002) provide the amu values for the reactants and products of the fusion reaction:

$$\text{mass of } {_1}H^2 = 2.014102 \text{ amu}$$

$$\text{mass of } {_0}n^1 = 1.008665 \text{ amu}$$

$$\text{mass of } {_2}He^3 = 3.016029 \text{ amu}$$

The masses of the reactants and products are

$$\text{Reactants} = 2 \cdot 2.014102 \text{ amu} = 4.028204 \text{ amu}$$

$$\text{Products} = 3.016029 \text{ amu} + 1.008665 \text{ amu} = 4.024695 \text{ amu}$$

The change in mass is

$$\Delta\text{mass} = 4.024695 \text{ amu} - 4.028204 \text{ amu} = -0.00351 \text{ amu}$$

and the energy released is

$$\Delta E = \Delta mc^2 = -0.00351 \text{ amu} \left(3 \cdot 10^8 \frac{m}{sec} \right)^2 = -5.246 \cdot 10^{-13} J = -3.274 \text{ MeV}$$

Fusion has the great advantages over fission of not requiring high-mass-number radioactive isotopes as a fuel and not yielding radioactive products (thus alleviating the spent fuel problem). Moreover, as the fusion reactions of Eq. (15-21) demonstrate, the "fuel" is deuterium ($_1H^2$) and tritium ($_1H^3$), an isotope of hydrogen in which the nucleus contains three protons. Although the individual fission of an atom of $_{92}U^{235}$ releases more energy (~200 MeV) than a single fusion reaction (3–18 MeV), on a mass basis the energy release ratio (fusion to fission) is about 7 to 1 (Sorensen 1983). The great appeal of fusion over fission and of fission over chemical reactions (burning) is well summarized by the Princeton Plasma Physics Laboratory on the website, www.pppl.gov/fusion_basics/pages/fusion_power_plant.html, from which Table 15.3 is adapted. Table 15.3 delineates the fuel requirements for a 1000-MW power plant

TABLE 15.3 Fuel consumption for various power plant types

Plant Type	Quantity of Fuel Used
Coal-fired	4.4×10^6 metric tonnes of black coal
Coal-fired	10.8×10^6 metric tonnes of brown coal
Fission	1.3 metric tonnes of $_{92}U^{235}$
Fusion	150 kg of deuterium and 500 kg of lithium

operating continuously for a year using coal (black or brown), nuclear fission, and nuclear fusion. While the fuel consumption of coal-fired power plants is measured in millions of metric tonnes and the fuel consumption of nuclear fission plants in metric tonnes, the fuel consumption of a fusion plant is measured in kilograms! The deuterium part of the fusion fuel requirement poses little problem since about 1 part in 5000 of sea water is deuterium. Since the oceans contain $\sim10^{15}$ tons of deuterium, the deuterium supply for fusion is essentially unbounded. The lithium is more problematic, but, compared with other energy sources (coal and uranium), is also plentiful.

Why, then, with all these possibilities, has fusion power not been realized? The requirements for fusion are very difficult to attain and sustain. Any of the fusion reactions listed in Eq. (15-21) requires the individual fuel nuclei to come together and then fuse. But all nuclei have positive charges; thus, the nuclei will repel each other unless the kinetic energy of each nucleus is sufficient to overcome the Coulomb repulsive force. For fusion to occur, the speed of each nucleus must be very high, and very high kinetic energies correspond to very high temperature since on average (Shultis and Faw 2002),

$$E_{\text{avg}} = k \cdot T \tag{15-22}$$

Depending on the fusion reaction, E_{avg} values of from a few keV to several hundred keV are needed in order to ensure that the Coulomb force is overcome and the fuel nuclei can be in position to fuse. The resulting temperature range is $10-300 \times 10^6$ K. At these elevated temperatures, the reactants are ionized with all electrons stripped from their orbits about the nucleus, and a plasma results. The plasma consists of electrons and the positively charged nuclei. The basic problems for a fusion reactor are threefold: (1) how to generate the high temperatures required (basically solved), (2) how to contain the hot plasma (in progress), and (3) how to achieve the required density of the plasma so that fusion takes place (in progress).

Of the fusion reactions listed in Eq. (15-21), the one of most interest is the deuterium-tritium reaction:

$$_1\text{H}^2 + {}_1\text{H}^3 \rightarrow {}_2\text{He}^4 + {}_0n^1 + 17.59 \text{ MeV} \tag{15-23}$$

Not only is the individual energy release per fusion high (17.59 MeV), but the temperature required is "only" $\sim50 \times 10^6$ K since 5 keV is required for the reaction. The deuterium-deuterium fusion reaction requires 50 keV or 500×10^6 K—an order of magnitude higher than the deuterium-tritium fusion reaction, with 4.03 MeV energy release. Since tritium, $_1\text{H}^3$, has a half-life of only 12.33 years, none occurs in nature. Thus, part of the problem with a fusion reactor using the deuterium-tritium reaction is providing the tritium. Fortunately, tritium is relatively easy to produce by means of neutron capture by lithium. The fission reactions of the two isotopes of lithium are as follows:

$$\begin{aligned} _3\text{Li}^6 + {}_0n^1 &\rightarrow {}_1\text{H}^3 + {}_2\text{He}^4 + 4.78 \text{ MeV} \\ _3\text{Li}^7 + {}_0n^1 &\rightarrow {}_1\text{H}^3 + {}_2\text{He}^4 + {}_0n^1 - 2.47 \text{ MeV} \end{aligned} \tag{15-24}$$

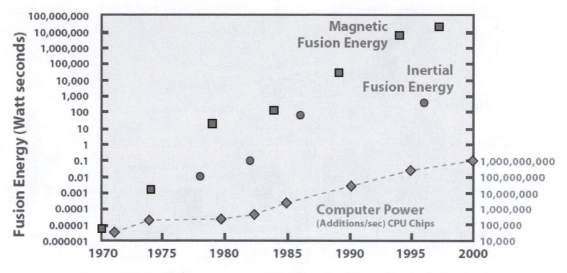

Figure 15.12 Fusion progress since 1970 (Princeton Plasma Physics Laboratory).

But how is a plasma at 50×10^6 K contained? Two methods have been proposed and extensively researched: (1) magnetic containment and (2) inertial containment. Since the electrons and ions are charged, the plasma can be confined by a magnetic field. In magnetic containment, the magnetic field forces the particles to follow spiral paths. Consequently, the charged particles in the plasma are confined by the magnetic field and do not strike fusion reactor walls. Inertial confinement is based on using high-energy lasers to vaporize, ionize, and ignite the fusion reaction in such a short period of time ($\sim 10^{-10}$ sec) that the fuel does not have an opportunity to expand—hence it is confined by its own inertia. Extensive research has been done using both techniques, but magnetic confinement is the more promising of the two. Figure 15.12 provides an indication of the progress of fusion power plant research. A measure of the progress made toward creating viable fusion reactors is the energy that has been attainable from fusion experiments. As the figure illustrates, since 1970 the energy (in watt-seconds) attainable has increased by 12 orders of magnitude— a truly remarkable advance considering that computer power over that time increased by 5 orders of magnitude. But much work is still required before an operational fusion power plant becomes a reality.

In what form might a fusion power plant appear? Figure 15.13, from Adomavicius and Lisak (2005), is a schematic of a fusion power plant. The overall shape of the fusion reactor is toroidal, a requirement that results from the magnetic containment and the helical path the charged nuclei must traverse. In earlier fusion research projects, toroidal fusion vessels were called "tokamaks." Tokamaks at the Princeton Plasma Physics Laboratory (PPPL) and the Joint European Torus (JET) are perhaps the best known of these research vessels.

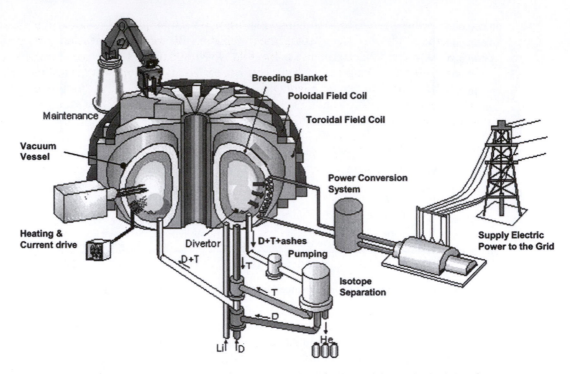

Figure 15.13 Schematic of a fusion power plant (Adomavicius and Lisak 2005).

The neutrons (14 MeV) are absorbed in the lithium-containing breeding blanket that surrounds the fusion chamber. Neutron capture by the lithium nucleus results in fission nuclear reactions that generate (or breed) tritium, a required fuel (see Eq. (15-24)). The tritium is separated and fed back into the fusion reactor for use as a fuel. As is evident from Figure 15.13, a fusion power plant is much more involved than a fission power plant. If a deuterium-tritium reaction is to occur, then Li must be injected into the breeding blanket so that tritium results. Unburned deuterium, tritium, and helium are recovered from the reactor; the He, which has commercial value, is removed and sold, and the deuterium and tritium are re-injected. Energy is recovered from the reactor vessel and used to generate electricity.

The JET is currently the largest tokamak in operation and has been known as the world's foremost tokamak machine for 20 years. In 1997 it successfully produced a peak power of 16 MW, with 10 MW being sustained over a 0.5-sec period (Wesson 2000). The most significant fusion research project of the next few decades will be the ITER (International Thermonuclear Experimental Reactor) program, a joint project funded by many countries, including the United States and the European Union. The ITER project (www.iter.org) has as its goals the development of equipment, techniques, and procedures required for commercial fusion power generation, and it is thus viewed as perhaps the last step before the realization of commercial fusion power plants. Specifically, the ITER goals are fourfold (www.efda.org): (1) produce more power than it consumes, (2) maintain a fusion plasma for up to eight minutes, (3) test technologies needed for fusion power plants, and (4) test and develop concepts for breeding tritium from a blanket surrounding the plasma. Figure 15.14 is a cutaway, same-scale rendering of the JET and ITER fusion reactor vessels. The ITER reactor will be located at Cadarache, France. The United States' participation and goals for the ITER program are detailed in "Planning for U.S. Fusion Community Participation in the ITER Program," available at www.ofes.fusion.doe.gov. Specific topics are as follows:

1. Integrated burning plasma sciences (fusion physics)
2. Macroscopic plasma physics (stability)
3. Waves and energetic particles (heating, control, and effects of products)
4. Multi-scale transport physics (confinement)
5. Plasma boundary interfaces
6. Fusion engineering science (fusion tools and technology)

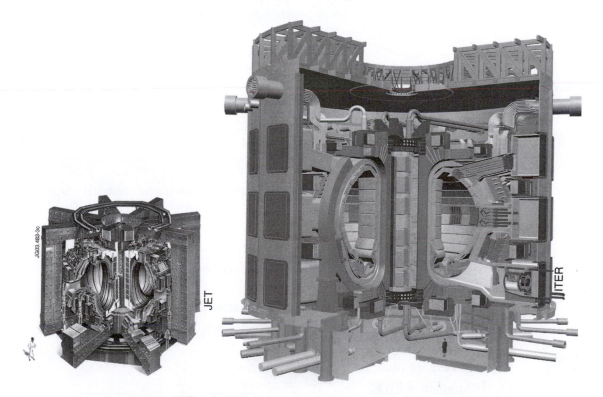

Figure 15.14 JET and ITER fusion vessel cutaway sketch (www.efda.jet.org).

Adomavicius and Lisak (2005) present a realistic assessment of the needed technology for commercial fusion power plant realization. The general industry consensus is that a commercial power plant may be a reality by 2050.

15.5 CLOSURE

Electricity from nuclear fission power plants represents a significant percentage of the total electricity generated. Fission power plant technology is relatively mature, but questions about safety issues, nuclear proliferation, and long-term radioactive waste disposal along with a generally adverse political climate have clouded the future of fission power. Nonetheless, energy security and greenhouse gas emission concerns have provided the impetus for renewed consideration of nuclear fission power plant construction, and many in the electrical power industry believe nuclear power will increase in importance and acceptance over the next few decades.

Fusion power represents a solution to the world's energy crisis and offers many advantages over both nuclear fission and conventional fuel for electricity and power generation. However, the technical problems are daunting, and even optimists in the fusion community suggest that 2050 may be the soonest that fusion can be commercially exploited for power generation.

REVIEW QUESTIONS

1. What is nuclear fission?
2. What is nuclear fusion?
3. What is deuterium?
4. What is heavy water?
5. What is the energy level difference between fission and chemical reactions on an atomic basis?
6. What is half-life?
7. What governs the operation of a nuclear reactor?
8. What is a thermal neutron? What is a fast neutron?
9. What are moderators?
10. What moderator is used in U.S. light-water reactors?
11. How many neutrons, on average, are produced by the fission of $_{92}U^{235}$?
12. What is a light-water reactor?
13. What are two types of light-water reactors?
14. Explain a PWR.
15. Explain a BWR.

16. Is nuclear power generation an important part of the power grid in the United States? Explain.

17. What is critical operation?

18. What part did graphite as a moderator play in the Chernobyl disaster?

19. Why are delayed neutrons important to the operation of a nuclear reactor?

20. Why is fusion so difficult to achieve?

21. How close are we to having power generated by fusion?

EXERCISES

1. Compute the energy resulting from the $_{92}U^{235} + _{0}n^{1} \rightarrow _{56}Ba^{137} + _{36}Kr^{97} + 2 _{0}n^{1}$ nuclear reaction.

2. Verify the energy release of each of the following fusion reactions:

$$_{1}H^{2} + _{1}H^{2} \rightarrow _{1}H^{3} + _{1}H^{1} + 4.03 \text{ MeV}$$
$$_{1}H^{2} + _{1}H^{3} \rightarrow _{2}He^{4} + _{0}n^{1} + 17.59 \text{ MeV}$$
$$_{1}H^{2} + _{2}He^{3} \rightarrow _{2}He^{4} + _{1}H^{1} + 18.35 \text{ MeV}$$
$$_{1}H^{3} + _{1}H^{3} \rightarrow _{2}He^{4} + 2 _{0}n^{1} + 11.33 \text{ MeV}$$

3. In the beta decay reaction $_{82}Pb^{214} \rightarrow _{83}Bi^{214} + _{-1}e^{0} + \nu$, determine the respective times required for the number of original atoms to be reduced by 25, 50, and 75 percent.

4. In the alpha decay reaction $_{88}Ra^{226} \rightarrow _{86}Rn^{222} + _{2}He^{4}$, determine the respective times required for the number of original atoms to be reduced by 25, 50, and 75 percent.

5. How much $_{94}Pu^{239}$ is required to produce 1 MWd if the average energy yield is approximately 190 MeV per fission (El-Wakil 1984) and the electrical conversion efficiency is 33 percent?

6. For $k_{eff} = 1.0005, 1.001$, and 1.002 and $0.1 \text{ sec} < t_{avg} < 0.0001 \text{ sec}$, plot n/n_{o}, the ratio of neutrons produced, for a time period of one second. Discuss the results with consideration of changing reactor power levels.

7. E_{avg} values from a few keV to several hundred keV are required for fusion. What is the required temperature range corresponding to $5 \text{ keV} < E_{avg} < 200 \text{ keV}$?

REFERENCES

Adams, M. L. 2005. "Sustainable Energy from Nuclear Fission Power," *The Bridge*, 32(4), Winter. Available at www.nae.edu.

Adomavicius, D., and Lisak, M. 2005. *A Conceptual Study of Commercial Fusion Power Plants*. EFDA-RP-RE-5.0, European Fusion Development Agreement.

Deutch, J. M., and Lester, R. K. 2004. *Making Technology Work: Applications in Energy and the Environment.* Cambridge, UK: Cambridge University Press.

Dorf, R. C. 2001. *Technology, Humans, and Society: Toward a Sustainable World.* San Diego, CA: Academic Press.

El-Wakil, M. M. 1984. *Powerplant Technology.* New York: McGraw-Hill.

Fay, J. A., and Golomb, D. S. 2002. *Energy and the Environment.* Oxford, UK: Oxford University Press.

Firestone, R. V., and Shirley, V. S., eds. 1999. *Table of Isotopes.* New York: Wiley.

Freidberg, J. P. 2007. *Plasma Physics and Fusion Energy.* Cambridge, UK: Cambridge University Press.

Harms, A. A. 2002. *Principles of Fusion Energy.* New Delhi: Allied Publishers.

LaMarsh, J. R., and Baratta, A. J. 2001. *Introduction to Nuclear Engineering,* 3rd ed. New York: Addison-Wesley.

McCracken, G., and Stott, P. 2005. *Fusion: The Energy of the Universe.* Amsterdam: Elsevier.

Murray, R. L. 1961. *Introduction to Nuclear Engineering,* 2nd ed. Englewood Cliffs, NJ: Prentice-Hall.

Nero, A. V. 1979. *A Guidebook to Nuclear Reactors.* Berkeley: University of California Press.

Shultis, J. K., and Faw, R. E. 2002. *Fundamentals of Nuclear Science and Engineering,* New York: Marcel Dekker.

Sorensen, H. A. 1983. *Energy Conversion Systems.* New York: Wiley.

Sweet, W. 2006. *Kicking the Habit: Global Warming and the Case for Renewable and Nuclear Energy.* New York: Columbia University Press.

U.S. Burning Plasma Organization. 2006. *Planning for U.S. Fusion Community Participation in the ITER Program.* Available at www.ofes.fusion.doe.gov.

Wesson, J. 2000. *The Science of JET.* JET Joint Undertaking, Abingdon, United Kingdom. Available at www.jet.efda.org.

Weston, K. C. 1992. *Energy Conversion.* St. Paul, MN: West.

White, R. S. 2005. *Energy for the Public: The Case for Increased Nuclear Fission Energy.* Santa Barbara, CA.

Nuclear Engineering International. 2007. *World Nuclear Industry Handbook.* Sidcup, Kent, UK: Nuclear Engineering International.

WEBSITES

www.ie.lbl.gov/toi2003/MassSearch.asp
www.uic.com.au
www.nrc.gov/reading-rm/basic-ref/teachers/0.4pdf
www.nucleartourist.com/
www.nukeworker.com/pictures/index-52.html
www.ofes.fusion.doe.gov
www.world-nuclear.org/info/chernobyl/inf07.html
www.eia.doe.gov/cneaf/nuclear/page/analysis/nucenviss2.html
www.iter.org
www.efda.org

Index